Harvesting Polluted Waters

Waste Heat and Nutrient-Loaded Effluents in the Aquaculture

Environmental Science Research

Volume 1 – INDICATORS OF ENVIRONMENTAL QUALITY
Edited by William A. Thomas ● 1972

Volume 2 – POLLUTION: ENGINEERING AND SCIENTIFIC SOLUTIONS
Edited by Euval S. Barrekette ● 1973

Volume 3 – ENVIRONMENTAL POLLUTION BY PESTICIDES
Edited by C. A. Edwards ● 1973

Volume 4 – MASS SPECTROMETRY AND NMR SPECTROSCOPY IN
PESTICIDE CHEMISTRY
Edited by Rizwanul Haque and Francis J. Biros ● 1974

Volume 5 – BEHAVIORAL TOXICOLOGY
Edited by Bernard Weiss and Victor G. Laties ● 1975

Volume 6 – ENVIRONMENTAL DYNAMICS OF PESTICIDES
Edited by Rizwanul Haque and V. H. Freed ● 1975

Volume 7 – ECOLOGICAL TOXICOLOGY RESEARCH: EFFECTS OF HEAVY
METAL AND ORGANOHALOGEN COMPOUNDS
Edited by A. D. McIntyre and C. F. Mills

Volume 8 – HARVESTING POLLUTED WATERS: WASTE HEAT AND NUTRIENT-LOADED
EFFLUENTS IN THE AQUACULTURE
Edited by O. Devik

A Continuation Order Plan is available for this series. A continuation order will bring
delivery of each new volume immediately upon publication. Volumes are billed only upon
actual shipment. For further information please contact the publisher.

Harvesting Polluted Waters

Waste Heat and Nutrient-Loaded Effluents in the Aquaculture

Edited by

O. Devik
The Chr. Michelsen Institute
Bergen, Norway

PLENUM PRESS · NEW YORK AND LONDON

Library of Congress Cataloging in Publication Data

Main entry under title:

Harvesting polluted waters.

(Environmental science research; v. 8)
Proceedings of a workshop sponsored by the Special Program Panel on Eco-Sciences of the NATO Science Committee.
Includes bibliographies and index.
1. Aquaculture–Congresses. 2. Sewage as fertilizer–Congresses. 3. Thermal pollution of rivers, lakes, etc.–Congresses. I. Devik, O. II. North Atlantic Treaty Organization. Special Program Panel on Eco-Sciences.

SH135.H37   630'.9162   75-40281
ISBN-13:978-1-4613-4330-1   e-ISBN-13:978-1-4613-4328-8
DOI: 10.1007/978-1-4613-4328-8

A Division of Plenum Publishing Corporation
227 West 17th Street, New York, N.Y. 10011

United Kingdom edition published by Plenum Press, London
A Division of Plenum Publishing Company, Ltd.
Davis House (4th Floor), 8 Scrubs Lane, Harlesden, London, NW10 6SE, England

Contributors

W. B. Brogden, Marine Science Institute, University of Texas, Port Aransas, Texas

Torkild Carstens, River and Harbor Laboratory, University of Trondheim, Trondheim, Norway

R. J. Conover, Marine Ecology Laboratory, Bedford Institute of Oceanography, Dartmouth, Nova Scotia, Canada

K. I. Dahl-Madsen, Water Quality Research Institute, Poppelgårdvej, Søborg, Denmark

Ole Devik, Christian Michelsen Institute, Bergen, Norway

Bent H. Fenger, Water Quality Research Institute, Poppelgårdvej, Søborg, Denmark

Curt Forsberg, Institute of Physiological Botany, University of Uppsala, Uppsala, Sweden

Arne Jensen, Institute of Marine Biochemistry, University of Trondheim, Trondheim, Norway

P. Korringa, Director of the Netherlands Institute for Fishery Investigations, IJmuiden, The Netherlands

Bo Møller, Water Quality Research Institute, Poppelgårdvej, Søborg, Denmark

Peter Mortensen, Danish Hydraulic Institute, Copenhagen, Denmark

C. H. Oppenheimer, Marine Science Institute, University of Texas, Port Aransas, Texas

J. E. G. Raymont, Department of Oceanography, University of Southampton, Southampton, England

John H. Ryther, Senior Scientist, Woods Hole Oceanographic Institution, Woods Hole, Massachusetts

Richard L. Saunders, North American Salmon Research Center, St. Andrews, New Brunswick, Canada

Hans Schrøder, Danish Hydraulic Institute, Copenhagen, Denmark

Carl J. Soeder, Gesellschaft für Strahlen- und Umweltforschung mbH, München, Abteilung für Algenforschung und Algentechnologie, Dortmund, Germany

Eberhard Stengel, Gesellschaft für Strahlen- und Umweltforschung mbH, München, Abteilung für Algenforschung und Algentechnologie, Dortmund, Germany

Kenneth R. Tenore, Assistant Scientist, Woods Hole Oceanographic Institution, Woods Hole, Massachusetts

P. R. Walne, Fisheries Experiment Station, Conwy, Gwynedd, U. K.

Preface

Waste heat from thermal power generation and waste minerals from domestic sewage are polluting rivers and coastal waters. It would thus be an important step forward if these *pollutants* could be made productive in aquaculture. The proposal to call a meeting of experts on *Waste Heat and Nutrient-Loaded Effluents in Aquaculture* came from Dr. Ole Devik.

His proposal was accepted and sponsored by the Special Program Panel on Eco-Sciences, a subsidiary body of the NATO Science Committee. NATO has extensive programs in the scientific and environmental fields, and I would like to take this opportunity to give you a short outline of these programs.

Collaboration and consultation between member countries of the Alliance have been of major concern to the North Atlantic Treaty Organization ever since it was established. In the mid-fifties a serious attempt was made to implement the collaboration in nonmilitary fields, and a report* from a committee of foreign ministers—Lester B. Pearson (Canada), Gaetano Martino (Italy), and Halvard Lange (Norway)—named scientific and technological cooperation as especially important. As a consequence of this report, a position as science adviser to the secretary-general of NATO (later changed to Assistant Secretary-General for Scientific and Environmental Affairs) and a science committee composed of one highly qualified scientist from each of the member countries of the Alliance was established in 1958.

The NATO science programs have changed during the years, but their predominant characteristics have remained an emphasis on cooperation and catalysis and a capacity for rapid response to new developments. Each of the programs has been conscientiously designed and deliberately implemented to improve the exchange of information. Over 50,000 individuals—of which some thousands come from countries outside the Alliance, including some hundreds

*Report of the Committee of Three: Non-Military Co-operation in NATO, NATO Information Service, B-1110 Brussels.

from Eastern Europe—have directly participated in these programs. The following programs have been in operation during 1973-74:*

The Senior Scientists Program. This is a small program awarding a few science lectureships, visiting professorships, and/or senior fellowships to outstanding scientists.

The Science Fellowships Program. This program, administered by the different member countries, awards about 600 NATO science fellowships each year. The program has allowed about 10,000 scientists to study for one year each in a foreign country.

The Advanced Study Institute Program. An ASI is primarily a high-level teaching activity at which a carefully defined subject is treated in considerable depth in a systematic and coherently structured program. About 50 institutes are supported each year.

The Research Grants Program. The main purpose of this program is to stimulate scientific research carried out in collaboration among scientists in the member countries of the Alliance. Grants are renewable for up to three years and 50 to 100 new grants are awarded each year.

The Science Committee Conference Program. The main purpose of these research evaluation conferences is to identify particularly fruitful areas for future research. The recommendations are directed both to those having a responsibility for selecting and supporting research programs and to the Science Committee itself. One or two conferences are held each year.

The Special Science Programs. In addition to the general and more permanent programs listed above, the Science Committee has frequently identified specialized scientific areas as deserving special encouragement or preferential support for limited periods. In 1974 there were special programs on: air-sea interaction; ecosciences, human factors; marine sciences; radiometeorology, stress corrosion cracking; and systems science.

The Science Committee Programs are guided by panels of scientists from the member countries and support is given to all fields of science, with emphasis on fundamental aspects rather than applications. Results from research projects are published in the literature, and scientific proceedings from ASIs and conferences are published in most cases.

I would also like to mention the Committee on the Challenges of Modern Society which, since 1969, has started and coordinated pilot studies in:† disaster assistance; environment and regional planning; road safety; air pollution; inland water pollution; advanced health care; coastal water pollution; advanced waste water treatment; urban transportation; disposal of hazardous substances; solar energy and geothermal energy.

Dr. A. Rannestad

*More information on the NATO science programs may be found in the booklet *Scientific Co-operation in NATO* or the book *NATO and SCIENCE, An Account of the Activities of the NATO Science Committee 1958-72*, NATO, Scientific Affairs Division, B-1110 Brussels.
†More information on the CCMS pilot studies may be found in the booklet *Man's Environment and the Atlantic Alliance*, NATO Information Service, B-1110 Brussels.

Contents

Large-Scale Culturing Systems

Food Chains and Their Use: Mussels, Mollusks, and Bivalves

Food Chains and Their Use: Fish and Fish Fry

Discussion and Index

Waste Heat and Nutrient-Loaded Effluents in the Aquaculture: The Setting of the Problem

Ole Devik

Christian Michelsen Institute
Bergen, Norway

INTRODUCTION

The obvious point of departure for our discussions is the concern about the way we are managing our resources. We can approach this problem from various angles, ranging from the trustful one that "God looks after those that look after themselves," to the downright pessimistic one that foresees a collapse of the human existence in the near future because we are overexploiting our resources in a way that will leave us to choke in our own waste.

We may begin our discussion with the statement that most of our resources are limited, but that the limits are still largely unknown, and that we still can stretch the approach to the limits with a combination of a judicious use of the resources and our own ingenuity.

I would prefer to describe our situation in terms of a simplified flowsheet (see page 2). In the center of the diagram is placed the unit group of the community, commonly the family, but it may also be any small group that fulfills his own needs and wishes. The family group will function when given a certain quantity of resources. The resources will exhibit a great variety in quantity and kind in the various parts of the world.

The flowsheet indicates a distinction between input in the form of energy, subsistence goods (food and clothing), and durable goods. To a certain degree, the distinction between energy and matter is arbitrary, but the distinction is an

1

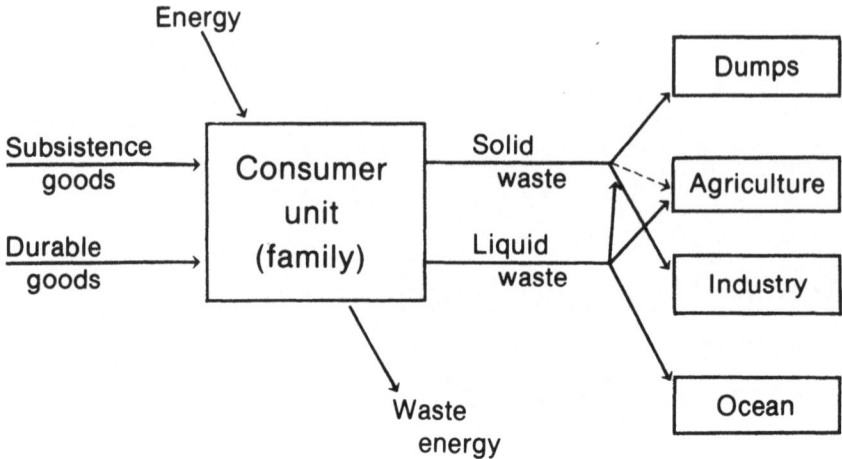

expression of the classification of the resources as to those that are not renewable and those that have to be recycled, whether we want it or not.

THE OBJECTIVE OF THE WORKSHOP

In this meeting the intention is to assess the possibility that we can achieve improvements in the present efficiency of recycling of materials through a judicious use of the photosynthetic process in combination with the heat produced in the course of the electricity generation.

We are dealing with output from the community in the form of liquid and solid waste which may be subdivided into domestic and industrial waste. The waste will consist of a mixture of organic and inorganic material, which will be broken down by microorganisms of various kinds so that the products of metabolism can be assimilated in the growth cycle of animals and plants. We may define a separate pathway for the various elemental components; in the diagram they are symbolized by the arrows designated agriculture and industry.

We may observe that the pathways as indicated are not closed; there is always a certain fraction of each elemental component that is drained off to the ocean. In other words, recirculation implies terrestrial cycles. The ocean may be regarded as a large sink, and once the components are absorbed in this sink, the cost of reclamation will as a rule be excessive.

GUIDELINES FOR THE RECIRCULATION

We can cite several guidelines for the optimum recirculation of the various elemental components.

1. In a recirculation, it is most economical to work in systems in which the concentration of the component is relatively large.
2. Reclamation in terrestrially based systems gives the best efficiency.
3. Restoration of materials to terrestrial use will improve the recycling efficiency.
4. In terms of quantity, the most important components to recycle will be PO_4, fixed N, K, possibly SiO_2.

For the domestic effluents, which have a relatively large concentration of fixed N, PO_4, K, and SiO_2, the most effective recycling schedule will probably include a treatment to remove in the form of sludge a major part of the mineral nutrient components, followed by the use of the runoff water for aquaculture in some form.

AQUACULTURE IN THE RECIRCULATION CONTEXT

The aquaculture will have its main value in the utilization of mineral nutrients in the runoff, which has a concentration of mineral nutrients too low to be reclaimed economically by other methods. From the point of view of the recycling of minerals, aquaculture methods may be ranked according to the fraction of matter returned to the terrestrial use.

The ranking will then reflect the number of links in the food chain. We may then rank the crop utilization in this order: (1) direct food use, (2) use in terrestrial animal husbandry, (3) use in marine animal husbandry or *ranching*, and (4) fish farming with intensive feeding. This ranking does not take into account the economy of the activity, which will be determined by such factors as the acceptability of the food produced, efficiency of the conversion of the feed, or investment costs.

The recent development of aquaculture reflects the influence of these factors. In our part of the world, we have seen that the age-old technique of raising such freshwater fish as carp in fish ponds has been supplemented by fish farming. Here, relatively high priced fish—such as salmon, rainbow trout, catfish, or sole—are produced from feed based on cheap protein, made from fish, or derived from suitable vegetable sources. These methods have been highly successful, and it now seems likely that the production of shrimp in large ponds will be commercially interesting.

The development in England to produce sole in ponds heated by cooling water from power plants utilizing atomic energy is in this context particularly interesting; it demonstrates the importance of additional heat to shorten the growth cycle of the fish.

I understand that the initial experiments also included alternative projects to utilize the primary production as a basis for the production of food animals. The results of these experiments, however, were not encouraging; they indicated that such an approach was not commercially attractive at the time. Since that time, an ever increasing research activity in fields related to aquaculture has provided us with additional data. We can point to investigations into the influence of various factors governing the rate of primary productivity, both in the oceans and in eutrophied areas. We have also made good progress in the mathematical modeling of such systems, and it therefore seems an opportune moment to ask whether these new experiences and research results may be utilized to achieve progress in the efforts to employ culture methods to control the productivity of selected areas.

The utilization of waste heat is of importance, because a moderate rise in the temperature will increase the turnover rate of the metabolic processes and lead to an increase in productivity. Besides, with the large amounts of heat available, the probability is that an increase in the productivity over large areas may make systems commercially attractive, even if under ordinary circumstances they are uneconomic.

The topics of the meeting have been chosen with these possibilities in mind, with the hope that the lectures will serve as starting points for discussions that should help to evaluate where we will meet the problems.

SIMILARITIES BETWEEN AGRICULTURE AND AQUACULTURE

It goes without saying that aquaculture and agriculture employ vastly different techniques, and that the commercially successful aquaculture owes little to the methods and principles of agriculture. All the same, it is worthwhile to ask whether these principles and methods are applicable. At any rate, I myself have found that a formulation of the principles serves as a useful frame of reference.

1. The area or volume under cultivation should be established and delimited physically and legally.
2. Physical and legal control should be established of the area or volume under cultivation.

3. Control over the exchange of matter (living or dead) between the ambient and the volume under cultivation should be effected.
4. Conditioning of the nutrient level by the addition of fertilizer or manure should be employed.
5. Control of species composition by the inoculation or implantation of key species and the weeding out of others should be maintained.

Using these principles, we may identify the various problems we have to solve.

DELIMITATION OF AREA OR VOLUME

The problems of delimitations of the volume under cultivation increase with the physical size and will be strongly influenced by the topography of the shore and the seabed. Grossly speaking, with coastal conditions as in Norway, aquaculture will comprise the exploitation of the euphotic layers, and the need is then to fence off the upper layers. The area for a production unit will be of the same order of magnitude that we find in the country's agriculture. In Norway, for example, the ordinary size farms have areas under cultivation from 100,000 m^2 upward, and we would expect similar areas to be indicated for aquacultural activities based upon the use of the primary production. We may compare this area to that of a standard supertanker.

PHYSICAL AND LEGAL CONTROL

The physical control of the volume will be dependent upon the local conditions and the construction of the fences, and will pose problems similar to the task of delimitation of the volume. I hope we will get an idea of the magnitude of these problems and how they can be tackled during the later discussions.

The legal control of the volume will be difficult as long as there exist only a few legal guidelines that can be applied. The rights of ownership are ill defined—which is only too reasonable, considering that aquaculture is a fairly new activity. During the planning of the program, it was considered whether a discussion of these problems should be included. The feeling was, however, that it is still somewhat premature to enter into a discussion on these aspects; it is better first to obtain more specific information regarding the feasibility of large-scale aquaculture and the specific problems such an activity will create.

CONTROL OF THE EXCHANGE OF MATTER WITH THE AMBIENT

A prerequisite for the aquaculture is to be able to exercise a certain control over the exchange of water between the volume under cultivation and the ambient. The objective should be to add the necessary nutrients at the same rate as they are assimilated and ultimately removed by cropping. We will probably need to know in more detail the interrelationships between the organisms cultivated, the nutrient level, and other factors such as light, temperature, and weather conditions. There will also be a need to monitor the changes, and we may imagine rather complex modeling of the systems, at least in an initial period. The ultimate aim is to eliminate as far as possible the need for such auxiliaries, and to try to get the control routines integrated into the know-how of the operators.

CONDITIONING OF THE NUTRIENT LEVEL

A consequence of the control of the exchange of matter is that we may be able to regulate the nutrient level within certain limits. It is convenient to consider the nutrient level separately, because the requirements will differ according to the organisms we intend to cultivate. We might also expect that selective control of one or more nutrients will facilitate the cultivation of chosen species.

There is obviously a lack of and need for characteristic growth data for the various species of interest, and this state of affairs will always exist, simply because improvement in the knowledge raises the need for supplementary data. To reach technical solutions, we have to seek pertinent information where it is available, and we hope that there is sufficient information to be found so that we can define the proper nutrient condition and point out the key factors for any one species.

As an added complication, we encounter the desire to utilize domestic effluents, which may be expected to contain both mineral nutrients and positive growth factors as well as inhibitors. Industrial effluents should not be considered in the first run, because of the added risk of harmful constituents that may be present, in spite of treatment of the effluent.

For the domestic effluent it seems realistic to assume that it has been through a certain amount of treatment which has removed the majority of the organic constituents, but that it still has a significant concentration of mineral nutrients and more or less known growth factors. We have a profusion of data that can be utilized in this context, but it is pertinent to ask what additional information is needed, and particularly what the consequences are for the public health when domestic effluent is extensively used.

CONTROL OF THE SPECIES COMPOSITION

The selection of species is, of course, dependent upon what kind of crop one intends to raise. At present, bivalve production seems the most imminent; only at a later stage may direct harvesting of the crop be practicable. Here we meet problems of inoculation and preconditioning of the selected species; one may raise the question whether genetic selection procedures are practicable. It is of importance to define the conditions for sufficient inoculum to secure the existence and growth of the desired organisms in a system that is as poorly controlled as those we are considering.

Furthermore, each specific use of the primary production will raise separate problems. In the production of bivalves, for example, a high temperature during the winter will lead to a high maintenance metabolism, and the animals may starve to death. Should we cool the system, or should we add suitable food during the dark months of the year, when the photosynthesis is negligible?

We may also mention the interrelation between grazers and the primary production, and among the grazers themselves. In the case of bivalves on a raft, for example, one may ask whether a heavy population of bivalves will squeeze out other grazers, with less interest for aquaculture.

I find then, that the dominant question is how we can make use of a eutrophied system, either in the aquacultural context, or as a ranching unit. Such a system could at least in principle be made to behave as a normal ecosystem, but with a higher level of productivity. This level may be raised even higher by the application of heat.

We may also expect that to the extent that organic materials sediment out of the system, there will result a gradient in nutrients, which will eventually lead to the accumulation of life proliferating on the system that is created.

These questions are not new; we find rather that they are asked again and again. In the course of time, some drop out when a sufficiently satisfactory answer has been reached, or when the question has been made obsolete by research results, but then new questions are raised.

No doubt, some of these questions will get an answer during our workshop, and additional questions will be raised. My hope is that we can use the deliberations here to define where the main problems are for the moment, and also to define a line of attack.

Controlled Primary Production in Marine Systems: Potentialities of Select Species/Control of Unwanted Grazers

The Technical Production of Microalgae and Its Prospects in Marine Aquaculture

Carl J. Soeder

Gesellschaft für Strahlen- und Umweltforschung mbH, München
Abteilung für Algenforschung und Algentechnologie
Dortmund, Germany

INTRODUCTION

The commercial utilization of natural vegetations of seaweeds (macroalgae) has an ancient tradition and is today the basis of an important industry (Levring *et al.*, 1969; Chapman, 1970; von Witsch, 1970). By contrast, microalgae have in only a few cases been exploited directly (Clement *et al.*, 1967; Clement and Van Landeghem, 1971; Aldave, 1969), since their dispersed standing crops are usually not accessible to economic harvesting techniques. Therefore, industrial production of microalgae is the logical approach toward making use of this group of minute, often single-celled water plants.

Whenever the mass culture of microalgae hit the headlines, freshwater species like *Chlorella* and *Scenedesmus* were concerned, or *Spirulina platensis*, a blue-green alga from warm brackish waters. This publicity probably stems from a futurological appeal of various projects which will eventually lead to the production of the aforementioned algae at the technical scale. Facts are, however, that the commercial utilization of marine microalgae is more widely distributed and established than that of their freshwater counterparts. In general, it will be possible to translate any technology developed for the mass culture of nonmarine microalgae into marine aquaculture.

Technologically we can distinguish two basic types of algae production systems, namely, (a) the clean processes, characterized by growth on defined substrates; and (b) the wastewater processes, characterized by the use of sewage or other liquid media of elevated BOD as the cultural media. In the wastewater processes we are dealing, in fact, with algae–bacteria systems.

From the energetical point of view a culture of algae may rely entirely on photosynthesis (autotrophic growth), on the utilization of organic substrates in the light as combined with true photosynthesis (mixotrophic growth), or on organic substrates alone (heterotrophic growth).

Until now the industrial production of microalgae as such has achieved some commercial establishment only in Japan (Takechi, 1971; Stengel, 1970) and in Mexico (Feldheim, 1973). The existing plants are still relatively small and render comparatively expensive products, the use of which is restricted to special applications, especially in the health food section and, to a very limited degree, in human nutrition. A number of technological as well as toxicological problems remain to be solved, before microalgal products can be expected to have a chance on bigger markets.

The aim of this chapter is to consider at first those systematic groups of microalgae which appear to be suitable for mass production, to describe the production technologies and the nutritive value of microalgae, and to discuss the aspects of microalgae production in the field of marine aquaculture.

TYPES OF CULTIVATED SPECIES OF MICROALGAE

An industrial production of microalgae will usually require a one-species culture system, i.e., an essentially pure culture of a given species. This aim is surprisingly easy to achieve, at least in the case of the cultures we have been working with so far. The greatest biological difficulty exists apparently in the fact that all types of microalgae are susceptible to infections by various types of parasites. These may be protozoa (Canter and Lund, 1968), unicellular fungi (e.g., Fott, 1967; Drebes, 1968; Soeder and Maiweg, 1969), bacteria (Shilo, 1971; Chet et al., 1971; Schnepf et al., 1974) or, in the case of blue-green algae, viruses (e.g., Saffermann and Morris, 1967). While the control of eukaryotic parasites is successful by the application of biocides (e.g., Kraut and Meffert, 1966; Soeder, 1971), the treatment of bacterial or viral infections of algae is still a problem. On the other hand, properly managed algal cultures are quite resistant to infections.

Maximal photosynthetic yields per unit culture surface require an optimal areal density (see below). At the optimal areal density the production is strictly light limited, and the overall growth rate of the algae is much lower than at light saturation of photosynthesis. We have recently compared three algae of very different potential autotrophic growth rate and found similar yields under light-limited conditions. This indicates that a selection of strains for fast growth at light saturation is not necessary (Soeder and Stengel, 1974).

In general, it has to be stated that neither has a thorough screening of the numerous types of microalgae been accomplished nor have the techniques of plant breeding been used to improve the quality or performance of cultivated microalgae, although these tasks are quite promising ones. For example, mutants of *Chlamydomonas* have been obtained which are lacking their normal cell wall and are certainly much more digestible than the wild type (Davies, 1971).

Unicellular green algae like *Chlorella*, which are widely used in laboratory work, have been preferred in most of the early studies on large-scale cultivation of microalgae (Burlew, 1953). Although they are highly productive (e.g., Payer *et al.*, 1973), it has to be emphasized that they are not very suitable for the rearing of larvae of aquatic animals like shrimp or oysters (Bardach, 1968). In some personal observations on the feedings of silver carp (*Hypophtalmichthys nobilis*) we also observed that the fish were able to filter *Chlorella* or *Scenedesmus* from the water but were unable to digest the algae. The rigid cell wall of chlorococcalean algae makes them also undigestible for monogastric mammals and humans, unless the structure of the cell wall is altered by a thermal shock (Kraut *et al.*, 1966) or other effective treatments. This difficulty does not exist in the case of naked chrysomonads such as *Isochrysis galbana* (fed, e.g., to oyster larvae; Rhodes, 1969), and is apparently not met with in diatoms (Sato and Serikawa, 1968) nor in blue-green algae. Of the latter, species of *Anabaena* were successfully fed to silver carp (Panov *et al.*, 1969), and the difference in digestibility between blue-green and green algae was generally reported from milkfish farms (Hickling, 1970).

A special feature of various blue-green algae is their ability to fix molecular nitrogen as amino N (Fogg, 1962). Open-air cultures of *Anabaena* and *Nostoc*, growing on atmospheric nitrogen as the only nitrogen source, gave yields of up to 7 g of algal dry matter per m^2 and day (Florenzano *et al.*, 1964).

An important criterion for the selection of strains suitable for industrial production is cell size, since only the larger-celled forms can be harvested by simple filtration methods. For this reason, the green alga *Coelastrum proboscideum* (von Witsch and Heussler, 1970) which forms porous, spherical colonies seems to be more attractive for mass culture than the smaller representatives of the same group which can only be harvested by more complicated techniques such as centrifugation or flocculation (Golueke and Oswald, 1965).

SOME REMARKS ON FACTORS CONTROLLING THE PRODUCTIVITY OF ALGAL MASS CULTURES

The physiological background data for the mass culture of microalgae are compiled in Burlew (1953), Lewin (1962), and Stewart (1974).

Light is the key factor for open-air production of algae. At suitable temperatures the yield is directly proportional to the daily amount of photosynthetically available radiation (PHAR) per unit culture surface. The theory of this light dependence and the prediction of yields from the light climate of a given site have recently been analyzed by Shelef *et al.* (1972).

It is one of the principles of autotrophic algal production to make optimal use of the incident PHAR. This is accomplished by adjusting optical density of the suspension to the *optimal areal density* in terms of algal biomass (Soeder *et al.*, 1970) or chlorophyll (Talling, 1972) per unit surface. By plotting the average daily yield against areal density, one will find a linear increase at low areal densities, approaching some kind of a saturation level (i.e., the optimal areal density) and a decline of yields, if self-shading (Tamiya *et al.*, 1953) becomes so strong that an increasingly large fraction of the population is no longer photosynthetically active (Nichiporovich, 1967). Since the optimal areal density is a function of PHAR ($kcal/m^2/day$) and light intensity changes during the day, it is only possible to adjust the areal density to an average optimum. The same holds for climates where the light energy input changes from day to day, and under such conditions light-adaptation effects may be a hitherto little understood complication of the situation (Steemann Nielsen *et al.*, 1964).

The *temperature* dependence of light-saturated, autotrophic algal growth displays an exponential increase of yields until the optimum temperature is reached (Tamiya 1957, Soeder *et al.*, 1967, Slobodskoi *et al.*, 1969). Important for the large-scale production of microalgae is the fact that the amplitude of the temperature dependence of yields decreases strongly with increasing optical density of the culture (Hegewald, unpubl., in Soeder, 1971), i.e., the Q_{10} is inversely related to the degree of light limitation. Nevertheless, the performance of dense cultures of thermophilic algae is a function of the daily changes in temperature (Stengel and Soeder, 1975).

Temperature requirements of algae vary over a wide range (e.g., Brock, 1969). Until now, only thermophilic or mesophilic strains have been cultivated at larger scale. The temperature of dense algal suspensions usually exceeds air temperature by several degrees. Thus, peak temperatures in mass cultures studied at Bangkok were on the order of 43°C (Payer *et al.*, 1973). Such transitory temperature extremes are surprisingly well tolerated even by algae of which the upper temperature limit at thermoconstant laboratory conditions is about 33°C. A very interesting question is how the yields of low-temperature algae—e.g., diatoms from the polar seas—compare with the yields of mesophilic strains below 15°C. This deserves an experimental evaluation.

For autotrophic production high *pH values* improve the efficiency of CO_2 utilization (e.g., Kraut and Meffert, 1966, Florenzano and Materassi, 1968). Clement and Van Landeghem (1971) see a special advantage of *Spirulina* in the fact that this blue-green alga grows well at pH 9.5 or more. Optimal pH values for green algae like *Scenedesmus* are definitely lower. Although these organisms are also able to utilize hydrogen carbonate ions as the carbon source, the strain tested by Stengel and Soeder (1975) prefers free carbon dioxide over hydrogencarbonate.

In order to make the best possible use of *water* and *nutrients* it is often desirable to feed the filtrate of the algal culture back into the culture basin after the harvesting of the biomass. This leads to a semicontinuous type of culture management and requires an adjustment of the nutrient proportions in the medium to the proportions of their uptake by the alga (Schultz, 1963). As nitrogen sources, ammonia or urea are cheaper and more suitable than nitrates, since the utilization of the latter leads to a large increase of culture pH.

Many microalgae are capable of utilizing *organic substrates* either in the light or in the dark. Various Japanese patents (e.g., Takechi, 1971) describe techniques for the heterotrophic cultivation of *Chlorella* in fermenters, and acetate is also used in Japan as a cheap carbon source for open-air cultures of green algae (cf. Stengel, 1970). Strains which are stimulated by the addition of sucrose to the medium seem to be fairly rare (Wiedemann, 1970; Zajic and Chiu, 1970).

TYPES OF CULTIVATION UNITS

Most of the work on mass culture of algae has been carried out in flat artificial ponds as the cultivation units. The first Japanese open-air system consisted of circular or oblong shallow basins into which CO_2-containing gas was injected in such a way as to create sufficient turbulence (Tamiya *et al.*, 1953). Turbulence is always required for keeping the algae in a homogeneously suspended state and to avoid a limitation of nutrient transport by diffusion gradients. At very high turbulences one can make use of the flashing-light effect (Fredrickson and Tsuchiya, 1970), which requires the algae to shift from high to low light intensities within about one millisecond. The Czechoslovakian system, an inclined surface with baffles (Prokes and Zahradnik, 1969), is perhaps the only one in which the production may take advantage of the flashing-light effect.

Recent open-air ponds for the cultivation of microalgae are either circular ponds or horizontal channels with mechanical or hydropneumatic generation of turbulence or inclined planes (cf. Stengel, 1970). Circular ponds are usually made of concrete, the stirrer units rotating around the center like the spokes of a cartwheel. The disadvantages of this type are: little or no stirring effect in the center of the pond, large interspaces between the individual ponds of a plant, and technical problems in injection of CO_2-containing gas. Various materials other than concrete (like plastic or thin metal) are suitable for building closed horizontal channels for the cultivation of algae. This type resembles closely the classic oxidation pond. The channels may have the form of a simple raceway-like loop or they may be meandering (Shelef *et al.*, 1972).

At low water depths, as desirable for the autotrophic clean-process units, paddle wheels are a good means for propulsion and turbulence generation. At water depths exceeding 30 cm, pumps can be used to serve the same purpose (Oswald and Golueke, 1968).

The French team developed a hydropneumatic propulsion for *Spirulina*. The rectangular channels are connected at their ends by sinks into which the diaphragm of the pond extends below the level of the channel level but not to the bottom of the sinks. Injection of CO_2-in-air mixture at one side of the first sink and at the other side of the second sink creates a turbulent flow of sufficient velocity according to the airlift principle.

For the cultivation of *Scenedesmus acutus* we consider a continuous flow velocity of 0.5 m/sec to be desirable. Oswald and Golueke (1968) are successfully working at lower current velocities on an intermittent pumping program with high-rate oxidation ponds.

An ingenious alternative to the abovementioned horizontal units has been developed in Czechoslovakia (Prokes and Zahradnik, 1969). The system consists of an inclined plane along which the algal suspension turbulently flows down. At the lower end the liquid is collected by some sort of a gutter and pumped back to the upper rim. Turbulence is created by numerous small baffles which have the function of microweirs, each being about 2 cm high.

YIELDS

In the early days of outdoor mass culture of microalgae, yields were usually on the order of $7g/m^2/day$ for autotrophic growth. Together with technology, the productivity of the cultures has been improved considerably. Some of the recent data on yields are summarized in Table I. An open question is whether open-air mass cultures of blue-green algae are actually somewhat inferior to potent chlorococcalean strains.

The yields of algal cultures growing on sewage are definitely higher than those of autotrophic culture systems. Oswald and Golueke (1968) obtained about 70 annual metric tons per hectare in California with cultures consisting mainly of *Scenedesmus*. A corresponding maximal value reported for Bangkok is about 170 annual metric tons per hectare (McGarry and Tongkasame, 1971). The greater yields of the sewage systems are probably due to a certain degree of mixotrophy. However, it can not be excluded that the algal substance was contaminated to a certain degree by bacterial biomass or detritus.

Table I. Theoretical Yields of Autotrophic Microalgal Cultures Growing under
Open-Air Conditions. Biomass Increase in Grams of Algal Dry Weight

	Maximum daily yield, g/m^2	Average yield for the vegetation period		Location	Authors
		$g/m^2/day$	Metric tons/ha/year		
Scenedesmus	28	10	25	Dortmund	Soeder et al., 1970
acutus (= Sc.	35	13.4	55	Bangkok	Payer et al., 1973
obliquus)	45	25	50	Rupite (Bulgaria)	Vendlova, 1969
Spirulina platensis	18	15	50	Mexico	Clement and van Landeghem, 1971

HARVESTING AND PROCESSING

Chlorella and other microalgae with average diameters below 10 microns are too small to allow filtration at the technical scale. Such minute organisms are, therefore, mostly harvested from the suspension by centrifugation. This makes the recovery of biomass a costly process (Golueke and Oswald, 1965), since suspension density is only on the order of 0.5 g of dry algal mass per liter. The alternatives in the harvesting of Chlorella-like microalgae are ion exchange (Golueke and Oswald, 1970) and froth flotation subsequent to the flocculation (McGarry and Tongkasame, 1971). Although the latter may require regenerative extraction of the flocculant from the algal slurry (McGarry, 1970), this seems to be the cheapest means of harvesting sewage-grown algae.

Larger microalgae are separated from the medium by filtration, which is especially easy in Spirulina (Clement et al., 1967; Clement and Van Landeghem, 1971). As filtration units (microsieves, etc.) are a less expensive and less energy consuming than are centrifuges, microalgae suitable for filtration are certainly of advantage as far as harvesting is concerned. The only industrial plant where harvesting of microalgae is routinely by filtration is the Spirulina factory of Sosa Texcoco in Mexico (Feldheim, 1973).

The raw product, an intensely colored slurry, is usually processed into a dry algal powder. It has been found by Bock and Wünsche (1967) that spray drying is insufficient in the case of green algae, since the rigid cell walls of such forms as Scenedesmus have to be denaturized by more effective means. This is accomplished by roller drying (Kraut et al., 1966, Soeder and Pabst, 1970). After having tested dry algal powders for various applications, we feel now that drying is probably not the most elegant way of making algal products both fully digestible and stable with regard to bacterial degradation. Especially if we consider the utilization of microalgal substances as proteinaceous ingredients for animal nutrition, the end product will usually be pelletized. In such cases we would prefer the direct mixing of the raw algal product with the other ingredients of the feed, thereafter processing the mixture under high pressure and heat by means of an extruder. We expect a perfectly digestible product which has the additional advantage of being more sterile than is the roller-dried product. As yet, however, we have no experience with the extruder or similar systems.

In general, and in agreement with the guidelines of the Protein Advisory Group of the FAO (PAG, 1971), we feel that bacteriological safety should be as great as possible in the processing of algal products.

End products containing the processed biomass of microalgae are intensely pigmented. In the field of human consumption of such products it has been feared that the blue-green or green color might create serious problems in their acceptance by the consumer. Mitsuda et al. (1966) have therefore developed methods for inexpensive pigment extraction. This would, however, add to the production costs and would forbid the utilization of the vitamins which are present in fairly high concentrations in the biomass of algae (Soeder et al., 1970). Moreover, the practical experience shows that the consumer in developing countries does indeed accept foods supplemented with green algae (Feldheim, 1971; Payer et al., 1973).

COMPOSITION AND NUTRITIVE VALUE

Of the many data on general chemical composition of algae (cf. Burlew, 1953; Strickland, 1960; Lewin, 1962; Levring et al., 1969) only a few examples for some microalgae powders are presented in Table II. For nutritional application their high protein content is in the foreground of interest.

Cultivated microalgae are also rich in many vitamins and essential fatty acids (Soeder et al., 1970; Takechi, 1971). The nutritional importance of their trace element contents has not yet been studied in detail.

Looking back on two decades of nutritional tests on microalgal substances in humans and animals (Soeder and Pabst, 1970), we can summarize that at least some microalgae are valuable sources of protein. For instance, the crude protein of strain 276-3a of *Scenedesmus acutus* has a biological value of 90% as compared to the casein standard in rats, a net protein utilization of 3.2 E and a protein efficiency rate of 67 (Kraut et al., 1966). For humans the biological value of the same alga is almost as high or higher, as found for soya crude protein (Kofranyi and Jekat, 1967; Müller-Wecker and Kofranyi, 1973), the digestibility being 78%.

Table II. From Chemical Composition of Two Selected Microalgae Powders
as Compared to the Edible Constituents of Soya Seed
(after Soeder et al., 1970). Values in Percent of Dry Substance

Constituent	Scenedesmus (276-3a)	Spirulina	Soya seed
Water	4–8	10	7–10
Crude protein	50–56	56–62	34–40
Lipids	12–14	2–3	16–20
Carbohydrates	10–17	16–18	19–35
Crude fiber	3–10	?	3–5
Minerals	6–10	4–6	4–5

A toxicological multigeneration test in mice has been carried out, again using
Scenedesmus 276-3a and following guideline No. 6 of the United Nations Protein Advisory Group (Pabst, unpublished). The histopathological data of this
test are still incomplete. Other microalgae have not yet been studied as extensively as *Scenedesmus*, so that a screening of the toxicological aspects of microalgae in nutrition is not yet accomplished. Of great importance, however, are the
mostly positive results of feeding sewage-grown microalgae to various animals
(Hintz and Heitmann, 1967).

MICROALGAE IN MARINE AQUACULTURE

Present Importance

As already indicated, the cultivation of microalgae has already gained some
importance for aquaculture in marine or brackish waters. Some examples shall
be presented to characterize the present situation.

First to be mentioned is the technique of rearing penaeid shrimp as origi-
nated in Japan (Bardach, 1968; Peres, 1968; Idyll, 1969; Westley, 1971). Two
types of microalgae are used in this process, namely diatoms and green algae.
The diatom, *Skeletonema costatum*, is cultivated in large open basins, and the
suspension is fed to *Penaeus* spp. from the second larval stage (zoea) onwards.
In the next stage (mysis) the young shrimp are fed zooplankton consisting
mainly of rotifers and brine shrimp. This zooplankton is produced by means of
special intensive cultures, relying in turn on mass cultures of the green alga,
Chlorella. In the postlarval stage the shrimp are either released to their natural
habitats (shallow coastal areas) or grown to market size on minced seafood
waste. Using the latter technique together with the artificial rearing of the larvae,
one obtains up to 15 metric tons of shrimp per hectare and year (Peres, 1968).

Brine shrimp (*Artemia salina*) is one of the favorite fish larvae feeds. It may therefore be of some commercial interest that the intensive cultivation of *Artemia* on either suspensions of artificially produced phytoplankton or on roller-dried green algae has been found to be quite successful (Sorgeloos, 1973).

Intensive culture of oyster and mussels relies to a large extent on the rearing of the mollusk larvae on microalgae which are produced in mass-culture setups, the dimensions of which are, however, much smaller than in the case of shrimp aquaculture. Various types of algae can be used for the primary steps of seed oyster production: chrysomonads, e.g., *Isochrysis galbana* (Rhodes, 1969); or diatoms, e.g., *Skeletonema costatum* (Westley, 1971), *Nitzschia closterium* (Sato and Serikawa, 1968), and *Chaetoceros simplex* (Kanazawa, 1969). Even the xanthophycean alga, *Monallantus salina*, has been proposed for the same purpose (Berland *et al.*, 1970). The hard clam (*Meretrix lamarckii*) is another mollusk which can be reared on artificially produced phytoplankton (Tanaka, 1969).

Although the production of seed oysters is currently the major purpose of growing microalgae as oyster feed, it is clear that one can indeed produce market-sized mollusks on the basis of microalgae cultivation. This is especially envisaged by Othmer and Roels (1973).

In terms of biomass production, benthic microalgae, especially blue-green, are of great economic relevance for the vast milkfish cultures in Southeast Asia. Hickling (1970) describes in detail what measures are taken to make the micro-algal pasture of *Chanos chanos* and some other fish as productive as possible. Green macroalgae like *Ulva* and *Enteromorpha* are also quasi-cultivated as milk-fish fodder. Their major counterpart in the Philippines is the benthic red alga, *Gracilaria confervoides* (Hickling, 1970).

Trends to the Future

Planktonic algae are the food for the larvae of many species of not exclu-sively carnivorous fish. This fact has been recognized by many authors but, as far as I know, has not yet been commercialized in practical aquaculture. Of special interest are those fish which are able to feed all their lives directly on planktonic algae. These fish are among the limnic species—the bighead and the silver carp—for which blue-green algae (e.g., *Anabaena*) are suitable (Panov *et al.*, 1969), and the brackish-water types of the genus *Tilapia* (Mironova, 1969), which are probably the only fish that can digest *Chlorella* and similar rigid-walled algae. With the technological progress of microalgae cultivation, its direct combination with pisciculture may become quite attractive in the near future. Pelletized food containing microalgae (Soeder *et al.*, 1969) would be a more costly alternative.

Considering the steep increase of prices for food and feed protein during the last years, I see a good chance for the further development of algal mass culture. They are highly productive and can yield proteinaceous matter of very high nutritional quality. There is no principal difficulty in the cultivation of marine microalgae with the same techniques that are nowadays available for the industrial production of limnic phytoplankton species. One will simply have to use more corrosion-resistant equipment. Corresponding to the situation in the mass culture of freshwater algae, the highest yields from artificially managed suspensions of marine microalgae are to be expected in warm climates, especially under conditions where cloudiness is scarce.

For some reasons the industrial autotrophic mass production of marine phytoplankton might be cheaper than in fertilized freshwater. If we grew the algae in some kind of a flow-through system which continuously receives fresh seawater, we would possibly be able to rely to a larger or lesser extent on the natural bicarbonate reserves and would not be bound to an extra carbon dioxide source.

Such a system would probably operate on periodical harvesting, since it requires a certain time until the mineral nutrients are used up to such a degree that the effluent of the algal plant does not create eutrophication nuisance when re-entering the coastal ecosystem. It we are indeed able to rely on the natural bicarbonate of seawater, the generation of a fairly high turbulence we need in our present systems can perhaps be replaced by gentle mixing as an energy-saving item.

The utilization of natural algae-nutrient resources is maximal if we come to some kind of artificial upwelling. Nutrient-rich water is pumped upwards from the bottom of the ocean to the surface, and there it enters the algal cultures. This ingenious system has recently been set to work by Othmer and Roels (1973) in the neighborhood of Puerto Rico, and large algal crops have been obtained. The authors are planning to use the temperature differences between the sea-bottom water and the surface for the operation of thermal pumps, thus generating much more energy than they possibly need for their aquaculture project.

The principal commercial implication of the system by Othmer and Roels (1973) is, however, the idea that the starting of an industrially managed marine food chain by the exploitation of the nutrient reserves in deep-sea water will lead to the production of a superseafood, free from environmental pollutants. Although no data have been published yet on the heavy metal and pesticide content of the mollusks and crustaceans produced in the pilot plant of Othmer and Roels, their expectations will probably be met by the practical results.

The accumulation of toxic metals and/or biocides by aquatic organisms is perhaps the greatest hazard in aquaculture; for example, it is well known that the nasty accumulation of DDT in the marine food chains shows its most conspicuous jump at the level of phytoplankton. If we grow microalgae in industrially polluted water, the product will contain the pollutants at much higher concentration. With respect to our mass cultures of green microalgae, this has been severely criticized by Wagner and Siddiqi (1973), and indeed, algae are potent pollutant collectors. On the other hand, we are able to demonstrate that we can produce the same algae in an unpolluted area in such a way that the final biomass contains neither heavy metals nor polycyclic hydrocarbons at concentrations worth mentioning (Payer et al., 1975). For example, *Scenedesmus* grown in the industrial city of Dortmund contains on the average 35 ppm of lead, whereas the corresponding value from *Chiengmei* in Thailand is below 1 ppm. Judging of microalgae will definitely yield products with acceptable and very low concentrations of toxic metals or other environmental pollutants. However, a regular check for pollutant concentrations is as essential in algae production as in all other branches of aquaculture.

REFERENCES

Aldave Pajares, A. 1969. Algas azul-verdes utilizadas como alimento en la region alto andina Peruana. *Bol. Soc. Bot. Libertad. Trujillo 1* (2): 5-43.

Bardach, J. E. 1968. Aquaculture. *Science 161*.

Berland, B. R., D. J. Bonin, R. A. Daumas, P. L. Labordeo, and S. Y. Maestrini. 1970. Variations du comportement physiologique de l'algue *Monallantus salina* (Xanthophycée) en culture. *Mar. Biol. 7:* 82-92.

Bock, H.-D., and J. Wunsche. 1967. Möglichkeiten zur Verbesserung der Proteinqualität von Grünalgenmehl. *Sitzungsber. Dtsch. Akad. Landw. Wiss. 16*(9): 113-119.

Brock, T. D. 1969. Microbial growth under extreme conditions. *Symp. Soc. Gen. Microbiol. 19:* 15-41.

Burlew, J. S. 1953. Algal culture from laboratory to pilot plant. Carnegie Inst. Wash., Publ. No. 600.

Canter, H. M., and J. W. G. Lund. 1968. The importance of Protozoa controlling the abundance of planktonic algae in lakes. *Proc. Linn. Soc. London 179:* 203-219.

Chapman, V. J. 1970. *Seaweeds and their uses.* Methuen & Co., London.

Chet, I., S. Fogel, and R. Mitchell. 1971. Chemical detection of microbial prey by bacterial predators. *J. Bact. 106:* 863-867.

Clement, G., and H. van Landeghem. 1970. Spirulina, ein günstiges Objekt für die Massenkultur von Mikroalgen. *Ber. Dtsch. Bot. Ges. 83:* 559-566.

Clement, G., C. Giddey, and R. Menzi. 1967. Amino acid composition and nutritive value of the alga *Spirulina maxima. J. Sci. Food Agric. 18:* 497-501.

Davies, D. R. 1971. Single cell protein and the exploitation of mutant algae lacking cell walls. *Nature (London) 233:* 143-144.

Drebes, G. 1968. Lagenisma coscinodisci gen. nov. spec. nov., ein Vertreter der Lagenidiales in der marinen Diatomee *Coscinodiscus. Veröff. Inst. Meeresforsch. Bremerhaven 3:* 67-70.

Feldheim, W. 1972. Untersuchungen uber die Verwendung von Mikroalgen in der menschlichen Ernährung. I. Ernährungsversuch mit algenhaltigen Kostformen in Thailand. *Int. Z. Vitaminforsch. 42:* 6-10.

Feldheim, W. 1973. Personal communication.

Florenzano, W. Balloni and R. Materassi. 1964. Indagini sulla coltura massiva non sterile delle alghe azotofissatrici. *Ann. Microbiol. (Milano) 14:* 115-127.

Fogg, G. E. 1962. Extracellular products. Pages 475-489 *in:* R. A. Lewin, ed. *Physiology and biochemistry of algae.* Academic Press, New York.

Fott, B. 1967. *Phlyctidium scenedesmi.* A new chytrid destroying mass cultures of algae. *Allg. Mikrobiol. 7:* 97-102.

Fredrickson, A. G., and H. M. Tsuchiya. 1970. Utilization of the effect of the intermittent illumination on photosynthetic microorganisms. Pages 519-541 *in: Prediction and measurement of photosynthetic productivity.* Centre of Agricultural Publ. and Doc., Wageningen.

Golueke, C. G., and W. J. Oswald. 1965. Harvesting and processing of sewage-grown algae. *J. Water Pollut. Control Fed. 37:* 471-498.

Golueke, C. G., and W. J. Oswald. 1970. Surface properties and ion exchange in algae removal. *J. Water Pollut. Control Fed. 42:* R304-R314.

Hickling, C. F. 1970. Estuarine fish farming. Pages 119-214 *in:* F. S. Russell and M. Yonge, eds. *Adv. Marine biol.,* Vol. 8. Academic Press, New York.

Hintz, H. F., and H. Heitmann. 1967. Sewage-grown algae as a protein supplement for swine. *Anim. Prod. 9:* 135-140.

Idyll, C. P. 1969. Status of commercial culture of crustaceans. Pages 55-64 *in:* H. W. Youngken, Jr., ed. *Food drugs from the sea: Proc. Symp. Marine Technol. Soc.* Washington, D. C.

Kanazawa, A. 1969. On the vitamin B of a diatom, *Chaetoceros simplex,* as the diet for the larvae of marine animals. *Mem. Fac. Fish. Kakoshima Univ. 18:* 93-97.

Kofranyi, E., and F. Jekat. 1967. Zur Bestimmung der biologischen Wertigkeit von Nahrungsproteinen. XII. Die Mischung von Ei, Mais, Soja, Algen. *Hoppe-Seyler's Z. Physiol. Chem. 348:* 84-88.

Kraut, H., and M.-E. Meffert. 1966. Über unsterile Grosskulturen von *Scenedesmus obliquus. Forschungsber. des Landes Nordrhein-Westfalen* Nr. 1648: 1-61. Westdeutscher Verlag, Köln.

Kraut, H., F. Jekat, and W. Pabst. 1966. Ausnutzungsgrad und biologischer Wert des Proteins der einzelligen Grunalge *Scenedesmus obliquus,* ermittelt im Ratten-Bilanz-Versuch. *Nutr. Dieta 8:* 130-144.

Levring, T., H. A. Hoppe, and C. J. Schmid. 1969. Marine algae—a survey of research and utilization. Walter de Gruyter & Co., Hamburg. 421 pp.

Lewin, R. A. 1962. Physiology and biochemistry of algae. Academic Press, New York.

McGarry, M. G. 1970. Algae flocculation with aluminium sulfate and polyelectrolytes. *J. Water Pollut. Control Fed. 42:* 191-201.

McGarry, M. G., and C. Tongkasame. 1971. Water reclamation and algae harvesting. *J. Water Pollut. Control Fed. 43:* 824-835.

Mironova, N. V. 1969. Comparison of growth of Tilapias *(Tilapia mossambica* Peters), when fed on *Chlorella* and other foodstuffs. *NASA Techn. Trnasl. TTF 529:* 478-484.

Mitsuda, H., K. Yasamoto, and H. Nakumura. 1966. Needs for new protein isolate techniques to utilize *Chlorella* and other unused resources. *Proc. 7th Int. Cong. Nutr. Hamburg 5:* 327-332.

Müller-Wecker, H., and E. Kofranyi. 1973. Zur Bestimmung der biologischen Wertigkeit von Nahrungsproteinen. 18. Einzeller als zusätzliche Nahrungsquelle. *Hoppe-Seyler's Z. Physiol. Chem. 354:* 1034-1042.

Nichiporovich, A. A. 1967. Aims of research and photosynthesis of plants as a factor in productivity. Pages 3-36 *in:* Q. A. Nichiporovich, ed. *Photosynthesis of productive systems.* Isr. Progr. Scient. Transl., Jerusalem.

Oswald, W. J., and C. G. Golueke. 1968. Harvesting and processing of waste-grown microalgae. Pages 371-389 in: D. F. Jackson, ed. Algae, man and the environment. Syracuse University Press, Syracuse.

Othmer, D. F., and O. A. Roels. 1973. Power, freshwater and food from cold, deep sea water. Science (Washington) Nov.

PAG Guideline No. 6. 1971. Guideline for preclinical testing of novel sources of protein. FAO/WHO/UNICEF Protein Advisory Group. United Nations, N. Y.

Panov, D. A., Yu. I. Sorokin, and L. G. Motenkova. 1969. Experimental study of the feeding of bighead and silver carp fry. Vopr. Ikhtiol. 9: 138-152.

Payer, H. D., C. J. Soeder, G. Feldheim, W. Feldheim, U. Gross, and R. Gross. 1973. Dortmunder Algen in Übersee. Umsch. Wiss. Tech. 73 (13): 484-485.

Payer, H. D., K. H. Runkel, H. Kunte, H. Gräf, E. Stengel, H. Mohn, and A. Polsiri, 1975. Die Kontamination von Mikroalgen mit einigen umweltbürtigen Schadstoffen. 1. Symposium Mikrobielle Proteingewinnung. Verlag Chemie, Weinheim, 191-200.

Peres, J. M. 1968. Aquaculture marine. Sci. Vie 86: 100-111.

Prokes, B., and J. Zahradnik, 1969. Outdoor cultures: Development of unit operations. Ann. Rep. Lab. Algol. Trebon 1969: 172-178.

Rhodes, W. W. 1969. Growth of oyster larvae (Crassostrea virginica) at various sizes in different concentrations of the chrysophyte, Isochrysis galbana. Proc. Nat. Shellfish. Assoc. 60: 10.

Saffermann, R. S., and M. E. Morris. 1967. Observations on the occurrence, distribution, and seasonal incidence of blue-green algal viruses. Appl. Microbiol. 15: 1219-1222.

Sato, T., and M. Serikawa. 1968. Mass culture of a marine diatom, Nitzschia closterium. Bull. Plankton Soc. Jap. 15: 13-16.

Schnepf, E., E. Hegewald, and C. J. Soeder. 1974. Elektronenmikroskopische Beobachtungen an Parasiten aus Scenedesmus-Massenkulturen. 4. Bakterien. Arch. Mikrobiol. 98: 133-145.

Schultz, G. 1963. Über eine zweckmässige Steuerung der Mineralstoffversorgung von Algengrosskulturen im Freiland. Z. Naturforsch. 18b: 946-950.

Shelef, G., M. Schwarz, and H. Schechter. 1972. Prediction of photosynthetic biomass production in accelerated algal-bacterial wastewater treatment systems. 6th Int. Water Poll. Res., Jerusalem. Pergamon Press, New York.

Shilo, M. 1971. Biological agents which cause lysis of blue-green algae. Mitt. Int. Ver. Theor. Angew. Limnol. 19: 206-231.

Slobodskoi, L. I., F. Ya. Sidko, V. I. Belyanin, V. F. Alypov, and G. F. Beresnev. 1969. Analytical expression of the effect of temperature on microalgae productivity. Biofizika 14: 196-199.

Soeder, C. J. 1971. Mikroalgenkultur im technischen Massstab. Biologie in unsere Zeit 133: 142.

Soeder, C. J., E. Hegewald, W. Pabst, H. D. Payer, I. Rolle, and E. Stengel. 1970. Zwanzig Jahre angewandte Mikroalgenforschung in Nordrhein-Westfalen. Jahrb. Landesamt fur Forschung, Nordrhein-Westfalen, 1-34.

Soeder, C. J., and D. Maiweg. 1969. Einfluss pilzlicher Parasiten auf unsterile Massenkulturen von Scenedesmus. Arch. Hydrobiol. 66: 48-55.

Soeder, C. J., and W. Pabst. 1970. Gesichtspunkte für die Verwendung von Mikroalgen in der Ernährung von Mensch und Tier. Ber. Dtsch. Bot. Ges. 83 (11): 607-625.

Soeder, C. J., G. Schultze, and G. Thiele. 1967. Einfluss verschiedener Kulturbedingungen auf das Wachstum in Synchronkulturen von Chlorella fusca Shihira et Krauss. Arch. Hydrobiol. (Suppl). 23 (Falkau-Arbeiten VI): 127-171.

Soeder, C. J., and E. Stengel. 1974. Action of external factors on growth and metabolism of algae. Pages 714-740 in: W. D. P. Stewart, ed. Algal physiology and biochemistry. Blackwell Publ., London.

Soeder, C. J., A. Strotmann, and E. Stengel. 1969. Aufzucht von Karpfen mit Grünalgen als Eiweissquelle. Umsch. Wiss. Tech. 11: 342.

Sorgeloos, P. 1973. High density culturing of the brine shrimp, *Artemia salina* L. *Aquaculture 1:* 385-391.

Stengel, E. 1970. Die Massenproduktion von Mikroalgen–Kulturverfahren und technische Anlagen. *Ber. Dtsch. Bot. Ges. 83*(11): 589-606.

Stengel, E., and C. J. Soeder, 1975. Control of photosynthetic production in aquatic ecosystems. Pages 645-660 *in:* J. F. Cooper (ed.), *Photosynthesis and Productivity in Different Environments.* Cambridge University Press, Cambridge.

Steemann Nielsen, E., V. K. Hansen, and E. Joergensen. 1964. The adaptation to different light intensities in *Chlorella vulgaris* and the time dependence or transfer to a new light intensity. *Physiol. Plant. 17:* 505-517.

Stewart, W. D. P., ed, 1974. *Algal physiology and biochemistry.* Blackwell Publ., London.

Strickland, J. D. H., 1960. Measuring the production of marine phytoplankton. *Bull. Fish. Res. Bd. Canada 122:* 1-172.

Takechi, Y. 1971. *Chlorella:* Its fundamentals and application. Gakken Co., Tokyo.

Talling, J. F. 1972. Generalized and specialized features of phytoplankton as a form of photosynthetic cover. *Proc. of the IBP/UNESCO Symp.* 6-12 May. Kazimerz Dolny, Poland.

Tamiya, H. 1957. Mass culture of algae. *Ann. Rev. Plant Physiol. 8:* 309-334.

Tamiya, H., E. Hase, K. Shibata, A. Mituya, T. Iwamura, T. Nihei, and T. Sasa. 1953. Kinetics of growth of *Chlorella.* Pages 204-234 *in:* J. S. Burlew, ed. *Algal culture.* Carnegie Inst. Wash. Publ. No. 600.

Tanaka, Y. 1969. Studies on propagation of a hard clam, *Merethrix lamarckii.* I. Artificial breeding. *Bull. Tokai Reg. Fish. Res. Lab. 58:* 163-168.

Vendlova, J. 1969. Outdoor cultivation in Bulgaria. *Ann. Rep. Algol. Lab. Trebon 1968:* 143-152.

Wagner, K.-H., and I. Siddiqi. 1973. Die toxischen Inhaltsstoffe der Mikroalge *Scenedesmus obliquus. Naturwissenschaften 60:* 109-110.

Westley, R. E. 1971. Observations on oyster shrimp culture in southern Japan. *Proc. Nat. Shellfish. Assoc. 61:* 14.

Wiedemann, V. E. 1970. Heterotrophic growth of algae. Pages 107-114 *in:* J. E. Zajic, ed. *Properties and products of algae.* Plenum Press, New York.

von Witsch, H. 1970. Mikro- und Makroalgen als Nahrungsmittel. *Ber. Dtsch. Bot. Ges. 83* (11): 519-526.

von Witsch, H., and P. Heussler. 1970. Erste Beobachtungen über die Eignung von *Coelastrum proboscideum* Bohlin zur Massenkultur. *Ber. Dtsch. Bot. Ges. 83* (11): 579-588.

Zajic, J. E., and Y. S. Chiu. 1970. Heterotrophic growth of algae. Pages 1-47 *in:* J. E. Zajic, ed. *Properties and products of algae.* Plenum Press, New York.

DISCUSSION

WALNE: When you start the *Scenedesmus* cultures in Dortmund, what proportion of failures do you get, and does the culture always come up to a thick culture? Do you have to use a very rich inoculum to do this?

SOEDER: We are growing the algae in a semicontinuous way. When we reach the appropriate density of the culture, we harvest every two days about one-half of the biomass. We have difficulties in growing thick suspensions in springtime, and under cloudy weather conditions with low light levels there is a certain degree of unreliability. In the seed culture this can be overcome by artificial light, and once we have the proper density, we have no problems.

KORRINGA: I have seen pictures of algal cultures from Siberia in round tanks similar to sewage purification tanks. The algae, I think were *Chlorella* and I was told that the algae were used for cattle.

SOEDER: Yes, the cattle drink the suspension and because of their intestinal rumen they can digest the algae to make good use of the nutrient content. I expect also that the bacteria present in the rumen are greatly stimulated by the algae. At present we are just working on stimulation of heterotrophic microorganisms by the organic trace materials of the algae.

CHRISTENSEN: Is waste heat being used in the experiments at Dortmund?

SOEDER: We have not actually used waste heat, but we have heated the ponds at times. There is equipment which enables us to heat the ponds, but in our climate this is probably not very useful. When it is cold at Dortmund, there is dull weather, and with insufficient light the production is low. As an example of heating I might mention the commercial mass culture plants in Bulgaria, where they are using volcanic heat to keep the algae around the optimum temperature of about 30°C.

I would also like to make the comment that it is possible that the marine phytoplankton from the northern seas and southern seas are able to photosynthesize at high rates at low temperatures, and this is something that should be looked at. Production figures on sea ice are impressive. Biomasses up to 1.5 kg/m^2 (dry weight) have been reported.

RAYMONT: Do you shut down the plant during winter with very low light intensity?

SOEDER: Yes.

RAYMONT: And there are no difficulties in starting the inoculation again?

SOEDER: No, it is possible to start large plants from conserved living materials, which can be stored at low temperature for half a year or longer.

KNUTSEN: How large is the dry weight loss during the night?

SOEDER: Nightly losses at Dortmund are on the order of 10%, and probably much larger in the tropics.

PERSOONE: Did you look for any special thermophilic strain or psychrophilic strains?

SOEDER: At first we were very much concerned about temperature, because we knew that *Scenedesmus* at a constant temperature has an optimum of about 32°C, while at 34°C it is killed. And we knew that in the tropics temperatures would exceed this upper limit. It was found by Dr. H. D. Payer that mesophilic strains are able to withstand the tropical temperatures for the reason that extreme temperatures do not last all the day as they do in regular lab cultures. Under natural conditions the enormous temperature stress occurs in the afternoon, and then the algae cool off during the night and relax.

FORSBERG: You mentioned the problem of parasites on the algae. Are the problems different in different climatic areas?

SOEDER: In Dortmund with drastic changes from fair weather to dull and vice versa, we indeed have problems. In cultures in Peru no parasites have shown up for one and a half years. W. J. Oswald told me that in California he grew *Scenedesmus* cultures on sewage for seven years without any problems.

Nitrogen and Phosphorus as Algal Growth-Limiting Nutrients in Waste-Receiving Waters

Curt Forsberg

Institute of Physiological Botany
University of Uppsala
Uppsala, Sweden

The utilization of waste heat and waste nutrients for primary production in eutrophied water systems may help feed the world population. However, many biological and technical problems require solution before a controlled aquaculture can be successfully handled.

The composition of sewage effluent from modern communities shows great variations. To illustrate this, Swedish data for suspended solids, COD, total-N, and total-P are listed which represent no treatment, biological treatment, and biological and chemical treatment.

The concept of limiting factors is briefly discussed. In many waste-affected waters nitrogen was the algal growth-limiting nutrient, especially when the degree of pollution was very high. In less polluted waters phosphorus played the corresponding role. In many of these waters, however, some factor other than N or P (light?, carbon?) was the primary growth regulator. From the results available it seems that waste-receiving lakes must have a comparatively low nutrient level if P or N are to act as the primary growth-limiting factors. A nutrient-controlled aquaculture should probably be most successful if phosphorus is relied upon as the principal growth-limiting factor at a nutrient level as high as possible. Among other things, waste heat and frequency of harvesting will influence the

Excessive supply of plant nutrients coming from the modern community has accelerated the eutrophication processes in slow-flowing rivers, lakes, reservoirs, and many coastal waters. This man-made eutrophication contrasts to the natural phenomenon which generally develops at a very slow rate. In order to maintain a desirable water quality, people in developed countries now are financing advanced waste water treatment to limit the nutrient flux to receiving waters. In parallel with efforts to reduce the outflow of nutrients, and thereby diminish the productivity, others are discussing methods to encourage biological productivity to help feed the world population. Fertilizing major areas of natural waters in order to increase productivity as a food resource will probably provide too low a yield for the cost.

INTEGRATED AQUACULTURE

The characteristics of the pollution that give rise to the undesired man-made eutrophication are the same as those needed to increase the productivity of waters and to augment the food supply. To a certain degree it might be possible to solve the two problems by combining efforts instead of attempting separate solutions. Another waste that affects aquatic productivity is cooling waters. Because hot water effluent can give rise to a marked increase in production (Backiel, 1971), receiving waters for both waste mineral nutrients and waste heat are of greatest interest as potential *food factories*. The utilization of waste heat and waste nutrients in eutrophied water systems can be looked upon as an integrated aquaculture.

Many biological and technical problems must be solved if the primary production in receiving waters is to be utilized for human consumption. A primary problem will be the determination of nutrient levels required for control of eutrophication. This problem involves the concept of growth-limiting factors and methods to reflect the nutrient status of receiving waters. This paper surveys the two key elements in eutrophication: nitrogen and phosphorus as growth-limiting nutrients in waste-receiving water and in effluent from waste water treatment plants.

Nitrogen and Phosphorus in Sewage Effluent

The composition of the sewage effluent from modern communities shows great variation from place to place. To illustrate the ranges for the central parameters, nitrogen and phosphorus values are given for representative Swedish effluents (Table I). Higher values for total-N and total-P in untreated sewage are given by other sources, ranging between 25 to 45 mg/liter for N and from 8 to

Table I. Characteristics of Sewage Effluent in Sweden. Residual Concentrations, milligrams per liter*

Treatment method	Suspended solids	COD	Total N	Total P	Approximate ratio, N:P
No treatment	100–200	200–400	15–25	4–8	3:1
Biological treatment	15–30	40–70	10–20	3–6	4:1
Biological and chemical treatment	5–15	20–50	8–18	0.3–0.8	25:1

*National Swedish Environment Protection Board, Water Protection Division, Treatment Section. Unpublished data.

15 mg/liter for P (OECD report; Rohlich and Uttormark, 1972). Within the Scandinavian countries, in spite of equivalent levels of economic development and similar pollution problems, quite different levels of nutrients in sewage have been noted (Lønholdt, 1973). Most striking is the phosphorus content of Danish sewage with values twice the corresponding ones in Finland and Sweden. High content of detergent phosphate seemed to be the explanation for that difference.

The ratio total N:total P in Danish sewage was 2.4:1, while the corresponding figure in Finland and Sweden was 4:1. Ratios of the same order of difference were also obtained in sewage effluent with biological treatment. After chemical treatment (phosphorus removal), the analyses of effluent gave an approximate ratio of 25:1. The effluent from the treatment plant at Åkeshov-Nockeby, Stockholm, operating with precipitation (Al-sulfate) and serving 230,000 people, gave an average N:P for 1972 of 30:1 (Forsberg and Hökervall, 1973). Since chemical sewage treatment plants have been operating for only a few years, most effluents still have high contents of both nitrogen and phosphorus in a ratio of about 4:1. Compared with the need of these nutrients for phytoplankton growth, this type of effluent contains an excess of phosphorus in relation to nitrogen.

Nitrogen and Phosphorus as Growth-Limiting Nutrients

The concept of growth-limiting factors in relation to the productivity of aquatic ecosystems has been treated and discussed several times from different points of view (see, e.g., Bringman and Kühn, 1965; Edmondson, 1970; Fuchs et al., 1972; Gerloff and Skoog, 1957; Goldman, 1965, 1972a, 1972 b; Lee, 1973; Lund, 1967; Massey and Robinson, 1971; Polisini et al., 1970; Samsel et al., 1972; Schindler, 1971; Sykes, 1973; The control of eutrophication, 1970). The concept of a limiting factor can be useful when total algal growth in culture is discussed, but because mechanisms of multiple responses exist in nature it should not be applied too rigidly when the conditions for natural phytoplankton populations are considered. Many different investigations performed indicate, however, that some nutrients play key roles in limiting the aquatic growth. Thus trace elements deficiencies were found more likely to occur in oligotrophic than in eutrophic waters (Goldman, 1972). Among macronutrients the general order of deficiency often stated is $P > N > C$. For the two most central nutrients, P and N, attempts have been presented to establish the critical levels in natural waters (Vollenweider, 1968). An OECD project for monitoring of inland waters, extending from 1973 to 1977, will give further data for this type of analysis.

In receiving waters having a comparatively high nutrient level, nitrogen often plays the primary role as a growth-limiting nutrient. To briefly illustrate this, results from marine, brackish, and fresh water recipients are listed in Table II. In heavily polluted waters—e.g., the New York bight, the Helsinki and Stockholm archipelago, and many enriched lakes—nitrogen was the primary limiting nutrient. In Frierfjord, where phosphorus limited the algal growth, special conditions caused by waste water from a fertilizer factory prevailed (Källqvist, 1972). The available results indicate that N-limited growth was noted in the heavily polluted waters, while P-limited growth or more equal effects of the two nutrients were observed in the less polluted ones. This is also illustrated by results that show a decreasing effect of N-enrichment in samples from stations further from the pol-

Table II. Algal Growth-Limiting Nutrients in Waters Receiving Sewage Effluent

Water area	Type of water	Limiting nutrient(s)	Reference
New York bight	Marine	N	Ryther and Dunstan, 1971
Oslofjord	Marine	N (P)	Skulberg, 1970
Trondheimsfjord	Marine	N	Sakshaug et al., 1972
Trondheimsfjord	Brackish	P	Sakshaug et al., 1972
Frierfjord	Brackish	P	Källqvist, 1972
Helsinki Archipelago	Brackish	N	Redogörelse, 1968
Stockholm Archipelago	Brackish	N (TE)	Lindahl and Melin, 1973
Lakes, West Berlin	Fresh	N (P)	Bringman and Kühn, 1965
Lake Erie, W. Basin	Fresh	N (P, TE)	Lange, 1971
Lake Superior	Fresh	P	Schelske et al., 1972
Lake Michigan	Fresh	P	Schelske and Stoermer, 1972
Virginia ponds	Fresh	N	Samsel et al., 1972
Shagava Lake	Fresh	P (N)	Powers et al., 1972
Lake Norrviken	Fresh	N, P	Ahlgren et al., 1973
20 recipient lakes	Fresh	N, P	Forsberg and Claesson, 1974

N = nitrogen.
P = phosphorus.
TE = trace elements.

lution source (Ryther and Dunstan, 1971; Redogörelse för undersökningar, 1968).

Nitrogen and Phosphorus in Swedish Waste-Affected Lakes

The role of N and P as growth-limiting nutrients in relation to the degree of pollution can be illustrated more in detail by results from research on the recovery of polluted lakes sponsored by the National Swedish Environment Protection Board (Forsberg, 1973; Forsberg and Claesson, 1974; Forsberg et al., 1972; Forsberg and Ryding, 1973). The water quality of these lakes can be illustrated with data for chlorophyll A and suspended solids, parameters which are also pertinent to study of aquaculture problems. Average values for surface water (0-2 m) from August 1972 are presented in Figure 1, which shows the correla-

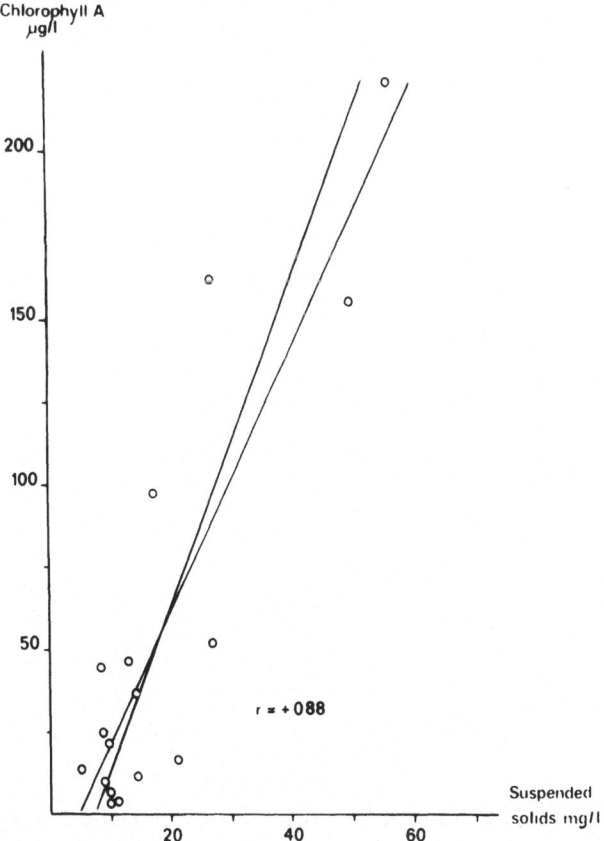

Fig. 1. The correlation between the chlorophyll A and the suspended solids. Average values for surface water, 0-2 m. August, 1972. 17 recipient lakes. Total number of samples = 334.

tion between chlorophyll A and suspended solids. The average water temperature in these lakes during August 1972 was 17°C. The total number of lake
water samples from this period was 334.

The role of N and P as growth-limiting nutrients in these lakes is being studied by using an algal assay procedure essentially the same as that of the bottle
test method (see, e.g., Maloney *et al.*, 1972). Enrichment experiments performed
with filtered lake water indicated that N limited growth of *Selenastrum capri-
cornutum* in two-thirds of the samples. In most samples the available nutrients
gave a marked algal growth without addition of extra nutrients. To obtain a
biological measure, a *green factor*, corresponding approximately to the total
amount of assimilable nutrients in lake water, the algal assay (*Selenastrum*)
chlorophyll was added to the initial lake water chlorophyll content. In Figure
2 the sum of these two fractions has been correlated to total-P. The values are

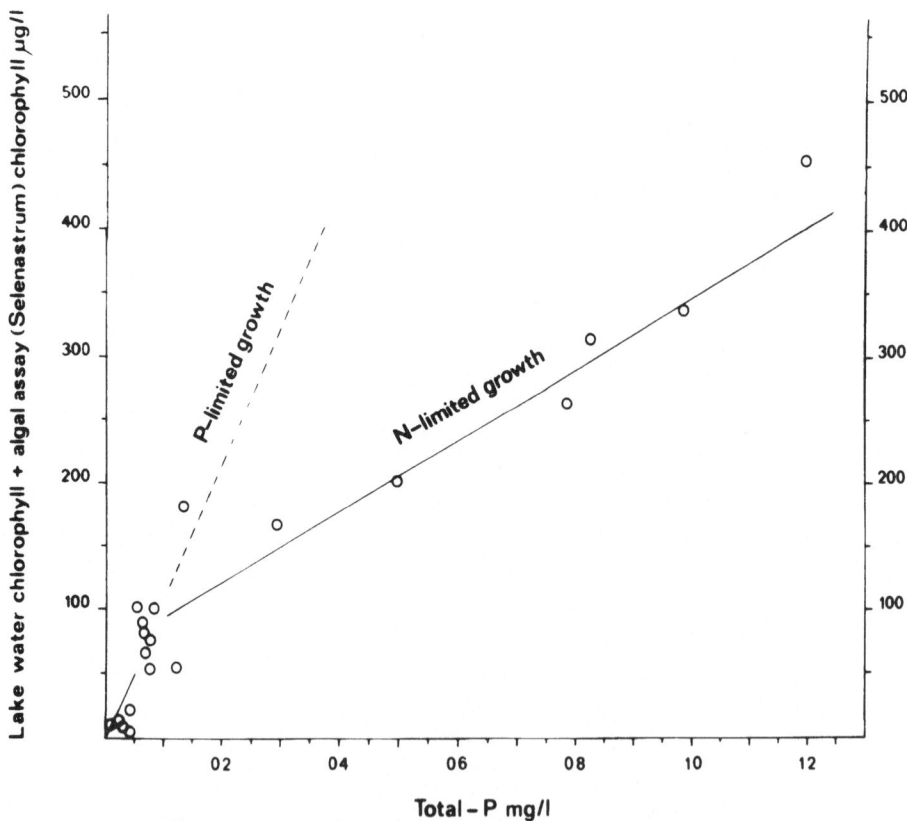

Fig. 2. The correlation between the lake water chlorophyll + the algal assay (*Selenastrum*)
chlorophyll and total-P. Average values for surface water, 0–2 m. August–October, 1972.
20 recipient lakes. Total number of samples = 896.

averages for surface water (0–2 m) from 20 waste-affected lakes. A total of 896 samples are represented during August-October, 1972. At total-P values less than 0.05 mg/liter, phosphorus was the growth-limiting nutrient. Above 0.1 mg P/liter nitrogen played the principal role. Between these values the growth was primarily limited by phosphorus or nitrogen or chelating agents.

In many waste-affected lakes, unenriched algal assays with filtered water gave chlorophyll values exceeding the natural levels by 50–75%. This indicates that in these lakes other factors than N or P (light?, carbon?) limited the algal growth.

DISCUSSION

In aquaculture, the determination of the required nutrient level will depend upon the type of production desired. In shallow waters, for instance, a high nutrient level would be expected to favor the growth of phytoplankton, while lower levels would likely result in a higher yield of vascular plants (Mulligan and Baranowski, 1969). In many waters the nutrient levels are very high, associated with a correspondingly high phytoplankton production. The approximate range of production in polluted lakes should range from 350 to 700 g C/m^2/year (Rodhe, 1969). For very polluted lakes a range of 400 to 1200 g C/m^2/year has been suggested (Gargas, 1972). Production levels of that magnitude represent a quantity of phytoplankton that is economically harvestable provided a suitable technique for algal removal (Parker et al., 1973) is available.

The production levels mentioned, however, are a consequence of eutrophication *running wild*, causing the well-known problems of productivity in the wrong place at the wrong time. In aquaculture a primary aim is to have a controlled eutrophication, which means that knowledge of the principal limiting factors will be of central importance to the problem.

In spite of the development of a large biomass, the water of many lakes still contains a lot of nutrients available for further algal growth. This can be exemplified by data from S. Bergundasjön, a very polluted lake belonging to the above-mentioned research program (Forsberg and Claesson, 1974; Forsberg and Ryding, 1973). In this lake with a production on the order of 650 g C/m^2/day (Bengtsson), ammonium, nitrate, and phosphate were always present in the water during the high productive periods. Thus when the monthly average for chlorophyll A during August 1972 reached 221 μg/liter, the corresponding average for NH_4-N, NO_3-N, and PO_4-P in filtered water was 0.78, 0.29, and 0.82 mg/l, respectively. Algal assay with this filtered water produced a large biomass and enrichment experiments indicated, as could be expected, that for further growth an added nitrogen supply was needed. In this type of lake the classic case of the nutrients P and N as the primary limiting factors did not prevail. Here, light or carbon or some other factor was the primary growth regulator. From the results available it seems that in order for P or N to act as primary growth-limiting factors, lakes must contain relatively low levels of those nutrients.

In aquaculture, as with water quality protection, it would be more favorable if P is the growth-limiting factor rather than N when attempting to control excessive fertilization. The latter is often derived primarily from natural sources that are more difficult to control than the P sources which are often more closely related to man's activity. In addition, many microorganisms possess the ability of N_2-fixation, which means that a nitrogen-controlled production will be difficult to achieve. Therefore it seems that in aquaculture a nutrient-controlled productivity may be best achieved with a P-limited growth at a nutrient level as high as possible. To attain a nutrient level necessary for an economic harvesting, a good understanding of the relationship between nutrients and algal growth as well as between productivity and harvestable crop must be developed. This development will require a combined chemical and biological approach. Bioassay techniques (see, e.g., Fitzgerald, 1968; Fitzgerald and Nelson, 1966; Fuchs *et al.*, 1972; Goldman, 1965; Jensen, 1972; Maloney *et al.*, 1972) can be expected to help with this by increasing the accuracy of predictions about the effects of different management programs.

In integrated aquaculture, waste heat will increase the production. But increasing temperature can, among other things, influence oxygen production, the size of the algal cells, and the species composition (see, e.g., Cairns, 1971; Eriksson and Forsberg, 1971; Margalef, 1954). Having this and other technical and economic problems in mind, the concept of limiting factors discussed above will be only a part of the problems necessary to study before a balanced and controlled integrated aquaculture can be successfully handled.

REFERENCES

Ahlgren, I., G. Alhgren, and B. Ulén. 1973. Limnologiska undersökningar i Norrviken, Edssjön och Oxundasjön. Miljöfaktorer och fytoplankton. Limnologiska institutionen, Uppsala. Meddel. nr. 12.

Backiel, T. 1971. Fishery-biological investigation of artificially heated lakes in Poland. *Proc. Fifth Br. Coarse Fish. Conf.*, Liverpool University, March 31–April 2.

Bengtsson, L. Institute of limnology, University of Lund. Personal communication.

Bringman, G., and R. Kühn. 1965. Nitrat oder Phosphat als Begrenzungsfaktor des Algenwachstums. *Gesund. Ing. 86:* 210–214.

Cairns, J., Jr. 1971. Thermal pollution– a cause for concern. *JWPCF 43:* 55–66.

Edmondson, W. T. 1970. Phosphorus, nitrogen, and algae in Lake Washington after diversion of sewage. *Science 169:* 690–691.

Eriksson, G., and C. Forsberg. 1971. Varmvattenutsläpp och alger. *Vatten 27:* 441-448.

Fitzgerald, G. P. 1968. Detection of limiting or surplus nitrogen in algae and aquatic weeds. *J. Phycol. 4:* 121-126.

Fitzgerald, G. P., and T. C. Nelson. 1966. Extractive and enzymatic analyses for limiting or surplus phosphorus in algae. *J. Phycol. 2:* 32-37.

Forsberg, C. 1973. Naturvårdsverkets RR-undersökning. 1. Naturvårdsverkets undersökning over återhämtningsförloppet hos sjöar som avlastas från föroreningar– RR-(Reningsverk-Recipient)-under-sökningen. Målsättning, organisation och omfattning. *Vatten 29:* 456-463.

Forsberg, C., and A. Claesson. 1974. Naturvårdsverkets RR-undersökning. 3. Algtest med vatten från RR-undersökningens sjoar, augusti-oktober, 1972. *Vatten 30:* 84-95.

Forsberg, C., and E. Hökervall. 1973. AGP-test av kommunalt avloppsvatten. III. Stockholms Kemiska reningsverk, 1972. *Vatten 29:* 281-289.

Forsberg, C., and S. O. Ryding. 1973. Naturvårdsverkets RR-undersökning. 2. Ytvattenkvaliteten RR-undersokningens sjoar, augusti-oktober, 1972. *Vatten 29:* 464-468.

Forsberg, C., B. Hawerman, and L. Ulmgren. 1972. A programme for studies of the recovery of polluted lakes. The effect of chemical sewage treatment and diversion of sewage effluent. *Vatten 28:* 156-161.

Fuchs, W. G., S. D. Demmerle, E. Canelli, and M. Chen. 1972. Characterization of phosphorus-limited plankton algae (with reflections on the limiting-nutrient concept). Nutrients and eutrophication. *Am. Soc. Limnol. Oceanogr. Spec. Symp. 1:* 113-133.

Gargas, E. 1972. Primaerproduktion. *Stads- og havneingeniøren 9:* 178-187.

Gerloff, G. C., and F. Skoog. 1957. Nitrogen as a limiting factor for the growth of *Microcystis aeruginosa* in southern Wisconsin lakes. *Ecology 38:* 556-561.

Goldman, C. R., 1965. Micronutrient limiting factors and their detection in natural phytoplankton populations. *Mem Ist. Ital. Idrobiol. 18* (Suppl.): 121-135.

Goldman, C. R. 1972. The role of minor nutrients in limiting the productivity of aquatic ecosystems. Nutrients and eutrophication. *Amer. Soc. Limnol. Oceanogr. Spec. Symp. 1:* 21-33.

Goldman, J. C., D. B. Porcella, E. J. Middlebrooks, and D. F. Toerien. 1972. The effect of carbon on algal growth– its relationship to eutrophication. Review paper. *Water Res. 6:* 637-679.

Jensen, A. 1972. Application of dialysis culture techniques to studies of water quality. Pages 41-45 *in: Algal assays in water pollution research.* Proceedings from a Nordic symposium, Oslo, October 25-26. Nordforsk.

Källqvist, T. 1972. Use of algal assay for investigating a brackish water area. Pages 111-123 *in: Algal assays in water pollution research.* Proceedings from a Nordic symposium, Oslo, October 25-26. Nordforsk.

Lange, W. 1971. Limiting nutrient elements in filtered Lake Erie Water. Water Res. 5: 1031-1048.

Lee, G. F. 1973. Role of phosphorus in eutrophication and diffuse source control. *Water Res. 7:* 111-128.

Lindahl, P. E. B., and R. Melin. 1973. Algal assays of archipelago waters. Quantitative aspects. *Oikos 24:* 171-178.

Lønholdt, J. 1973. Råspildevands indhold af BI_5, N og P. *Stads- og havneingeniøren 7:* 1-7.

Lund, J. W. G. 1967. Planktonic algae and the ecology of lakes. *Sci. Prog. Oxf. 55:* 401-419.

Maloney, T., W. E. Miller, and T. Shiroyama. 1972. Algal responses to nutrient additions in natural waters. I. Laboratory assays. Nutrients and eutrophication. *Am. Soc. Limnol. Oceanogr. Spec. Symp. 1:* 134-140.

Margalef, R. 1954. Modifications induced by different temperatures on the cells of *Scenedesmus obliquus (Chlorophyceae) Hydrobiologia 6:* 83–94.

Massey, A., and J. Robinson. November 1971. A review of the factors limiting the growth of nuisance algae. Water and Sewage Works. Pp. 352–355.

Mulligan, H. F., and A. Baranowski. 1969. Growth of phytoplankton and vascular aquatic plants at different nutrient levels. *Verh. Int. Ver. Limnol. 17:* 802–810.

National Swedish Environment Protection Board, Water Protection Division, Treatment Section. Unpublished data.

OECD: Report of the expert group on treatment processes. NR/ENV/72.24.

Parker, D. S., J. B. Tyler, and T. J. Dosh. 1973. Algae removal improves pond effluent. *Water and Wastes Engineering.* Pp. 26–29, January.

Polisini, J. M., C. E. Boyd, and B. Didgeon. 1970. Nutrient limiting factors in an oligotrophic South Carolina pond. *Oikos 21:* 344–347.

Powers, C. F., D. W. Schultz, K. W. Malueg, R. M. Brice, and M. D. Schuldt. 1972. Algal responses to nutrient additions in natural waters. II. Field experiments. Nutrients and eutrophication. *Am. Soc. Limnol. Oceanogr. Spec. Symp. 1:* 141–156.

Redogörelse för undersökningar av Helsingfors och Esbo havsområden. 1968. Helsinki City Municipal Services Department, Construction Division, Water Protection Laboratory. Mimeographed.

Rodhe, W. 1969. Crystallization of eutrophication concepts in Northen Europe. Pages 50–64 in: *Eutrophication: Causes, consequences, correctives.* Proceedings of a symposium. Washington.

Rohlich, G. A., and P. D. Uttormark. 1972. Waste water treatment and eutrophication. *In:* Nutrients and Eutrophication. *Limnol. Oceanogr. Spec. Symp. 1:* 231–245.

Ryther, J. H., and W. M. Dunstan. 1971. Nitrogen, phosphorus, and eutrophication in the coastal marine environment. *Science 171:* 1008–1013.

Sakshaug, E., A. Haug, A. Jensen, and S. Myklestad. 1972. Phytoplankton ecology of the Trondheimsfjord. Biological Station, N-7001, Trondheim. Mimeographed.

Samsel, G. L., J. R. Reed, and H. J. Winfrey. 1972. Investigations on nutrient factors limiting phytoplankton productivity in two central Virginia ponds. *Water Resour. Bull. 8:* 825–833.

Schelske, C. L., L. E. Feldt, M. A. Santiago, and F. Stoermer. 1972. Nutrient enrichment and its effect on phytoplankton production and species composition in Lake Superior. *Proc. 15th Conf. Great Lakes Res.* Pp. 149–165.

Schelske, C. L., and E. F. Stoermer. 1972. Phosphorus, silica and eutrophication of Lake Michigan. Nutrients and eutrophication. *Am. Soc. Limnol. Oceanogr. Spec. Symp. 1:* 157–171.

Schindler, D. W. 1971. Carbon, nitrogen, and phosphorus and the eutrophication of freshwater lakes. *J. Phycol. 7:* 321–329.

Skulberg, O. M. 1970. The importance of algal cultures for the assessment of the eutrophication of the Oslofjord. *Helgol. Wiss. Meeresunters. 20:* 111–125.

Skulberg, O. M. Personal communication.

Summary report of the agreed monitoring projects on eutrophication of waters. 1973. OECD, Paris.

Sykes, R. M. 1973. Identification of the limiting nutrient and specific growth rate. *JWPCF 45:* 888–895.

The control of eutrophication. 1970. Canada Centre for Inland Waters. Technical Bull. No. 26.

Vollenweider, R. A. 1968. Scientific fundamentals of the eutrophication of lakes and flowing waters with particular reference to nitrogen and phosphorus as factors in eutrophication. OECD, Paris.

DISCUSSION

SOEDER: The ratios for N/P refer to concentrations on a weight basis?

FORSBERG: Yes.

CHRISTENSEN: What is the amount of heat needed in this form of aquaculture? Have you made any estimate of the increase of yield that an increase in temperature will bring about?

FORSBERG: Only in general terms. For more specific information I would refer to the Konyn project in Poland; no doubt Dr. Nyman can furnish more details.

SOEDER: If we include the primary production in the aquaculture systems, it is important to consider the nutrient relations discussed by Forsberg. In a sewage effluent the nutrient proportions are usually not equivalent to the proportions of nutrients taken up by the primary producers. The effluents from such a culture would usually contain large amounts of residual nutrients which have not been assimilated because some other factors were limiting. In order to make the best of the nutrients of the effluent, we then either have to remove part of the phosphorus or improve the sewage by efficient fertilization. By both methods the objective is to achieve a system where all added nutrients can approximately be used to such an extent that the final effluent is no hazard to the natural environment.

FORSBERG: A new situation is arising with the chemical sewage treatment plants, which the government wants to operate so that the outflow of nutrients will be as low as possible. In particular, the objective is to lower the phosphorus content to below 0.5 milligrams per liter, which might lead to a situation where phosphorus might be limiting.
   In cases where the effluent of the sewage treatment plant is to be utilized for primary production, we might have to decrease the dosage of, say, aluminium sulphate in order to obtain an effluent having 15 to 10 parts of nitrogen to one part of phosphorus by weight, which will more or less correspond to the uptake ratio of the plants cultivated. Of course, the exact value will depend upon a number of factors, for instance the frequency of harvesting, the type of plants being cultivated, and so forth.

PILLAY: I think the point raised by Dr. Soeder is very important, particularly the use of oxidation ponds for aquaculture purposes. The primary oxidation ponds are normally used for the production of algae, while the secondary ones very often are used for various other types of culture.

WALNE: Do sewage effluents contain an excess of phosphorus and is this excess solely due to detergents?

FORSBERG: In the present situation it is considered that 30 to 50% of phosphate in sewage effluent comes from detergents. The objective of the sewage treatment is to remove at least 90% of the phosphate. By chemical treatment 95 to 99% will be removed. The present proportions in raw sewage, I believe, is still to the favor of nitrogen.

PILLAY: In ordinary pond fertilization it has been observed that the requirements of fertilizers depend upon the nature of the water rather than the soils in contact with the water. Generally in such waters phosphorus seems to be the ultimate limiting factor, except where sewage is used. To my knowledge, most of the sewage effluents presently used for aquaculture certainly contain more phosphorus than can be utilized beneficially.

OPPENHEIMER: The reason for the large amount of phosphorus in the process of primary and secondary treatment is that phosphorus is released from organic matter directly and by microbiological activity, whereas the nitrogen compounds to a larger extent are conserved and taken out as solid sludge.

It is an interesting philosophical point that the natural population of a stream effected a proper balance of nutrients before man began to manage global organic material. So if one attempts to reconstruct systems which are more natural and to increase productivity, perhaps then we should eliminate the primary and secondary treatment processes and try to use the direct sewage to achieve a total balance between nitrogen and phosphorus, which might be less costly than to try to use N and P. But this is just a philosophical point when we are discussing the deficiency of artificial aquaculture using sewage treatment nutrients.

WALNE: What you are saying is that in a complete biological treatment where you finish up with everything as nitrate and phosphate, and with no biomass, you will have a correct ratio if you ignore the detergent situation.

OPPENHEIMER: This is what happens in nature in the actual environment.

CONOVER: This might not apply to the sea, where the ratio that one finds in the water generally is far different from what one finds in organisms.

OPPENHEIMER: Yes. It is true that the balance is not perfect in seawater. The reason is that there is a steady drain of phosphate as apatite, which in shallow waters is replenished by producing an anaerobic environment in one part of the system, the bottom. Under anaerobic conditions the apatite is resolubilized to complete the cycle.

Effects of Cooling Water Discharge on Primary Production and Composition of Bottom Fauna in a Fjord

K. I. Dahl-Madsen, Bo Møller,
and Bent H. Fenger

Water Quality Research Institute
Poppelgårdvej, Søborg, Denmark

INTRODUCTION

The increasing number and size of power stations result in an increase of discharges of cooling water. It is, therefore, necessary to develop instruments for calculating ecological effects of these cooling water discharges. When such instruments are developed, it is possible to make rational decisions on topics like location of power stations and their discharges and limits to the amount of discharges. In this chapter, two types of instruments for calculating the ecological effects of cooling water discharges are described.

The first instrument is a theoretical, mathematical model of the causal relationships between the phytoplankton primary production and the forcing functions of light, temperature, water exchange, and waste water discharges. The second instrument is an empirical relationship between composition of bottom fauna and excess temperatures in the discharge area.

The data used in the primary production model and for establishing the empirical bottom fauna relationship were obtained through investigations in the Kalundborg Fjord (Water Quality Research Institute, 1973). A detailed description of the hydraulic part of the primary production model is shown in the following chapter, by Schrøder and Mortensen.

PRIMARY PRODUCTION, SYSTEM ANALYSIS

The dependence of primary production on light, temperature, and discharges of nutrients is of a very complex nature and it must, therefore, be formulated through a mathematical model. The model is established through expansion of a box model of water exchange with equations for photosynthesis and respiration, after Dahl-Madsen (1974). The basis of the model is the analysis of the primary producing system, which is given in the following parts of this chapter.

Box Model of a Fjord

A fjord can hydraulically be described as a series of completely mixed boxes. In Figure 1 the number and size of the boxes in Kalundborg Fjord are shown. For computer-technical reasons the number of boxes is cut down to 10 instead of 16 in the hydraulic box model.

State Variables

State variables and their interrelations are shown in Figure 2. The state variables are:

Algae biomass	: AB	g C/m^3
Detritus phosphorus	: DP	g/m^3
Inorganic phosphorus	: CP	g/m^3
Detritus nitrogen	: DN	g/m^3
Inorganic nitrogen	: CN	g/m^3

The model does not take into account the production of the submerse vegetation. Owing to this, the model will underestimate the primary production in the shallow areas of the fjord, where light can penetrate to the bottom. Furthermore, zooplankton is not included as a state variable.

Processes

The state variables are related through the following processes:

Photosynthesis.

$$PROD = MY(T) \cdot AB \cdot f(I) \cdot \frac{CN}{CN + MN} \cdot \frac{CP}{CP + MP}$$

where MY is maximal specific growth rate, day^{-1}; I is light intensity, kcal/m^2/

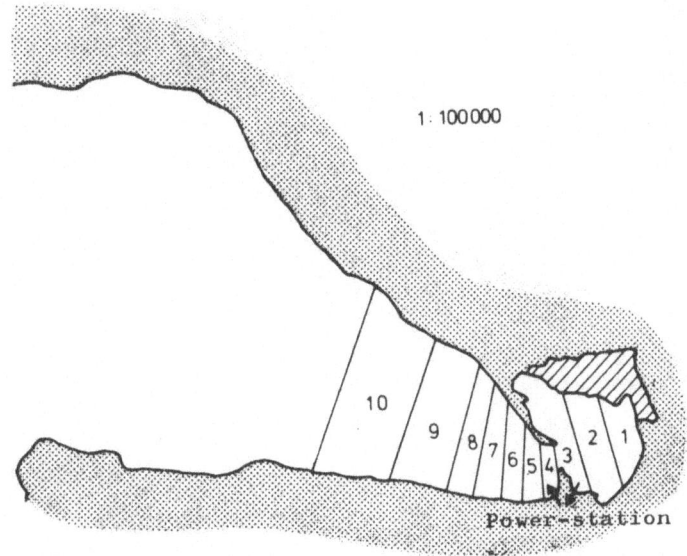

Fig. 1. Boxes in Kalundborg Fjord.

day; T is temperature, $°C$; and MN and MP are half saturation parameters for nitrogen and phosphorus, g/m^3.

Inactivation of Algae.

$$INAK = K_1(T) \cdot AB$$

where K_1 is inactivation rate parameter, day^{-1}. The inactivation is caused, e.g., by the grazing of zooplankton.

Decomposition of Detritus.

$$DECN = K_2(T) \cdot DN \qquad DECP = K_2(T) \cdot DP$$

where K_2 is decomposition rate parameter, day^{-1}.

Sedimentation.

$$SED = K_3 \cdot AB$$

where K_3 is sedimentation rate parameter, day^{-1}.

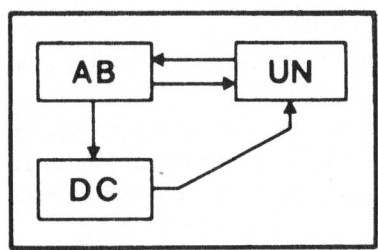

AB: algae biomass

UN: inorganic nutrients

DC: detritus carbon

Fig. 2. Relations between state variables.

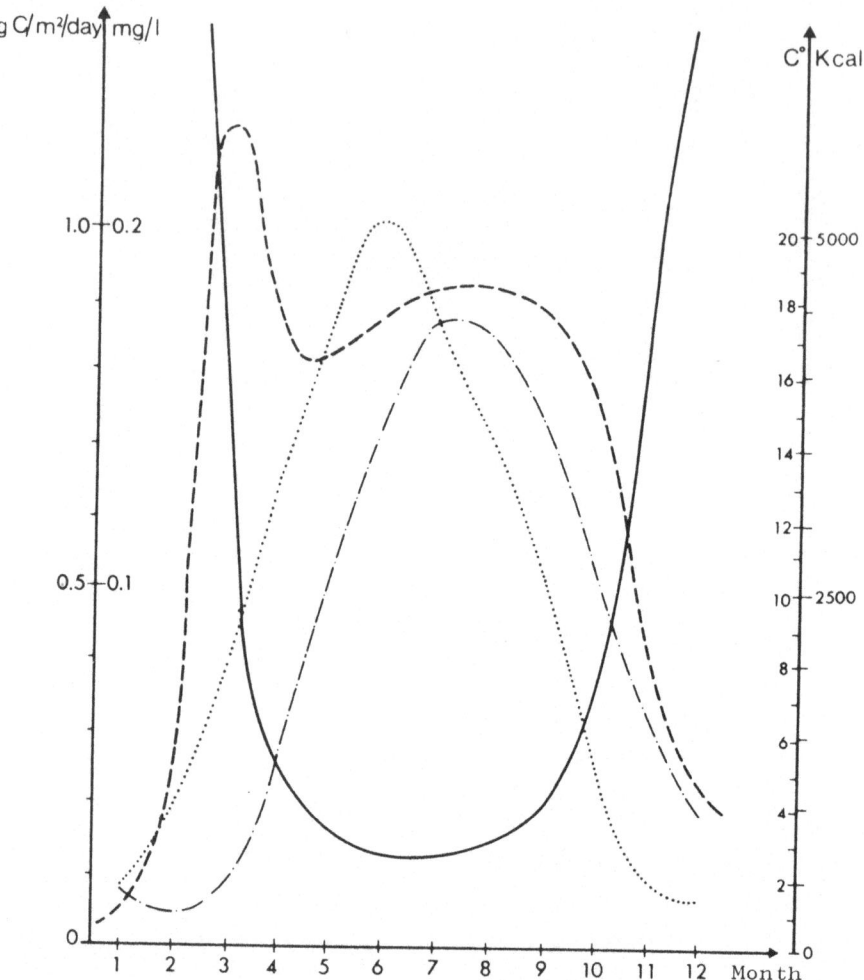

Fig. 3. Calculated yearly variation of primary production. (‐‐‐‐) Primary production, g C/ m²/day; (——) inorganic nitrogen, mg/liter; (—·—) temperature, C°; (·····) light intensity, kcal.

Forcing Functions

The forcing functions are: light, temperature, water exchange, and discharge of nutrients.

Light. The calculations are carried through with values of a mean yearly variation of light intensity at the surface. See Figure 3. The light intensity at a given depth is calculated by the formula

$$I = I_0 \cdot \exp\left(- \alpha - \beta \cdot AB\right) \cdot D$$

where I_0 is light intensity at surface, kcal/cm²day; α is the extinction coefficient of water, m⁻¹; β is the extinction coefficient of algae, m²/g C; D is depth.

Light intensity at the given depth is used in calculations of production by the formula:

$$f(I) = I/I_K \quad \text{for } I \leqslant I_K$$

$$f(I) = I \quad \text{for } I > I_K$$

where I_K indicates the crossing between initial and horizontal parts of curve, kcal/cm²/day. The production is calculated at 5 depths and is integrated per m².

Temperature. The calculations are carried through with values obtained by measurements of the yearly variations of temperature in Kalundborg Fjord for a number of years preceding the start of discharge of cooling water. See Figure 3. The temperature is used in the following formula for rate parameters:

$$MY(T) = MY(20) \cdot 1.07^{(T+OT-20)}$$

$$K_1(T) = K_1(20) \cdot 1.07^{(T+OT-20)}$$

$$K_2(T) = K_2(20) \cdot 1.07^{(T+OT-20)}$$

where OT is excess temperature calculated as a function of the discharge of cooling water.

Water Exchange. Variations in water exchange influence the primary production through change in gradients of nutrient concentrations and excess temperatures. The water exchange variations in Kalundborg Fjord are shown in a simplified form in Figure 4.

Calculations are carried through for two situations. First is complete vertical mixing with small exchange volumes. In this situation, gradients of nutrient

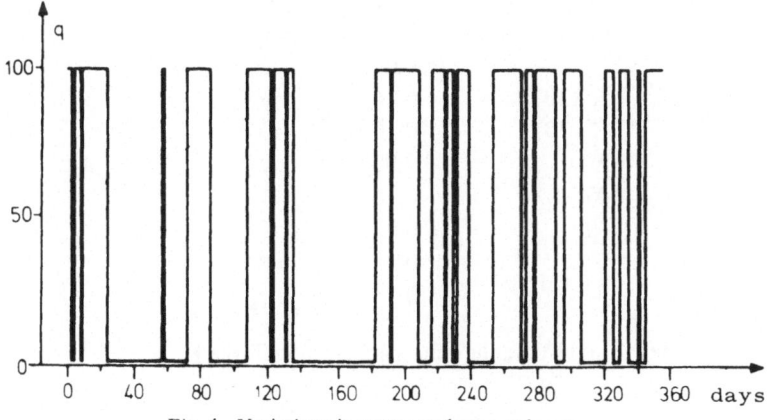

Fig. 4. Variations in water exchange volumes.

concentrations and excess temperatures are created. The steepness of the gradients depends on the length of the period. Second is the halocline situation with 100 times greater exchange volumes. In this situation, the gradients are smoothed out in a short period of the time.

Discharges of Nutrients. The values used for the calculations are 243 kg N/day and 83 kg P/day discharged to box No. 1.

Mass Balances

Without mixing transport, the mass balance for the state variables can be formulated in the following systems of equations for one box:

$$\frac{dAB}{dt} = PROD - INAK - SED - Q \cdot AB/V$$

$$\frac{dDN}{dt} = YN \cdot INAK - DECN - Q \cdot DN/V$$

$$\frac{dCN}{dt} = N/V - YN \cdot PROD + DECN - Q \cdot CN/V$$

$$\frac{dDP}{dt} = YP \cdot INAK - DECP - Q \cdot DP/V$$

$$\frac{dCP}{dt} = P/V - YP \cdot PROD + DECP - Q \cdot CP/V$$

where YN and YP are the relation between N, P and C in algae; Q is advective transport, m^3/day; V is box volume, m^3; N and P are discharges of nutrients, g/day.

Values of parameters and constants are:

$MY(20)$	1.0	day^{-1}
MN	0.005	g/m^3
MP	0.200	g/m^3
I_K	1.000.0	kcal/cm^2/day
$K_1(20)$	0.06	day^{-1}
$K_2(20)$	0.10	day^{-1}
K_3	0.002	day^{-1}
YN	0.167	
YP	0.024	
α	0.1	m^{-1}
β	0.2	m^2/g

The geometric and hydraulic values are given in Table I.

Table I. Geometric and Hydraulic
Variables and Constants

Box	Volume, $10^6 \ m^3$	Depth, m	Volume of water exchange, m^3/s
1	1.2	1	90
2	4.3	3	238
3	4.4	5	156
4	2.5	6	85
5	2.8	6	126
6	3.5	6	180
7	5.0	7	276
8	6.3	7	260
9	17.1	8	400
10	36.4	10	430

Solution of Equations

Figure 3 shows a solution of the equations. Furthermore, yearly variations of light and temperature are shown.

COMPOSITION OF BOTTOM FAUNA

The dependence of the composition of bottom fauna on temperature, waste water discharges, and water exchange is of an even more complex nature than the dependence of primary production on these forcing functions.

Considerably more basic work of describing bottom fauna composition, and especially secondary production, has to be done before the causal relationships of the bottom fauna system can be formulated by methods of mathematical modeling. Predictions of effects of cooling water discharges on bottom fauna must, therefore, follow lines of purely empirical and statistical reasoning. The use of the investigation results from the Kalundborg Fjord may show how to make such empirical predictions.

Sampling Stations and Methods of the Bottom Fauna Investigation

In Figure 5 an example of the choice of sampling stations is shown. At stations 1-12 the samples were taken with an 0.1 m^2 Van Veen bottom sampler. Two samples were taken at each station. At stations A, B, C, D, and 0 an 0.025 m^2 Van Veen bottom sampler was used. Three samples were taken at each station.

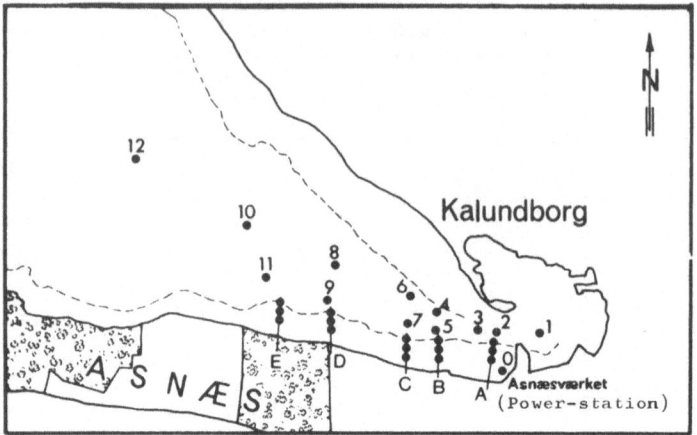

Fig. 5. Sampling stations in Kalundborg Fjord for the bottom fauna investigation.

The samples were passed through a sieve with a 1 mm mesh size. After preservation with formaldehyde, the samples were sorted, the species were determined, the number of individuals were counted, and the weight of this sum of individuals of each species was found.

Results

A complete description of all the results of the investigation has been reported in WQRI report to Isefjord Power Company. Results especially interesting in connection with predictions of effects are shown in Figure 6.

Figure 6 is based on calculations of the percentage of cold water species, e.g., species with an arctic, boreal distribution and with a distribution border south of the English Channel. These percentages are related to excess temperatures for a period with stationary, low water exchange and with the present cooling water discharge.

From Figure 6 a steady increase can be seen in the percentage of cold water species in the direction of the fjord entrance. There are several possible explanations of this phenomenon.

Occurrence of Cold, Saline Bottom Layer. The frequency of occurrence of a cold bottom layer with high salinity increases in the direction of the fjord entrance. Because of the increasing depth, results of samplings from the transections show that the variation of bottom fauna composition related to depth is very pronounced.

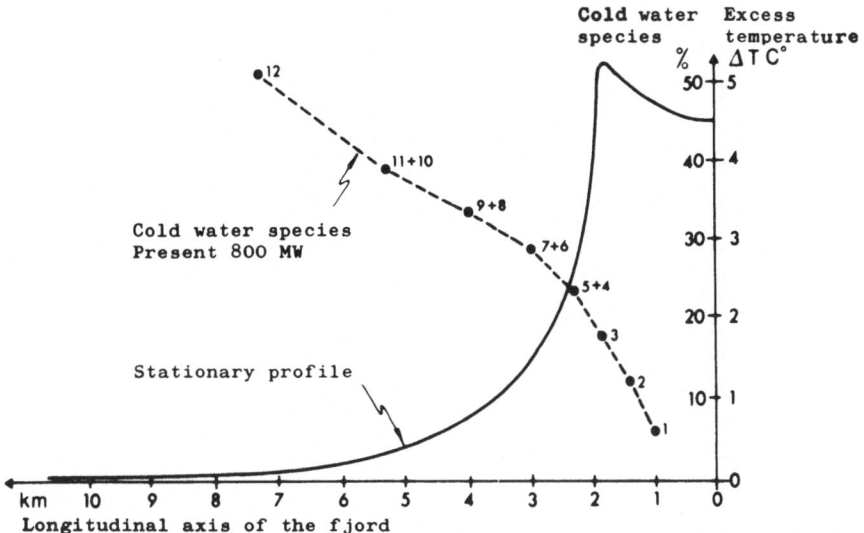

Fig. 6. Percentage of cold water species in Kalundbord Fjord related to excess temperatures.

Discharge of Waste Water. The inner part of the fjord is heavily polluted with discharges of untreated waste water. The steep decline in number of cold water species in the inner part of the fjord must be seen in relation to these discharges combined with the fact that most of the cold water species are intolerant to pollution.

It can be summarized that the decline of cold water species in the inner part of the fjord is caused by a combination of waste water discharge and cooling water discharge, whereas the cooling water discharge is most likely responsible for the decline in the outer part of the fjord.

CALCULATIONS OF THE EFFECTS
OF CHANGED DISCHARGES OF COOLING WATER

Calculation results from the primary production model are shown in Figure 7, in which comparisons can be seen between different amounts of cooling water discharges and a standard situation with no discharge. The uncertainty of the calculations is essential to evaluation of the usefulness of the calculations. This uncertainty is found in three areas.

First, there are uncertainties in the model itself. For example, submersed vegetation is not incorporated in the model. The primary production in the

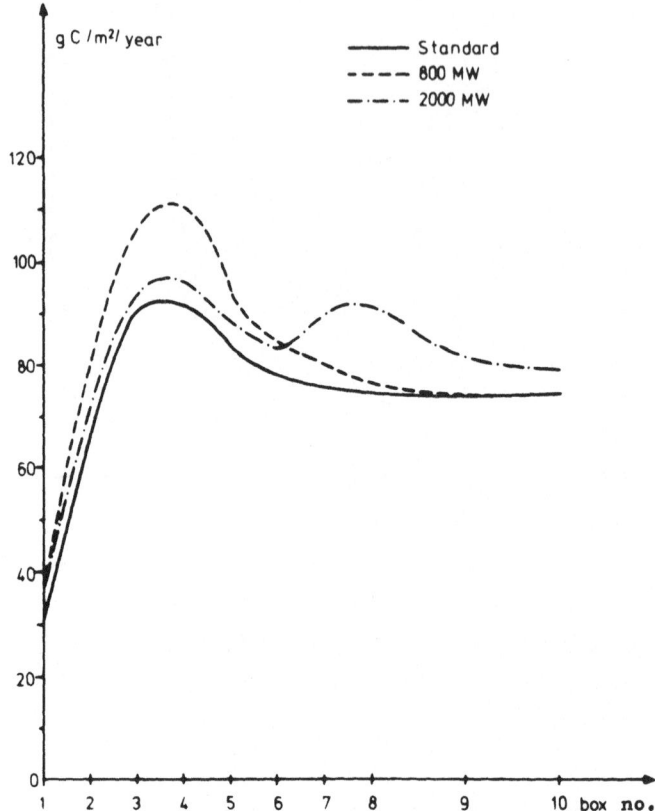

Fig. 7. Calculation of effects of cooling water discharges.

three inner shallow boxes is, therefore, considerably underestimated. Besides, zooplankton and its grazing is not explicitly incorporated. Second, some of the model parameters, especially the temperature parameters, are uncertain. Third, the values for discharges and boundary values are uncertain because of lack of measurements.

These uncertainties indicate that the *absolute*, calculated values must be considered critically. The relative values can, on the other hand, be considered with more confidence. The calculated level of yearly primary production in the outer part of the fjord is 75 g $C/m^2/day$. This level can be compared with measurements from the Great Belt which show a level of 100 g $C/m^2/day$.

Assuming that cooling water discharge is a primary factor in reducing the percentage of cold water species in the outer part of the fjord, the percentage of cold water species after a change in the discharges can be calculated. Results of these calculations are seen in Figure 8, which shows that the expected reduction in the number of cold water species is 10–15% after an increase of cooling water discharge corresponding to 2000 MW.

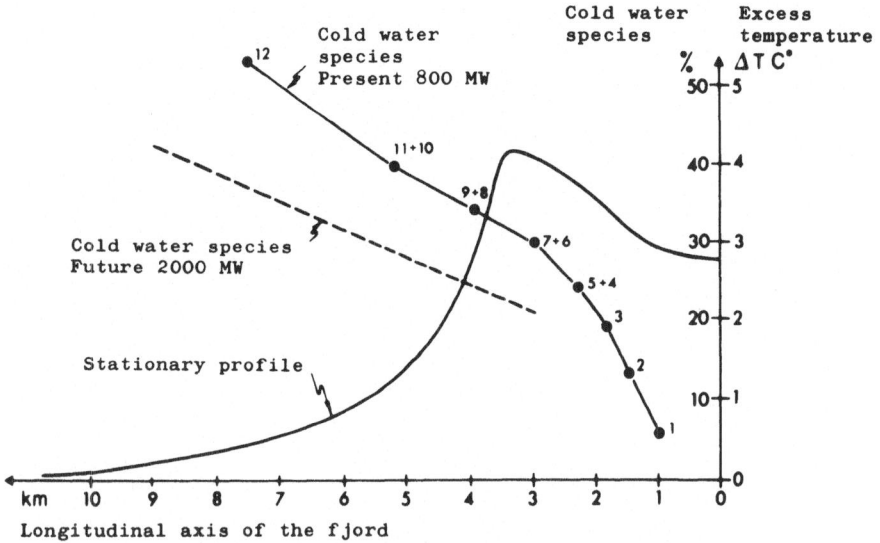

Fig. 8. Results of calculations of future bottom fauna composition.

CONCLUSIONS

Having worked with calculations of the ecological effects of cooling water discharges we feel that it is reasonable to draw up the following conclusions:

1. Calculations of ecological effects of cooling water discharges on the Kalundborg Fjord have shown that the relative changes of primary production and composition of bottom fauna is on an order of magnitude of 0–25% as a function of alternative amounts of cooling water discharges.
2. An increase in *correctness* of the primary production model must be established partly through incorporation of more state variables (as, e.g., macrovegetation and zooplankton), partly through collection of more complete data for calibration of the model.
3. The value of calculations of changes of bottom fauna composition will be increased if they are based on causal instead of on empirical relationships. This would, however, require allocation of more resources into theoretical as well as applied analysis of the bottom fauna system.

REFERENCES

Dahl-Madsen, K. I., and E. Gargas. 1974. A preliminary eutrophication model of shallow fjords. IAWPR conference, Paris, September 9–13.
Water Quality Research Institute report to Isefjord Power Company. 1973. Copenhagen, Denmark.

A One-Dimensional Model for Prediction of Excess Temperatures in a Fjord

Hans Schrøder and Peter Mortensen

Danish Hydraulic Institute
Copenhagen, Denmark

A one-dimensional mathematical model for direct calculation of the longitudinal mixing parameters and distribution of excess temperatures is presented. The model has been applied for prediction of excess temperatures due to an increase in cooling water discharge in a Danish estuary, Kalundborg Fjord. The model calibration is direct in the sense that explicit calculations of the longitudinal mixing parameters are obtained on the basis of observed longitudinal excess temperature profiles caused by the present use of cooling water. The effect of discharging the cooling water of the plant extension closer to the mouth of the estuary is investigated. The influence of occasional density circulations is discussed.

INTRODUCTION

Waste heat disposal from power plants may give rise to technical problems as well as problems concerning the influence on the aquatic environment. In both cases, a reliable method for prediction of the excess temperature is needed whether the problem is limited to a region close to the outfall, or concerns a large body of water in which the heat disposal gives rise to excess temperatures.

The present treatise deals with a mathematical tool for prediction of excess temperatures in an estuarine environment. The methods were developed and applied in connection with a feasibility study for a planned extension of a thermal power plant situated in a Danish fjord. In this chapter the hydraulic part of the investigation is presented, while in a succeeding article the application of the mathematical model by biological experts as the basis for a water quality model is treated.

Mollowney (1972) used the box model concept in combination with an oscillating reference frame in a similar study of heat distributions in the Thames estuary. In this discussion a method based on a fixed reference frame is utilized to calculate the longitudinal mixing parameters from observed longitudinal profiles of excess temperatures and to predict future distributions in an essentially nontidal estuary with occasional density circulations.

THE TRANSPORT-MIXING MODEL

General

The calculation of the distribution of excess temperatures or pollutants in most estuaries can be treated as a problem in one dimension, since the variations in temperatures or concentrations of substance over the cross sections are small compared to the variation in the longitudinal direction. In doing so, the following two transport mechanisms appear: (a) an advective transport and (b) a longitudinal mixing transport.

In a model based on the one-dimensional concept, average cross-sectional values of velocities and concentrations are calculated, and the fact that deviations over the cross sections do exist is accounted for in the transport parameters.

Suppose that current velocities and concentrations were evenly distributed over the cross section. If this were the case, the advective transport carried by the net flow would dominate over the longitudinal mixing transport, which then would be a transport generated by the turbulent diffusion only. However, current velocity variation in the transverse direction, and notably in the horizontal direction, gives rise to a longitudinal mixing and mass transport, which very often is dominating over the advective transport.

Let x denote the coordinate in the longitudinal direction perpendicular to the cross section with the time dependent area $A(x, t)$, and let y and z be the transverse horizontal and vertical coordinates. Then the instantaneous current velocity and concentration of substance can be written as

$$u(x, y, z, t) = U + u'$$ (1)

$$c(x, y, z, t) = C + c'$$ (2)

where U and C are the mean values of u and c, and the departures from the

means are u' and c'. It may be noted that U and C further can be divided into a
tidal fluctuating part, and a part which appears as the average value over one or
more tidal cycles.

The instantaneous mass transport of substance through the cross section is
then given by

$$\int_A ucdA = \int_A (U + u')(C + c')dA = UCA + \overline{u'c'}A \qquad (3)$$

where the bar denotes the average value over the cross section. The first term is
the advective transport due to the average movement of water particles with the
net flow, and the second term expresses the longitudinal mixing transport.

It is traditionally assumed that the longitudinal mixing transport can be expressed as

$$F = - AD \frac{\partial C}{\partial x} \qquad (4)$$

where $D = D(x, t)$ is a dispersion or mixing coefficient which now may be defined as follows:

$$D(x, t) = \frac{\overline{u'c'}}{\partial C/\partial x} \qquad (5)$$

Note that the mixing coefficient is assumed to consist of a tidal fluctuating part
and a mean value which can vary with time. It is furthermore stressed that the
mixing coefficient is strongly related to the distribution of concentration of the
considered substance, such that a mixing coefficient for one substance is not
necessarily applicable for prediction of distributions of other substances with
different concentration distributions over the cross sections.

The Box Model Concept

Keeling and Bolin (1958) used the box model concept in oceanographical
studies. Since then this concept has been applied widely in pollution studies,
see for instance Mollowney (1972). It is believed that the box model concept
offers an excellent alternative to the application of the commonly used transport
diffusion equation, since it gives a rather simple analysis and provides means by
which the mixing parameters may be expressed explicitly.

The basic assumptions are that the concentrations vary linearly between box
centers and that the longitudinal mixing transport can be simulated by assuming
a continuous exchange of water between adjacent boxes by equal and opposite
flows.

The advective transport between adjacent boxes ($i - 1$ and i) is found simply as

$$Q_{i-1,i} \frac{C_{i-1} + C_i}{2} \tag{6}$$

where $Q_{i-1,i}$ is the instantaneous flow (m³/s) over the cross section of area $A_{i-1,i}$, and C_{i-1} and C_i are the concentrations in the box centers.

The longitudinal mixing transport is

$$q_{i-1,i}(C_{i-1} - C_i) = -A_{i-1,i} D_{i-1,i} \frac{C_i - C_{i-1}}{\Delta X_{i-1,i}} \tag{7}$$

where $q_{i-1,i}$ is the exchange volume (m³/s) over the boundary and $\Delta X_{i-1,i}$ is the distance between box centers. The relation between the mixing coefficient and the exchange volume is then

$$q_{i-1,i} = \frac{A_{i-1,i} D_{i-1,i}}{\Delta X_{i-1,i}} \tag{8}$$

Equations of Conservation of Excess Heat

The balance of excess heat in box i of Figure 1 is considered, and it is assumed that the rate of heat loss to the atmosphere can be expressed as

$$R_i = -\frac{f A_{si}}{\gamma p} \Delta T_i \tag{9}$$

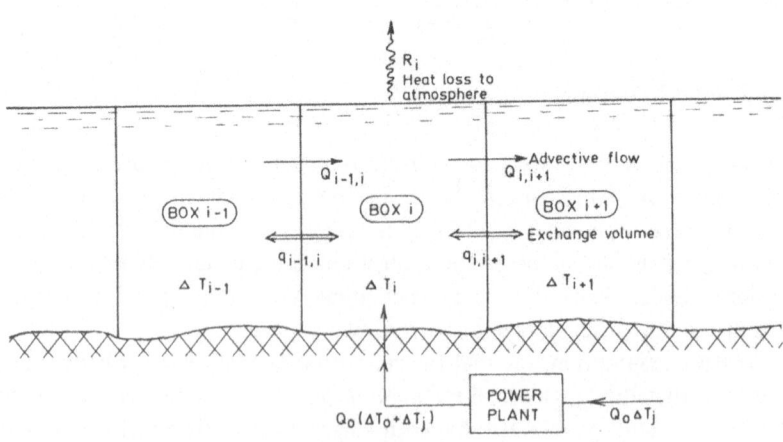

Fig. 1. Parameters in the heat balance of box i.

where f is a heat exchange coefficient (cal/m^{2}s$°$C); A_{si} is the surface area of box i (m^2); γ is the specific weight of sea water (kg/m^3); and p is the specific heat capacity of sea water (cal/kg$°$C). Furthermore, excess heat is discharged to box i at a rate of $Q_0(\Delta T_0 + \Delta T_j)$ where Q_0 is the condenser water flow and ΔT_0 is the temperature rise through the power plant's condensers. The equation of conservation of heat is then

$$\frac{d(V_i \Delta T_i)}{dt} = Q_{i-1,i} \tfrac{1}{2}(\Delta T_{i-1} + \Delta T_i) - Q_{i,i+1} \tfrac{1}{2}(\Delta T_i + \Delta T_{i+1})$$

$$+ q_{i-1,i}(\Delta T_{i-1} - \Delta T_i) + q_{i,i+1}(\Delta T_{i+1} - \Delta T_i)$$

$$+ Q_0(\Delta T_0 + \Delta T_j) + R_i \tag{10}$$

where ΔT_j is the excess temperature of the intake water.

Time-Averaged Model

Provided that the tidal fluctuations of ΔT_i are small compared with the long-term variations, it appears useful to consider values averaged over a tidal period. The left-hand side of equation 10 then becomes

$$\frac{1}{T} \int_t^{t+T} \frac{d(V_i \Delta T_i)}{dt} \, dt = V_i \frac{d\langle \Delta T_i \rangle}{dt} \tag{11}$$

where $\langle \Delta T_i \rangle$ denotes the excess temperature in box i averaged over a tidal cycle. This result is an approximation which is good if the change of volume from one tidal cycle to the next is small. In periods with irregular changes in the water level as a consequence of wind set-up, errors may be introduced using this method. However, if the aim of an investigation is to provide a picture of seasonal changes, a time step of, say, one week may be used and the errors will then be insignificant.

The same integration can be applied to the right-hand side of equation (10). It can be shown that calculations can be performed with values considered as averaged over the time step, and that the approximation is good if the deviations from one time step to the next are small.

The finite difference approximation for box i takes the following form:

$$A_i \Delta T_{i-1}^{n+1} + \left(B_i - \frac{2 V_i^{n+\frac{1}{2}}}{\Delta t} \right) \Delta T_i^{n+1} + C_i \Delta T_{i+1}^{n+1}$$

$$= -A_i \Delta T_{i-1}^n - \left(B_i + \frac{2 V_i^{n+\frac{1}{2}}}{\Delta T} \right) \Delta T_i^n - C_i \Delta T_{i+1}^n - D_i \tag{12}$$

where

$$A_i = q_{i-1,i}^{n+\frac{1}{2}} + \tfrac{1}{2} Q_{i-1,i}^{n+\frac{1}{2}}$$

$$B_i = -q_{i-1,i}^{n+\frac{1}{2}} - q_{i,i+1}^{n+\frac{1}{2}} + \tfrac{1}{2} Q_{i-1,i}^{n+\frac{1}{2}} - \tfrac{1}{2} Q_{i,i+1}^{n+\frac{1}{2}}$$

$$C_i = q_{i,i+1}^{n+\frac{1}{2}} - \tfrac{1}{2} Q_{i,i+1}^{n+\frac{1}{2}}$$

$$D_i = 2 Q_0^{n+\frac{1}{2}} (\Delta T_0 + \Delta T_j^{n+\frac{1}{2}})$$

in which the upper indices $n + 1$ and n denote values pertaining to time $t = (n + 1) \Delta t$ and $t = n \Delta t$ respectively. The index $n + \frac{1}{2}$ is used for values centered in the time step.

To start the calculation of this set of algebraic equations, the initial values of excess temperature in all boxes are given. Furthermore, the boundary values are assumed to be known. The values obtained for the first time step (in which the initial values are applied) are used for calculation of the right-hand sides of the succeeding set of equations, and so on.

In case conditions can be assumed to be at a steady state, only one set of equations is solved:

$$A_i \Delta T_{i-1} + B_i \Delta T_i + C_i \Delta T_{i+1} = -\tfrac{1}{2} D_i \qquad (13)$$

Calibration of the Model

Consider the ideal situation in which a complete time series of excess temperatures averaged over a tidal period for each box is given. Equation (10) can then easily be solved with respect to the exchange volumes centered in the time steps.

Under steady state conditions, the calibration becomes extremely simple. The exchange volumes can in this case be found separately by considering a control volume extending from the boundary between box $i + 1$ and box i, by expressing that the inflow of excess heat to the control volume balances the outflow:

$$q_{i,i+1} = \frac{\tfrac{1}{2} Q_{i,i+1} (\Delta T_{i+1} + \Delta T_i) - Q_0 \Delta T_0 - \sum\limits_{i=1}^{i=i} R_i}{\Delta T_{i+1} - \Delta T_i} \qquad (14)$$

Extreme care should be taken that the ΔT values used in such a calibration actually represent values of a steady state condition, since very small changes in the excess heat within the control volume may cause the results to be greatly in error.

The Kalundborg Fjord Model

Kalundborg Fjord is a 10-km-long indentation about 6500 m wide at the entrance and only 500 m wide at the entrance to the inner fjord. The maximum depth is about 15 m. The general topography, the model segmentation, and the position of the existing 760 MW thermal power plant, Asnæsværket, is shown in Figure 2.

An extension of the power plant is presently under consideration, for which reason it was deemed necessary to investigate the marine aspects of the increased cooling water discharge. The emphasis of the investigation was placed on the question of the biological impact of increased water temperatures in the fjord.

The cooling water discharge at maximum load is about 30 m^3/s taken in at the inner part of the fjord, heated about 8°C through the plant's condensers, and discharged on the west side of the plant as shown in Figure 2.

The basis for a biological assessment of the thermal effects is, of course, a reliable prediction of future excess temperatures. This was done by means of a box model calibrated on the basis of the temperature distribution caused by the

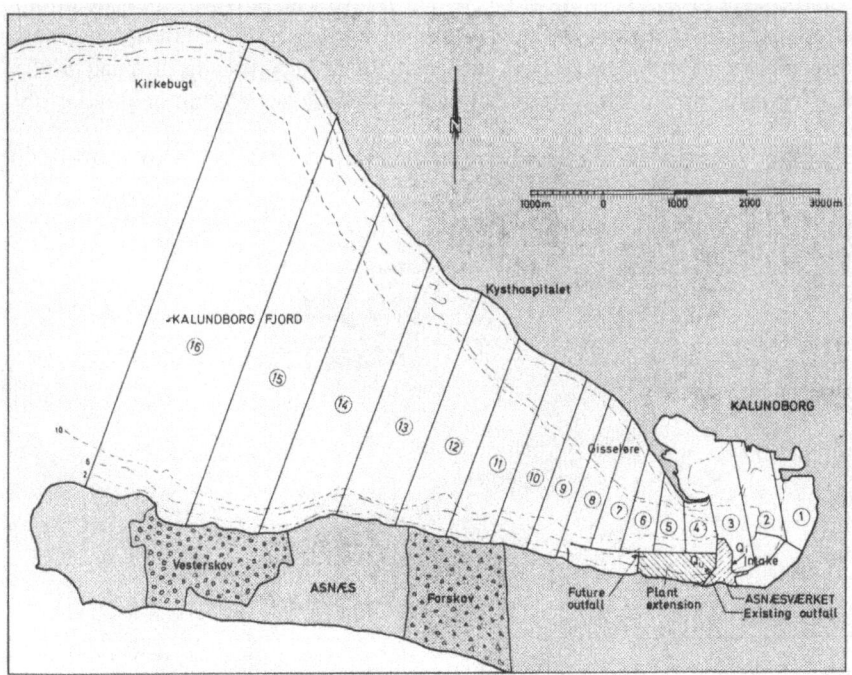

Fig. 2. Model segmentation of Kalundborg Fjord.

present power plant's disposal of excess heat (see below). As a part of the biological assessment, the mathematical model was used as the framework for the
establishment of a water quality model by means of which the effect of increased
temperatures on the primary production can be evaluated. The methods applied
by biological specialists and the results of this part of the study have been presented previously (cf. p. 39 *ff*).

Hydrography

The hydrography of the Kalundborg Fjord shows features which are characteristic of the conditions of the Danish belts and sounds. The mouth of the fjord
is situated at the northern part of the Great Belt, which is the main branch of
the large stratified to partly mixed estuary connecting the brackish water of the
Baltic Sea and the oceanic waters of the North Sea. Considerable fluctuations in
salinity (and temperature) are encountered mainly as a result of meteorological
conditions.

The level of the interface between the salty bottom layer ($20-30\%_{00}$), and the
surface layer ($10-20\%_{00}$) in the Great Belt is highly sensitive to wind conditions
and variations in barometric pressure. The average level of interface is most
often found 15 m below surface, such that the waters of the fjord is part of the
surface layer in the Great Belt. However, meteorological conditions frequently

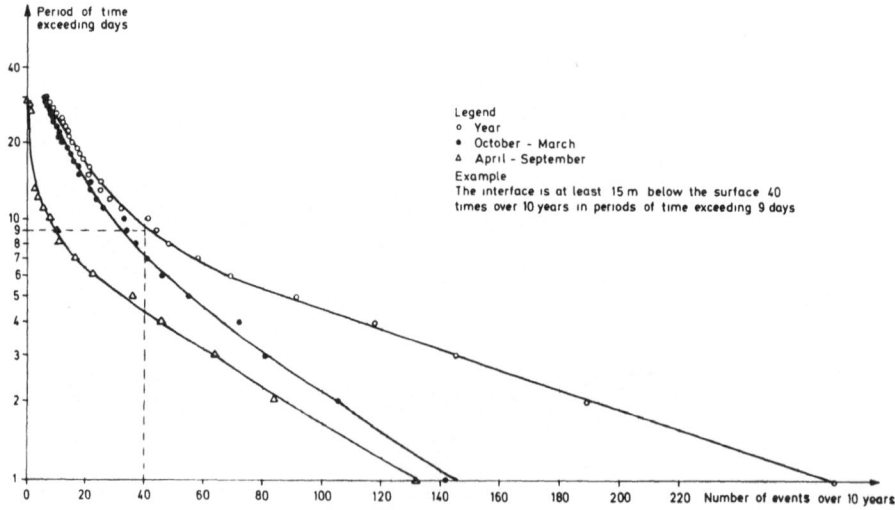

Fig. 3. Number of events and corresponding periods of time in which the interface is at
least 15 m below the surface.

cause the level to rise as a whole, or cause the interface to tilt. Under such situations, a rapid increase in the salinity at the mouth of the fjord is found to give
rise to a density circulation in the fjord. A saltwater wedge protrudes into the
fjord with a considerable velocity (up to 10 cm/s), and gives rise to an extra-
ordinary rate of water renewal.

On the basis of long-term measurements of salinity and temperature performed from a light ship in the southern part of Kattegat, an analysis of the
frequency of occurrence and length of time of situations with the interface at
different levels have been made. The results are depicted in Figure 3.

Continuous measurements of the temperature in two verticals in the fjord
by means of recording temperature chains have shown that internal waves (generated by the surface tidal wave) of heights up to 4 m are present in periods with
stratified water. This gives rise to considerable current velocities in both layers
and greatly enhances the longitudinal mixing. The tidal range is only about
40 cm, which means that the tide itself contributes insignificantly to the longitudinal mixing under density homogeneous conditions.

Calibration and Present Temperature Distributions

The continuous temperature measurements, and occasional measurements
using direct reading instruments from a boat cruising from one end of the fjord
to the other, gave rise to the following conclusion regarding the excess temperature field:

> In periods with density homogeneous conditions, the excess temperature created
> by the cooling water discharge corresponding to a daily average load of about 300
> MW may rise to about 2°C above normal in the inner part of the fjord. After the
> occurrence of a density circulation, no excess temperature can be found, save
> a region in the vicinity of the cooling water outlet directly influenced by the
> jet (Danish Hydraulic Institute, 1973).

A box model valid for density homogeneous conditions only, with a segmentation as shown in Figure 2, was calibrated by means of the observed longitudinal
profiles of excess temperatures. Steady state conditions were assumed for the
times of measurements. One example of an observed excess temperature profile
and the mixing coefficients calculated from the model exchange volumes and
equation (8) are shown in Figure 4. A heat exchange coefficient of 5 cal/m²s°C
was used in the calculations. It is emphasized that the mixing coefficients are
valid for periods with density homogeneous conditions only, and therefore
represent minimum values corresponding to the most adverse conditions from
the point of view of recirculation.

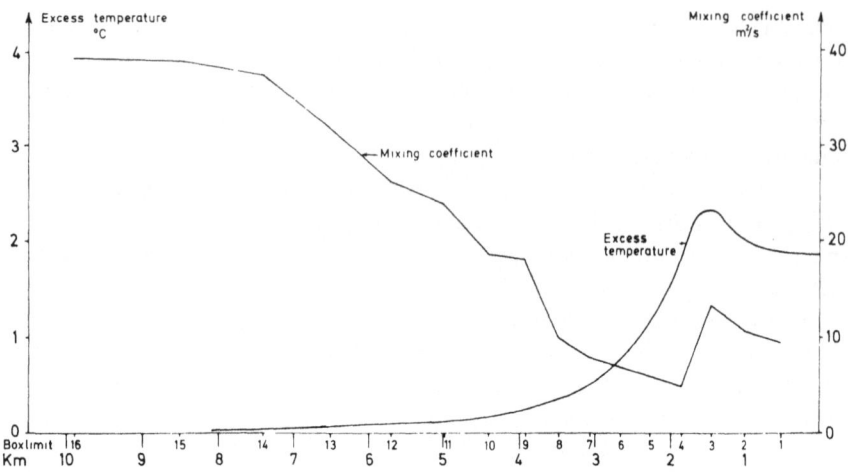

Fig. 4. Observed longitudinal excess temperature profile and mixing coefficients.

Predicted Temperature Distributions

A suppression of the level of heat accumulation can be obtained by discharging future cooling water closer to the mouth of the fjord. The effect of such an arrangement has been studied by means of the calibrated model. The combined cooling water of the existing and the new plant is discharged through

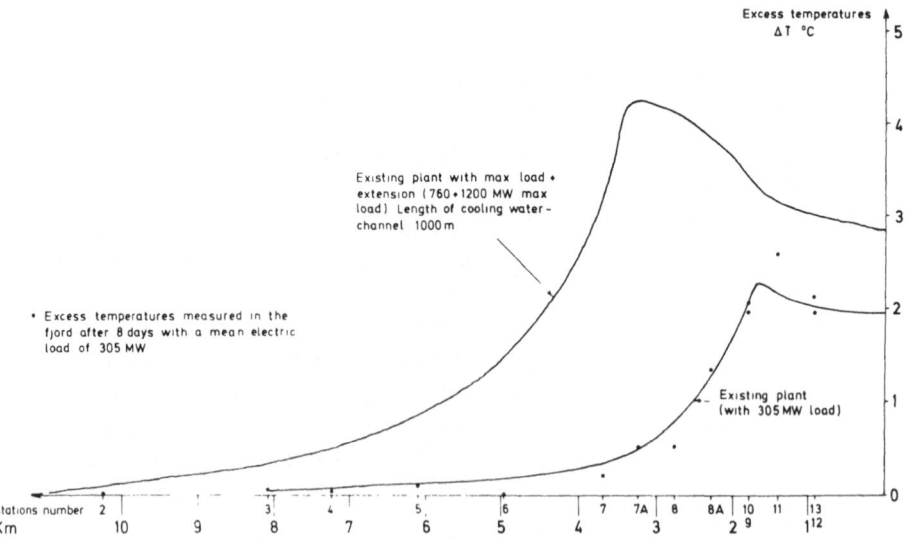

Fig. 5. Excess temperatures computed with steady state model.

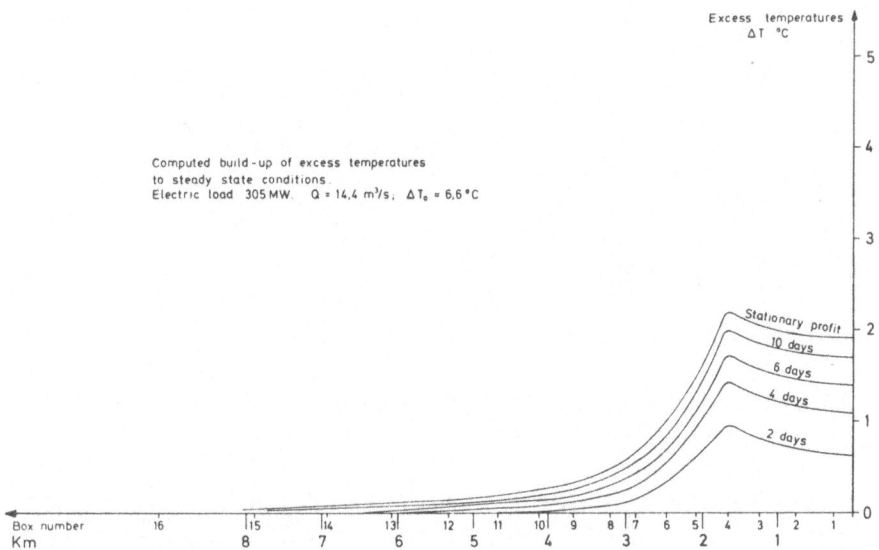

Fig. 6. Computed excess temperatures in the time-averaged model.

a channel running between the coastline and a sea wall placed at 4 m water depth. The existing intake is maintained, and the intake for the plant extension is placed about 1000 m west of the existing intake.

Results of the calculations with the steady state model are depicted in Figure 5.

In order to acquire an impression of the period of time needed to build up a steady state excess temperature distribution, runs with the time-averaged model were performed. Initial excess temperatures were chosen at zero throughout the fjord. The results are shown in Figure 6. Steady state conditions are reached after about two weeks.

Statistical Considerations

The analysis of the frequency of occurrence and period length of situations corresponding to different levels of the interface has shown that the fjord is stratified in about 40% of the time. Decisive for the distribution of excess temperatures through time is the distribution of length of periods with density homogeneous conditions only. In these situations, heat will be accumulated in the fjord and the level of heat accumulation is determined by the length of the uninterrupted periods of time with density homogeneous conditions.

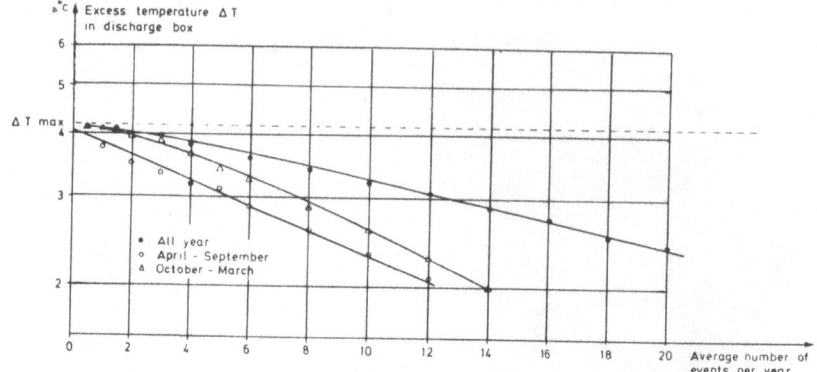

Fig. 7. Frequency curves for excess temperature in discharge box. Electric load 2000 MW.
$Q = 80 \ m^3/s$. $\Delta T_0 = 8°C$. Length of cooling water channel 1000 m.

Using the frequency curves for the interface position shown in Figure 3, one
can get the distribution of length of periods with homogeneous conditions in
the fjord in an average year. Together with computed excess temperatures with
the time-averaged model, it is possible to represent frequency curves for excess
temperatures in the fjord. The results, represented in terms of frequency curves
(number of occurrences) for the highest excess temperature, are shown in
Figure 7.

The length of periods with excess temperatures above a certain level can be
determined by using Figures 3 and 7.

It is possible to take into account the statistical occurrence of excess temperatures in a water quality model. This will be treated in a following article.

REFERENCES

Keeling, C. D., and B. Bolin. 1958. The simultaneous use of chemical tracers in oceanic
    studies. II. A three-reservoir model of the North and South Pacific Oceans. *Tellus 20:*
    17–54.
Mollowney, B. M. 1972. One-dimensional models of estuarine pollution. Water Pollution
    Research. Technical Paper No. 13. London.
Danish Hydraulic Institute. 1973. Kalundborg Fjord—a receiving water for waste heat. Re-
    port to the electricity board Isefjord voerket, Nov. 1973. (In Danish).

DISCUSSION

CONOVER: In your model I gather that you are considering primary production in the
    pelagic communities only. In a model of this sort, I think you should consider the primary

production in the benthic zone as well. In this kind of an inlet in northern America, the benthic primary production can exceed the planktonic primary production. I do not know how this will affect your model or whether it does affect it at all. However, if eutrophication is a problem there, one serious aspect could be a tremendous proliferation of attached forms in the littoral zone.

FENGER: Are you referring to the effect of grazing?

CONOVER: No, actually I am discussing the growth of both attached algae and various types of spermatophytes, marsh grasses, and pond weeds of one kind and another, which would be growing in the coastal zone.

FENGER: This fjord is not so shallow; it is very deep and during summer the light you can get down into the fjord is very low. With a visibility in the summertime of 2–3 meters, the production of macrophytes is very small. In the shallow areas, the productivity from the benthic area for the coastal zone may be just as high as of the phytoplankton itself.

CONOVER: The estuary that I am referring to is St. Margaret's Bay, Nova Scotia, which has an average depth of perhaps 40 or 50 meters, and the macrophyte production there on an annual basis is larger than the primary production.

FENGER: Yes, but is is very weakly eutrophicated.

CONOVER: Yes, light is perhaps not a serious limiting factor.

FENGER: In contrast the Kalundborg Fjord is so heavily eutrophicated that you get almost no macrophyte vegetation.

SKULBERG: Can you give the figures for the volume of cooling waters used today and the future volume?

FENGER: I don't have the exact figures, but they will correspond to the effect of the power plant installations. The present installed effect is 800 megawatts and with the new installation there will be a total of 2000 megawatts.

FORSBERG: Is nitrogen possibly a limiting factor in this system, and have you considered the possibility of nitrogen fixation?

FENGER: A weak point of the present model is that nitrogen fixation is not included. We plan, however, to remedy this situation in the future investigations. We intend to measure the extent of nitrogen fixation as a routine analysis, because in many of our eutrophicated lakes and fjords nitrogen fixation is probably of major importance. Another weak point is that so far we do not have experimental data to corroborate the parameters of the exchange of nutrients between the bottom sediments and the water. The parameters of the model were evalued from experiments in other Danish fjords, where recent results were available. Moreover the effect of grazing has not been included in the model.

KORRINGA: In the effect on the environment, the primary production is only one aspect. Another aspect is the effect of the hot water effluent on the movements of the fish. In the warm water effluent from a large power plant in Holland, we find that in winter many fish will congregate around the water outlet. In winter they will not find food in this area, but as the metabolic rate of the fish is increased by the increase in temperature, the fish will lose considerable weight with corresponding loss of quality.

FENGER: The Kalundborg Fjord is not a fishing area, but such a situation might be of interest in the vicinity where previously we had quite a good fishery.

RAYMONT: Is there a salinity gradient with depth in this fjord?

FENGER: The exchange of water in this fjord takes place through two different mechanisms. The wind from the west will turn the water around in the bottom of the fjord, while the cold bottom water from the Kattegat sometimes will enter. The influx of the Kattegat water is a relatively rare occurrence, and it will not always advance to the same distance up fjord. These situations will create a salinity gradient, which has been included in the hydraulic model.

RAYMONT: The point of the question is whether the intake to the power station is so deep that the cooling water will have a higher salinity than the water at the outlet.

FENGER: The intake for the future power plant is placed so far in the bottom of the fjord that the more dense Kattegat water will never advance so far.

RAYMONT: The question was raised because of a similar problem in the Southampton area, where the Marcawood power station is making use of cooling water with a moderately high salinity. It was hoped that with a surface discharge the cooling water would give off sufficient heat to be cooled at a satisfactory rate, but under certain circumstances the warm water will sink down under the surface. This gives a lens of relatively warm water which literally moves with the tide up and down the estuary, and which will lose its heat relatively slowly. This can have effect on the estuary, not necessarily bad, but is a point that is often forgotten.

FENGER: In our case the situation is different, because the cooling water is taken in from about 12 to 13 meters. At this depth there is a difference in temperature between the upper and lower levels, but as the salinity gradient is very, very small, we believe that the discharged cooling water will stay on the surface and cool off sufficiently rapidly.

CONOVER: What about sedimentation in this system? Biological sedimentation might be a very considerable factor, if you have a persistent phytoplankton bloom in this type of environment. Are the sediments anaerobic? And is the water column above the sediment still oxygenated?

FENGER: At times the bottom sediment is anaerobic, while the columns above are oxygenated. In the model the sedimentation is not taken into account, but in the case of sedimentation taking place the organic nutrient will be fixed to the bottom and the inorganic nutrients released by mineralization.

CONOVER: In a well aerated system mineralization would be roughly proportional to the rate of sedimentation. However, the sedimentation rate will be too high for complete mineralization to occur if the system produces more than it exports.

FENGER: You mean that the rate of accumulation will be increased during the future condition?

CONOVER: That could very easily be. Moreover, heavy blooms can affect the density of the water quite independent of the salinity and temperature conditions. They can in fact contribute to an increase in their own sedimentation role.

SKULBERG: Are there plans for using the locality for aquaculture?

FENGER: At present the land resources are very limited, but for the new nuclear power plant the situation might be different.

OPPENHEIMER: Will you define the primary productivity as net or gross? Is it corrected for maintained respiration?

FENGER: It is defined as the total productivity, and measured by the C-14 method as the production of organic carbon per square meter and year.

OPPENHEIMER: When the productivity data are based on the C-14 method, they might be compared with the natural productivity in our area, which is about 400 grams per square meter per year.

FENGER: No waste waters?

OPPENHEIMER: No. This was our hope, that the bay systems will have waste water introduced from a multiple environment which would account for its high productivity. Productivity is quite high.

FORSBERG: Will the future situation be one where you have a discharge of waste nutrients and waste heat in the same area, or will they try to find another solution to the problems of this area?

FENGER: This point has not been clarified yet. This particular study has just been finished, and the authorities have not allowed the plant to go on with their plans. However, it seems that the most likely solution to the problem is one where the city of Kalundborg will discharge its sewage by pipeline into the open sea.

SOEDER: What is the nuisance of this mixed waste heat and waste sewage?

FENGER: The main objection is that we have anaerobic conditions at the bottom at certain times of the year, and moreover that the productivity in the summertime is so high in the shallow areas that you get excessive bloom and bad odors. The drawbacks are mainly esthetic.

The Role of Filter Feeders in Stabilizing Phytoplankton Communities with Some Considerations for Aquaculture

R. J. Conover

Marine Ecology Laboratory
Bedford Institute of Oceanography
Dartmouth, Nova Scotia, Canada

Looking at the hydrosphere globally, there is little doubt that most primary production is carried out by phytoplankton. It follows logically that most of the animals in the sea are fine particle feeders, either directly on the phytoplankton itself or on detrital particles largely derived from plant sources following death and partial decomposition of the primary producer. The fine particle feeders that concern us in planning marine propagation experiments will be primarily of two types: free-swimming planktonic animals and attached benthic forms. While there is considerable diversity of feeding mechanisms among the groups, they too can be broadly classified into two types: cirral feeding mechanisms and ciliary–mucous systems. To be able to predict something about the effects of fine particle feeding on suspensions of phytoplankton, we should first consider how these systems work.

Of the cirral feeding animals, the most widely studied—but not necessarily the best understood—are the copepods. In the classical concept of filter feeding, as described by Cannon (1928) for *Calanus finmarchicus*, lateral gyres, which are created by movement of the mouth parts, pass under the forward-directed tips of the swimming feet and out through a filter formed by the second maxillae. Particles are then strained from the water by the setules perpendicular to the setae and forming the meshes of the filter (Figure 1). Marshall and Orr (1956)

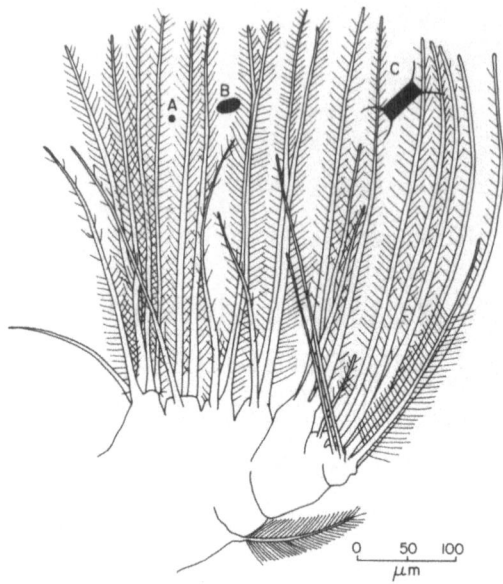

Fig. 1. Left maxilla of *Calanus helgolandicus* ♀ from the right. A, B, and C represent the sizes of food organisms used. (A) *Nannochloris oculata;* (B) *Syracosphaera elongata;* (C) *Chaetoceros decipiens* (from culture) (after Marshall and Orr, 1956).

found the minimum distances between setules to be about 1 μm but, as is clear from the relative sizes of particles and setae, there is small chance of a 2-4 μm particle, such as *Nannochloris* (A in Figure 1), being retained efficiently. Indeed, numerous experimental studies on feeding by several species of copepods indicate considerably reduced efficiency when particles are smaller than 5 μm.

Among the Crustacea only the branchiopods, such as *Artemia salina* and some cladocerans, seem capable of efficient fine particle filtration. Perhaps this is facilitated by glandular secretions in the vicinity of the food groove (Jørgensen, 1966). In any event, *Daphnia longispina* can make efficient use of bacteria in its nutrition (Monakov and Sorokin, 1961) and the marine cladoceran, *Penilia avirostris*, does not use particle sizes greater than 8 μm (Pavlova, 1959).

There are a few ciliary-mucous feeders in the planktonic community such as the salps and other pelagic tunicates. These animals have tremendous capacities for growth (Heron, 1972) and they feed with great rapidity, taking particles from less than 1 μm to greater than 1 mm (Madin, 1974). As "colonizing species" they could constitute a serious nuisance in mariculture experiments, but they seem to be confined to open ocean situations and are rather difficult to maintain in healthy condition in the laboratory.

The larvae of many benthic invertebrates are planktonic and use their cilia for both swimming and feeding. These mechanisms again fall into two general types: the single band system and the opposed band system (Strathmann *et al.*, 1972). In the single band system, a current of water passes through the ciliary row and if a particle is detected, there is a reversal of ciliary beat locally so that particles are collected on the upstream side of the row. In the opposed band system the long preoral cilia create the feeding current and also assist the particles into the food groove, where they are transported toward the mouth by short cilia. The postoral cilia aid in retaining the particles in the food groove and also help to clear unwanted material. Neither the single-band nor the opposed-band system is highly efficient for the smaller-sized particles, probably because alteration in beat pattern of cilia is necessary for capture and very small particles are not "sensed" soon enough for the necessary adjustments in behavior pattern. Quite obviously, large particles would be excluded from the food groove in opposed-band animals as well.

Of the benthic organisms likely to be important in aquacultural projects, only the barnacles are cirral feeders and, like the copepods, they are not particularly efficient at capturing small cells (Crisp and Southward, 1961), although they do have fine cirri near the base of the larger ones which filter the water leaving the mantle cavity.

Three other benthic groups are probably of concern in aquaculture. The sponges are not exactly ciliary mucous feeders, but rather create feeding currents by means of special flagellated cells called choanocytes. The choanocytes are surrounded by collarlike structures which are in reality miniature filters capable of removing particles down to 0.1 μm. The sponge is made up of different-sized pores and spaces, some of which are lined with choanocytes. Particles too big to enter these flagellated chambers are phagocytized by amoeboid cells, and even larger particles may be ingested by external dermal cells (Rasmont, 1968).

Tunicates feed by means of a constantly renewed mucous net of very high retention. Recent studies by Holmes (1973) show that certain species, at least, can filter water at about the same rate as mollusks, which are doubtless the most efficient and commercially the most important of the benthic suspension feeders.

In the common edible mussel, *Mytilus edulis*, an incurrent stream of water, generated by the lateral cilia, brings food and oxygen to the animals from which particles are removed by the latero-frontal cirri, which screen the space between the individual gill filaments (ostium). The food particles are then transported toward the mouth by the frontal cilia. The effective minimum aperture size was stated to be 2-3 μm $\times$ 5-6 μm by Dral (1967) but electron micrograph studies by Moore (1971) have shown that, indeed, the latero-frontal cirri have lateral cilia of their own reducing the aperture size to 0.6 $\times$ 2.7 μm. Experimental feeding studies have shown *Mytilus* to be one of the most efficient fine-particle feeders retaining 1-2 μm objects with nearly maximum efficiency (Vahl, 1972a). The queen scallop, *Chlamys opercularia*, does not reach maximum retention until particles are around 7-8 μm, however (Vahl, 1972b).

The point of this brief description of feeding mechanisms is merely to emphasize that there are grazers capable of handling the entire range of particle sizes likely to be encountered in natural or cultured suspensions of photosynthetic cells.

A second point to be made is that there are no truly automatic filter feeders simply passing a stream of water through a fixed-meshed sieve until it clogs. Historically, it was assumed that a suspension feeder would remove particles at a rate proportional to the number of particles such that

$$dP/dt = kP$$

which leads to the familiar formula, "volume swept clear," for filtering rate

$$F = (v/t) \ln (P_t/P_0)$$

where v is the volume available to the filterer, and P_0 and P_t are concentrations of particles at the start and after time t. Considering only observations on *Calanus*, for instance, the most studied copepod, a glance at the literature suggests values for F ranging from about one ml to over a liter per day. Therefore, it seems unlikely that we can predict the effect of grazing by assuming that a constant proportion of the volume of the system will be swept clear.

There is current argument about what kind of model does best describe the behavior of the grazer in a suspension of particles. Most agree that if the concentration of particles offered is increased, the amount ingested will increase to some limiting concentration beyond which there will be no further increase in the amount ingested. But it is argued by some that the filtering rate will remain constant up to that maximum concentration and thereafter decrease, while ingestion increases in a linear fashion to a maximum and thereafter remains constant, as is shown in Figure 2 taken from Reeve (1963). The animal studied by Reeve was *Artemia salina*, which probably comes nearest to fitting the stereotype of a "filtering automaton" of any pelagic grazer. On the other hand, copepods feed independently of their locomotory activities and show considerable control over their feeding behavior. They have been shown to select cells only 50% larger in volume in a mixture of single and dividing cells (Richman and Rogers, 1969). Figure 3 shows the effect of increasing cell concentration on filtering rate and ingestion for *Calanus hyperboreus* (Mullin, 1963). In such experiments the scatter of points is usually high, but there is no evidence for zero slope to the filtering rate curve. Ingestion, on the other hand, increases in a nonlinear fashion to a maximum and then declines at higher concentrations. Frost (1972) has carried out a similar series of experiments using different-sized diatoms of similar shape (pillbox) in rather larger volumes of water (3.5 liters) for longer periods (2-5 days) than in earlier experiments. As shown in Figure 4, he

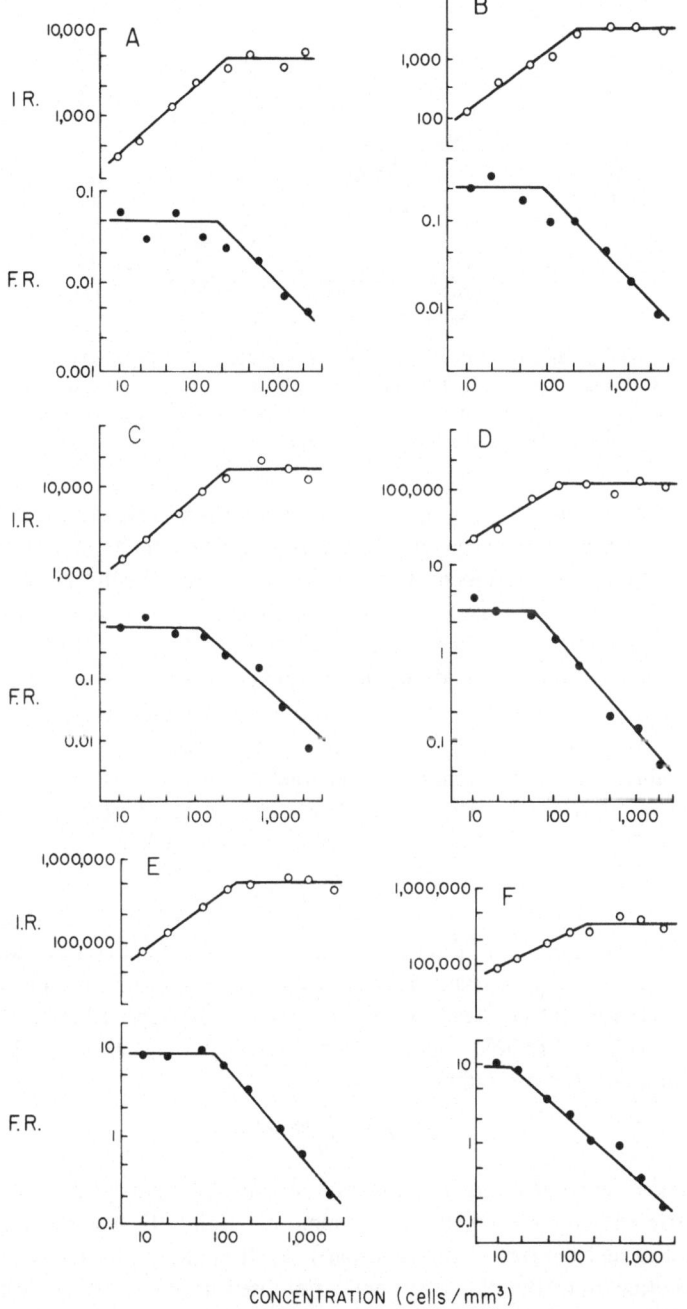

Fig. 2. Filtration and ingestion rate (F.R. and I.R., respectively) with increasing cell concentration, for *Artemia* at six different ages (increasing from A to F) feeding on *Phaeodactylum*, both axes logarithmic (after Reeve, 1963).

72   R. J. Conover

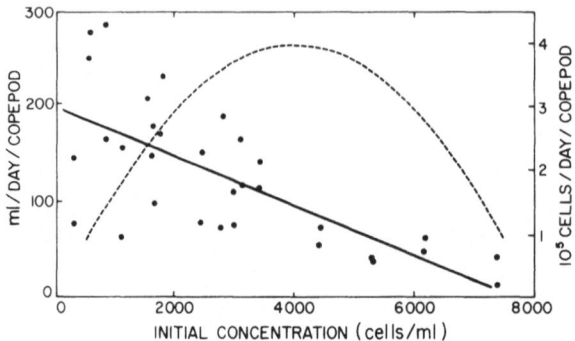

Fig. 3. Grazing by female *Calanus hyperboreus* on various concentrations of *Thalassiosira fluviatilis* (four experiments). Left ordinate and solid line (——), grazing rate. Right ordinate and broken line (---), rate of intake of cells (after Mullin, 1963). Regression equation: $\overline{Y}_x = -0.246x + 197$. Correlation coefficient: $r = -0.696$.

has fitted a constant-filtering rate type of model (like Reeve) to his data but I, personally, am not convinced that his filtering rate data are better described this way than by a single regression line such as Mullin used. He did show, however, that the "plateau" of maximum consumption is higher for previously starved individuals than for those already feeding. It could be, then that the humpbacked ingestion curve observed by Mullin and others could be some sort of overshoot phenomenon resulting from use of experimental animals not previously sufficiently conditioned to the experimental feeding conditions.

Probably the most common model used to describe feeding in copepods just now is a development from the exponential used by Ivlev (1945) to describe feeding ecology of fish. In his original format the ration taken

$$r = R\left(1 - e^{-kp}\right)$$

where R is the maximum ration at saturation, p is the food concentration, and k is a proportionality constant. However, Parsons *et al.* (1967) found that copepods apparently did not feed at food concentrations approaching zero. There appeared to be a threshold concentration of p which they called p_0. They then wrote the IVLEV curve so that

$$r = R\left(1 - e^{-kp}\,e^{kp_0}\right)$$

Some typical curves are shown in Figure 5. An absolute value of p_0 has yet to be demonstrated and it very well may be some function of the volume available to be searched and the size of the food particles. It could be an experimental artifact resulting from the closed systems often used in laboratory grazing experiments, but, as I will show shortly, a p_0-type threshold is very important in regulating grazing on natural populations.

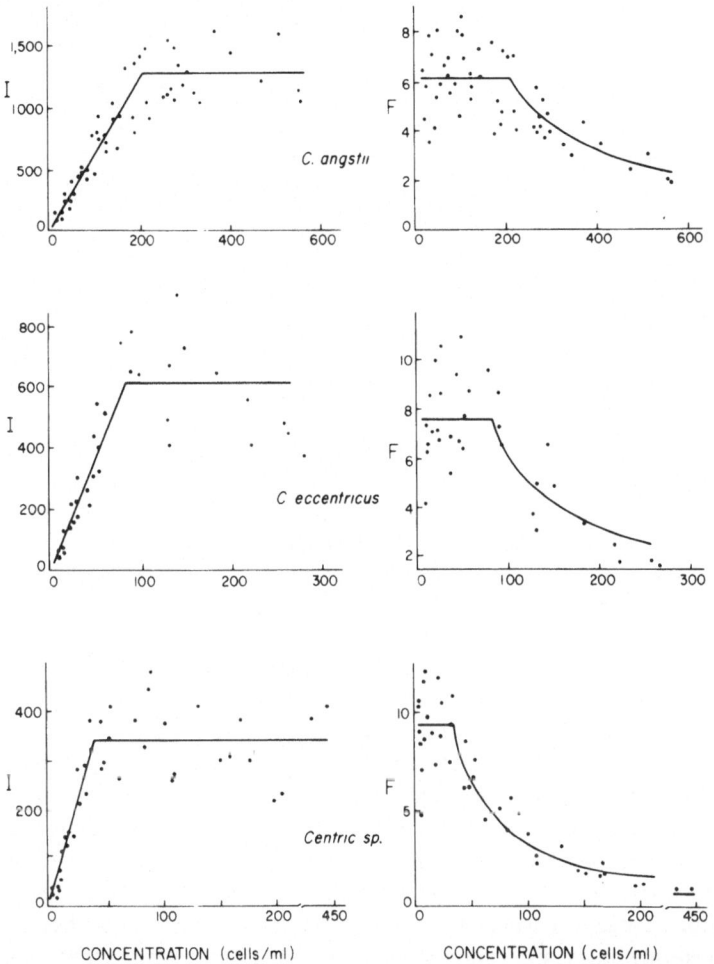

Fig. 4. Effect of cell concentration on ingestion rate, I, and volume swept clear, F, of adult females of *Calanus* feeding on *Coscinodiscus angstii* (top graphs), *Coscinodiscus eccentricus* (center graphs), and centric sp. (lower graphs) (after Frost, 1972).

The response of other grazers to increasing food concentration has been little studied, but there has been argument about whether filtering rate slows before the gill clogs in bivalves. Recent work by Winter (1969, 1973) and Tenore and Dunstan (1973) indicate that it probably does.

Winter describes seven stages of response to increasing particle supply in bivalves. With little or no food present, there is no filtration and reduced metabolism. With the addition of some food, the latero-frontal cirri first take up a position completely screening the ostium and a thin mucous filter is employed to further increase retention, but with further increases the cirri are gradually

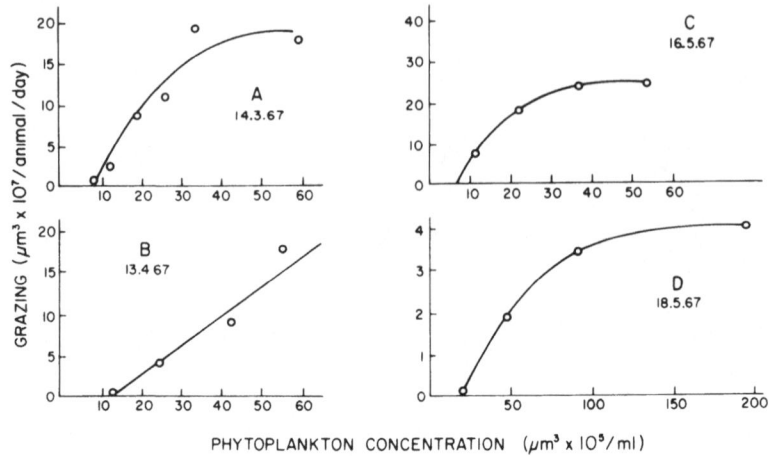

PHYTOPLANKTON CONCENTRATION (μm³ x 10⁵/ml)

Fig. 5. Zooplankton grazing at different concentrations of phytoplankton. (A) *Calanus pacificus* grazing on *Thalassiosira* spp. (B) *Calanus plumchrus* III and IV grazing on *Skeletonema costatum* and μ-flagellates. (C) *Calanus plumchrus* V grazing on *Skeletonema costatum* and μ-flagellates. (D) *Pseudocalanus minutus* and *Oithona* grazing on *Skeletonema costatum* and μ-flagellates (after Parsons *et al.*, 1969).

raised to open the ostium wider and the mucous filter is discarded. Eventually, pumping is reduced and pseudofeces formed. In the extreme case pumping ceases altogether and the valves may close.

In my opinion, the key tool for studying feeding relations of grazing animals is the Coulter Counter, especially some of that later models that incorporate pulse height analysis so that the size spectrum can be displayed almost instantly. The pioneering work here was also done by Parsons *et al.* (1967, 1969). We have been using this tool in the study of feeding by neritic copepods and mussels on the natural range of particulate matter in the sea. On the basis of these studies I will make my third major point, which is that filter feeders, while capable of selective feeding, select their food, not on the basis of size or energy content, but on availability. Hence, in nature selective feeding may normally be of relatively minor importance.

In studies with cultured populations of cells it has been shown numerous times that the copepod selects the larger-sized cell. In this way it inevitably obtains its maximum ration faster and the decrease in filtering rate with time has a greater slope (Mullin, 1963). But if one studies the feeding of a copepod on a natural suspension of cells, it may not be the large cell that is selected but rather the cell which has the highest concentration in biomass (volume) units. For instance, my colleague Poulet (1973) has shown that the common neritic copepod, *Pseudocalanus*, feeds predominantly on a 10-μm peak in the surface water of Bedford Basin, Nova Scotia, Canada (Figure 6). However, in deeper water the particle size spectrum is flattened and the distribution of biomass more uniform among the different size categories. *Pseudocalanus* adapts by shifting grazing emphasis to larger particle sizes as they become relatively more important.

Somewhat similar behavior is shown by the same copepod when studied in successively less neritic water (Figure 7). Note, however, that while the 10-μm peak is still present as we go further offshore, the total concentration is lower than in the Basin and there is little feeding upon it. In other words, there is probably a p_0 for different particle sizes and we surmise that it is higher for smaller particles than for large. On a seasonal basis *Pseudocalanus* continues to be quite opportunistic, feeding with relish on precisely the peak which is most abundant (Poulet, 1974). By shifting grazing pressure from one peak to another, presumably as concentration slips below p_0 for one size category and increases above it for another, *Pseudocalanus* exerts a stabilizing influence on the phyto-

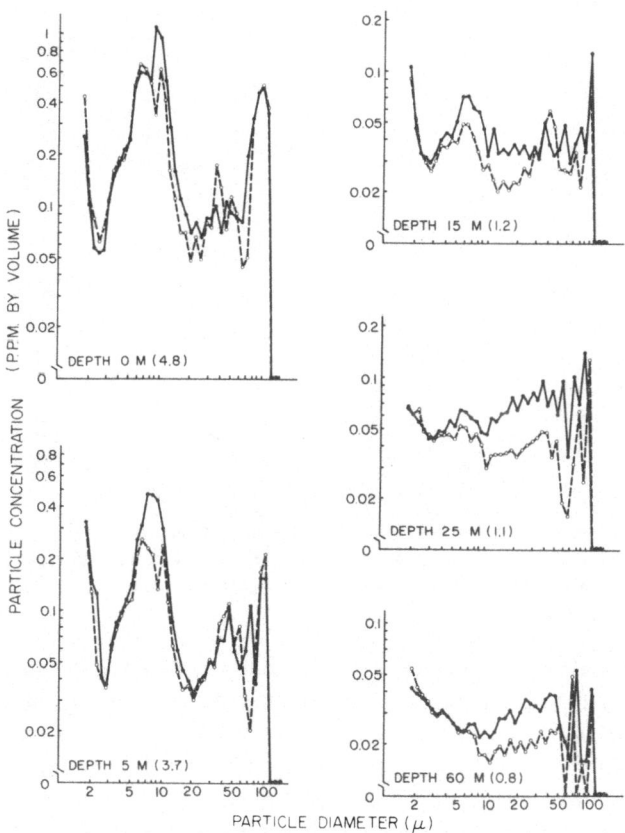

Fig. 6. Particle size spectra in the controls (●——————●) and after grazing (O-------O) by *Pseudocalanus minutus* on water samples from 5 depths at station 2. Total particle concentration (ppm) in the controls are in parentheses. Date of experiments: April 27, 1971 (after Poulet, 1973).

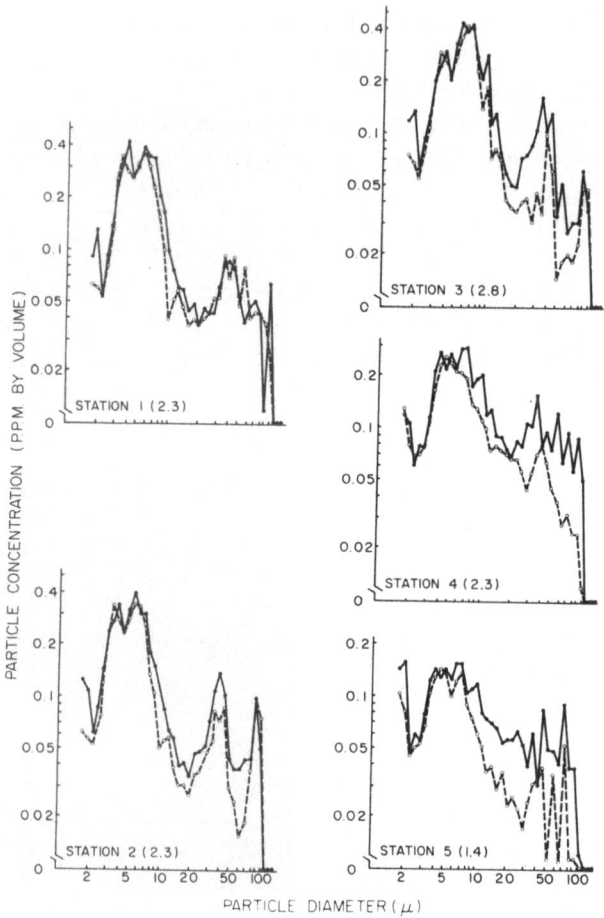

Fig. 7. Particle size spectra in the controls (solid circles) and after grazing (open circles) by *Pseudocalanus minutus* on water samples from five closely spaced stations. Total particle concentrations (ppm) in the controls are in parentheses. Date of experiments: May 11, 1971 (after Poulet, 1973).

plankton community. Thus, no single size category is grazed to extinction nor is any permitted to proliferate to the extremes characteristically associated with *eutrophication*.

The extreme case of a balanced or stable planktonic community would seem to be that found in so-called blue-water or oligotrophic locations such as the Sargasso Sea. In these localities the concentration of particles and C^{14} primary production measurements are generally low, but if grazing pressure is released, very rapid proliferation of certain-sized peaks occurred (Figure 8, Sheldon *et al.*,

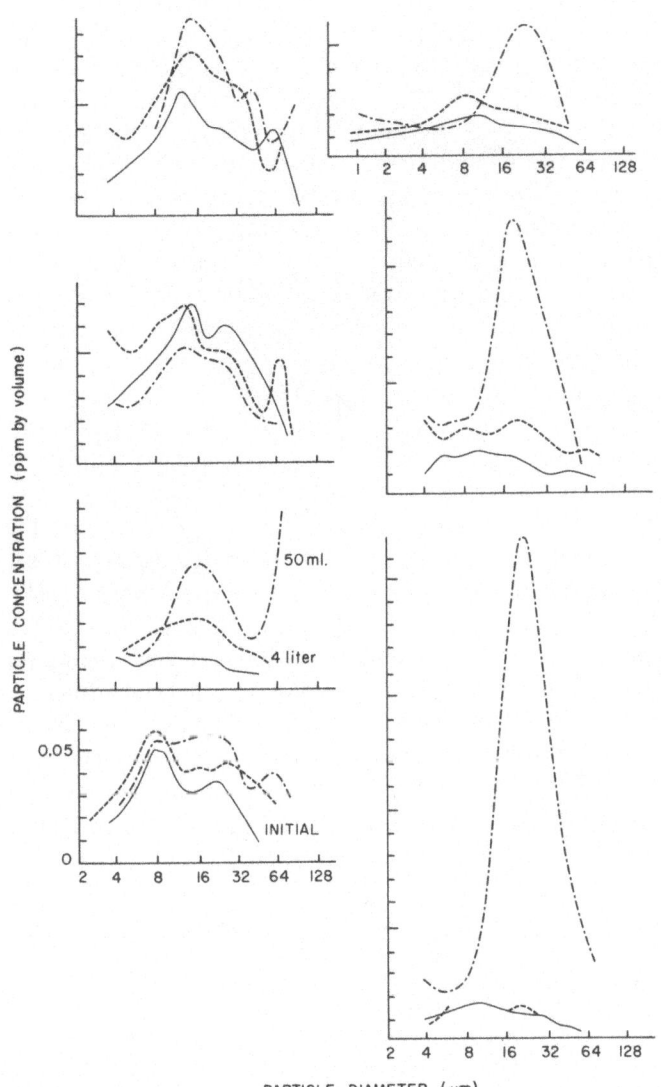

Fig. 8. The growth of particles in the Sargasso Sea (July, 1972). (——) initial standing stock; (—·—·) after incubation in a 50-ml bottle; (- - -) after incubation in a 4-liter bottle. All incubations were for 7–9 hr in the light starting at 0900. Locations were, reading from top to bottom: left-hand side—42° 17′N, 63° 23′W; 41° 44′N, 61° 36′W; 39° 05′N, 62° 27′W; 42° 09′N, 62° 55′W; right-hand side—33° 30′N, 64° 00′W; 34° 36′N, 63° 44′W; 39° 28′N, 62° 45′W (after Sheldon *et al.*, 1973).

1973). Another recent paper by Taniguchi (1973) makes a strong case for such communities having a higher ecological efficiency than northern food chains where productivity is greater and biomass of plankton tends to accumulate.

I also mentioned that there has been recent work at our laboratory using the Coulter Counter as a tool for studying the grazing of *Mytilus edulis*. These experiments were carried out in a tidal marsh at Petpeswick Inlet, about 35 miles east of Dartmouth, Nova Scotia. Initially, we tried to measure changes in particle spectra as water passed over a natural mussel bed, and though it was possible to demonstrate differences, the results were highly variable and proper controls difficult to achieve. Subsequent experiments which I mention here were carried out by Franziska Knips as part of her graduate research program (Knips, 1973). To make the experiments more controllable, a portable pumping system was built and natural tidal creek waters were pumped directly over mussels recently removed from the bed to special grazing chambers. Experiments were carried out in the fall and spring only, so really we cannot generalize too extensively, but there appeared to be no major shift in the particle sizes at peak concentration with season, even though the spring concentrations were an order of magnitude higher. As shown in Figure 9, which describes an October experiment, *Mytilus* grazed in proportion to concentration. It behaved in the same fashion with the much higher spring peak (Figure 10). When current speed was varied, maximum filtering rates and ingestion were obtained at 100 ml/min but were always less at higher flow rates. In contrast Walne (1972) found maximum filtering rates for *Mytilus* at 400–500 ml/min. We are not certain why this should be but probably the geometry of our vessels resulted in excessive turbulence near the animals at higher flow rates. Walne found this to be the case with smaller vessels he used.

While these two figures give no evidence that *Mytilus* selects any particular

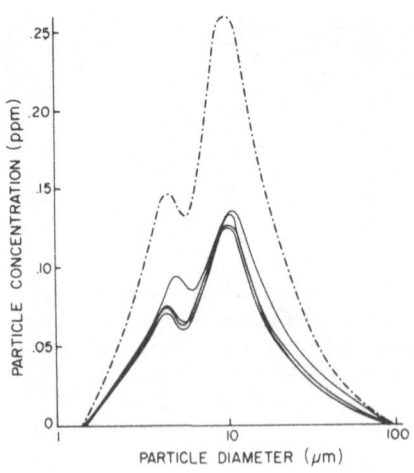

Fig. 9. October 6, 1972, size–frequency distribution for 50 ml/min flow rate; particle concentration is by volume. Broken line is from control chamber and solid lines from four separate chambers containing mussels (after Knips, 1973).

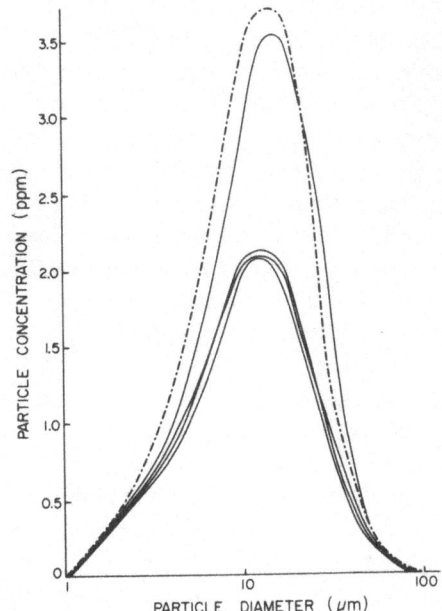

Fig. 10. May 8, 1973, size–frequency distribution for 50 ml/min flow rate. Remaining legend as for Fig. 9 (after Knips, 1973).

size range of particles, when Knips examined the percentage removed in each size category as shown in Figure 11, she found nearly uniform removal at low flow rates and low food concentrations; but as food concentration and flow rates were increased, a greater percentage of larger-sized particles were taken.

It has frequently been suggested that animals cannot afford to feed selectively when food resources are low (e.g., Ivlev, 1945). As the amount of food increases, selectivity takes place perhaps even in such an unselective filterer as the bivalve *Mytilus*.

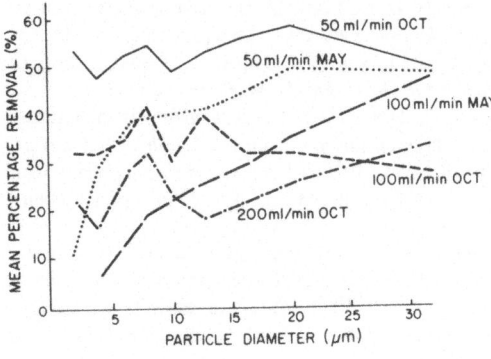

Fig. 11. Percentage removal of particles as a function of particle size (diameter) at 3 flow rates in October, 1972, and two flow rates in May, 1973 (after Knips, 1973).

What then are the implications of these studies toward planning for large-scale culture of aquatic organisms? As I have tried to show, in nature grazing tends to stabilize phytoplankton communities, and as long as there is a well-balanced population of pelagic and benthic filterers and as long as food supply limits growth, selective feeding is of minor importance. The ultimate examples of such stabilized communities are found in the tropics and they yield virtually nothing for export. Hence, population stability, per se, is not desirable if our aim is high yield of some particular food chain component. On the other hand, if we increase food supply by increasing total productivity, as by uncontrolled nutrient enrichment for instance, grazers can afford to be selective, which means that grazing pressure will no longer be balanced over the entire phytoplankton community. The effect is to release grazing pressure on less desirable forms and eutrophication results. This is not to say that a little enrichment is not a good thing, for clearly the Great Central Lake experiments have demonstrated an increase in juvenile salmon production with fertilization (Barraclough and Robinson, 1972), but careful monitoring will be required in aquaculture experiments and, unwanted species will be a necessary by-product, if gross instabilities are to be avoided. What we really want to achieve is controlled instability in which the food chain imbalances are in the species we wish to harvest. The alternative seems to be monoculture or, at best, joint culture of a selected few species that interact in a symbiotic fashion mutually beneficial to all. Only this way can we channel the benefits of increased nutrients, higher temperatures, and the like, into producing the desired species. Of course, agriculturalists have recognized this since the Bronze Age, so I really am not telling anyone anything new.

REFERENCES

Barraclough, W. E., and D. Robinson. 1972. The fertilization of Great Central Lake. II. Effect on juvenile sockeye salmon. *Fish. Bull. 70:* 37–48.

Cannon, H. G. 1928. On the feeding mechanism of the copepods *Calanus finmarchicus* and *Diaptomus gracilis. J. Exp. Biol. 6:* 131–144.

Crisp, D. J., and A. J. Southward. 1961. Different types of cirral activity of barnacles. *Phil. Trans. Roy. Soc. Lond. B. 243:* 271–307.

Dral, A. D. G. 1967. The movements of the latero-frontal cilia and the mechanisms of particle retention in the mussel (*Mytilus edulis*). *Neth. J. Sea Res. 3:* 391–422.

Frost, B. W. 1972. Effects of size and concentration of food particles on the feeding behavior of the marine planktonic copepod *Calanus pacificus. Limnol. Oceanogr. 17:* 805–815.

Heron, A. C. 1972. Population ecology of a colonizing species: The pelagic tunicate *Thalia democratica*. I. Individual growth rate and generation time. *Oecologia 10:* 269–293.

Holmes, N. (1973). Water transport in the ascidians *Styela clava* Herdman and *Ascidiella aspersa* Müller. *J. Exp. Mar. Biol. Ecol. 11:* 1–13.

Ivlev, V. S. 1945. The biological production of waters. *Uspekhi Sovrem. Biol. 19*(1): 98–120.

Jørgensen, C. B. 1966. *Biology of suspension feeding.* Pergamon Press, Oxford.

Knips, F. 1973. *Field observations of filter feeding in* Mytilus edulis *populations in Petpeswick Inlet, Nova Scotia.* Masters dissertation, McGill University, Montreal.

Madin, L. P. 1974. Field observations of the feeding behavior of salps (*Tunicata: Thaliacea*). *Mar. Biol. 25:* 143–147.

Marshall, S. M., and A. P. Orr. 1956. On the biology of *Calanus finmarchicus.* IX. Feeding and digestion in the young stages. *J. Mar. Biol. Ass. U.K. 35:* 587–603.

Monakov, A. V., and Yu. I. Sorokin. 1961. Experimental investigation of *Daphnia* nutrition using C^{14}. *Doklady (Biol. Sciences) 135:* 925–926.

Moore, H. J. 1971. The structure of the latero-frontal cirri on the gills of certain lamellibranch molluscs and their role in suspension feeding. *Mar. Biol. 11:* 23–27.

Mullin, M. M. 1963. Some factors affecting the feeding of marine copepods of the genus *Calanus. Limnol. Oceanogr. 8:* 239–250.

Parsons, T. R., R. J. LeBrasseur, and J. D. Fulton. 1967. Some observation on the dependence of zooplankton grazing on the cell size and concentration of phytoplankton. *J. Oceanogr. Soc. Japan 23:* 10–17.

Parsons, T. R., R. J. LeBrasseur, J. D. Fulton, and O. D. Kennedy. 1969. Production studies in the Strait of Georgia. Part II. Secondary production under the Fraser River plume, February to May, 1967. *J. Exp. Mar. Biol. Ecol. 3:* 39–50.

Pavlova, Ye. V. 1959. On grazing by *Penilia avirostris* Dana. *Tr. Sevast. Biol. Sta. 11:* 63–71. (Fish. Res. Bd. Can., Trans. No. 967).

Poulet, S. A. 1973. Grazing of *Pseudocalanus minutus* on naturally occurring particulate matter. *Limnol. Oceanogr. 18:* 564–573.

Poulet, S. A. 1974. Seasonal grazing of *Pseudocalanus minutus* on particles. *Mar. Biol. 25:* 109–123.

Rasmont, R. 1968. Nutrition and digestion. Pages 43–51 *in:* M. Florkin and B. T. Scheer, eds. *Chemical Zoology*, 2. Academic Press, New York.

Reeve, M. R. 1963. The filter-feeding of *Artemia.* I. In pure cultures of plant cells. *J. Exp. Biol. 40:* 195–205.

Richman, S., and J. N. Rogers. 1969. The feeding of *Calanus helgolandicus* on synchronously growing populations of the marine diatom *Ditylum brightwelli. Limnol. Oceanogr. 14:* 701–709.

Sheldon, R. W., W. H. Sutcliffe, Jr., and A. Prakash. 1973. The production of particles in the surface waters of the ocean with particular reference to the Sargasso Sea. *Limnol. Oceanogr. 18:* 719–733.

Strathmann, R. R., T. L. Jahn, and J. R. C. Fonseca. 1972. Suspension feeding by marine invertebrate larvae: Clearance of particles by ciliated bands of a rotifer, pluteus and trochophore. *Biol. Bull. Mar. Biol. Lab., Woods Hole 142:* 505–519.

Taniguchi, A. 1973. Phytoplankton-zooplankton relationships in the western Pacific Ocean and adjacent seas. *Mar. Biol. 21:* 115–121.

Tenore, K. R., and W. M. Dunstan. 1973. Comparison of feeding and biodeposition of three bivalves at different food levels. *Mar. Biol. 21:* 190–195.

Vahl, O. 1972a. Efficiency of particle retention in *Mytilus edulis* L. *Ophelia 10:* 17–25.

Vahl, O. 1972b. Particle retention and relation between water transport and oxygen uptake in *Chlamys opercularis* (L.) (Bivalvia). *Ophelia 10:* 67–74.

Walne, P. R. 1972. The influence of current speed, body size and water temperature on the filtration rate of five species of bivalves. *J. Mar. Biol. Ass. U.K. 52:* 345–374.

Winter, J. E. 1969. *Über den Einfluss der Nahrungskonzentration* und anderer Faktoren auf

Filtrierleistung und Nahrungsausnutzung der Muscheln *Arctica islandica* und *Modiolus modiolus*. *Mar. Biol. 4:* 87–135.

Winter, J. E. 1973. The filtration rate of *Mytilus edulis* and its dependence on algal concentration, measured by a continuous automatic recording apparatus. *Mar. Biol. 22:* 317–328.

DISCUSSION

RAYMONT: I understood you to say that when food is scarce, the animal—say, a copepod—will take more or less everything it can get. This then assumes that it is filtering at random. If the animal is actually sighting and grabbing it, from an energetic point of view the animal will get far more by selecting the large particle. I thought there was considerable evidence for such a selection.

CONOVER: Yes, there is, perhaps. I think we have to look here, not at a single grazing animal, but at the community of grazing animals. And as I tried to show in the experiments with copepods in the Bedford Basin, as food becomes scarcer there is a tendency for the copepods to feed on the larger-sized particles, but they tend to ignore this trend when a peak gets below a specific level, p_0. This raises a point which perhaps I did not emphasize earlier. We feel that there are a series of p_0's for different particle sizes in the sea and presumably these vary for different filterers. They will also vary for different kinds of particles as well. The threshold concentration of one cell may be quite a lot higher or lower than is the p_0 for another type of cell. When you have, in a fluctuating natural environment, a concentration of small particles, reduced a certain amount by grazing, then you start getting a shift in the general feeding pattern. This is the way individual organisms work; in a natural grazing situation you will have a lot of different kinds of organisms, and they will each tend to operate within their own range of p_0's. We are talking really of communities. We are going from the particular to a community on the basis of what we know about a few individuals. Much of it is surmise, of course.

KORRINGA: You spoke about the mucous feeding sheets on the gills of mussels. That was the concept of McGinitie at the time, which was seriously criticized. Have you any recent information that it is really true that mucous sheets are used in the filtering system of bivalves?

CONOVER: We have not examined this ourselves, but Winter thinks it is important in the feeding on the relatively low food concentrations. And if you look at the gill, the lateral cirri have motility. They can be raised and lowered to allow greater amounts of water and larger-sized particles to pass through. This is presumably the way the mussel regulates its feeding and probably the way it is selecting for size. We are getting this tendency for larger-size selection by the mussel when it is feeding on dense natural particle suspensions. In this case the mucous sheet is probably dispensed with and you are getting some ostial opening by raising the lateral frontal cirri to allow more water and particles of a smaller size range through.

KORRINGA: I have myself often thought that the mucous could be important in collecting the ions the mussels need for building up their shells. The second point that interests me very much is your reference to the size of the container. I have myself found in working with mussels that if you put them in too small containers, they won't feed normally, but if you put in large containers, the same mussel in the same fluid starts to feed. I have no explanation for this phenomenon.

CONOVER: This is a problem we have with feeding experiments on all grazers. If you put them in a sufficiently natural environment you cannot detect any grazing, and if you put them in environments that restrain them sufficiently so that you can measure changes in the actual concentration of particles in the water, then you have serious doubts about the naturalness of the experiments. We are now trying to work with flowing systems to provide a constant renewal of the environment and we can increase filtering rates this way. For the smaller grazers we still have not had too much experience with such systems. Dr. Walne has shown that there is quite an effect of the flow rate on the filtering rate of bivalve mollusks.

MATTHEWS: We are talking about animals which we can demonstrate fairly clearly are filtering. Parsons' suggestion, if I remember correctly, mostly concerned salmon fry and other carnivorous organisms. But there is certainly a lot of evidence to show that there is selectivity among true filter feeders, as you demonstrated very clearly today. I wonder, however, whether this example of *Pseudocalanus* is not exaggerating it a little bit: The deeper one goes, the bigger the copepods tend to be; the deeper *Pseudocalanus* might be adapted for feeding on bigger organisms.

CONOVER: These experiments were done effectively with the same copepods and different water. We took the copepods from one depth and compared their feeding on the particle spectra from different depths. With *Pseudocalanus* there is quite a size range of individual copepods possible, but I doubt very much that the actual feeding capability of the organism is greatly altered over this range of differences. The size of the filtering apparatus shows relatively minor changes with increasing copepod size, even if there is quite a large change in the volume or weight of the organisms themselves. Where you are comparing an organism of the size of *Pseudocalanus* to one many times larger or to a different kind of filterer like a euphausiid, that is something else. The euphausiids are obviously filtering over a different part of the particle spectrum in most cases.

CARSTENS: May I have a question on this filtering mechanism? Do these animals distinguish between plankton and clay and silt, or do they run it all through the intestinal tract?

CONOVER: I do not think we know how well they are able to select between nutritious and nonnutritious particles, or how they do it if they do achieve it. In experiments with *Artemia salina* Reeve showed that there was an upper limit to the amount of food which could be shoved through the gut, feeding the animal on cells alone, but if you put in something which was less nutritious – such as sand – on a volume basis, you could increase the total amount of material passed through the gut of the animal. The animal just continued to feed until it apparently had obtained enough to eat. A number of other organisms, e.g., bivalves, may shut down completely and cease to filter feed when confronted with a heavy suspension of silt and not enough organic matter in it to make feeding worthwhile. The copepod *Calanus* has been shown to select organic from inorganic also, but just how they do it I don't know. Apparently the organism, after collecting material on its filter and bringing it to the general region of the mouth, has the ability to decide at that point whether it is going to ingest this bolus or whether it is going to kick it away.

OPPENHEIMER: I return to the question of mucous feeding. I like to support the thought that one of the mechanisms resulting from mucous feeding is selection; that small particle size is influenced by surface chemistry and this effect becomes much more important than metabolic requirements and so surface potential or the zeta potential may be bases for selection.

If you watch filter feeding organisms such as rotifers or copepods as the particles enter the mouth area of the animals, they will select and discard materials. One of the theories is that the mucous has a surface potential which will attract opposite charged particles in the water, and thus you have a surface chemistry sieve in addition to a mechanical sieve.

PERSOONE: There is apparently a much higher selection in some animals like ciliates and rotifers as compared to crustaceans. Zillioux demonstrated in a recent paper that copepods can eat anything. He has shown beautiful photographs of latex particles in the gut, and you will never see a rotifer eating such a thing.

CONOVER: In experiments of this sort, using inert particles, you can demonstrate that a copepod will take and ingest them all right—inadvertently. They cannot avoid doing this but they can restrict the amount of feeding. You cannot get a representative feeding rate for a copepod by feeding it latex particles, but you can get the animal to take a few. This was done by Wilson when he demonstrated how the copepod effects size selection. Such experiments may possibly demonstrate how selection may work. When you are feeding on individual particles, as a rotifer does, at the moment of contact it makes a decision as to whether it will ingest the particle or not. A filtering copepod usually has to make a decision as to whether to ingest a bolus of particles which it has collected from its filter and which represent an average of particles passing through the filtering mechanism.

WALNE: A fact of great importance in the study of filter feeders is that in laboratory studies it is easy to demonstrate filter feeding of particles, e.g., by putting an animal in a glass beaker and you can stir some graphite into the water. This makes a nice demonstration, but from the animal's point of view it will be of limited significance. Instead, we must do much more in studying filter feeding in situations in which we know the animals are living, growing, reproducing, and it is only by these studies that we can get some idea whether the selection observed in a laboratory in reality means anything. One example is the study of *Phaeodactylum* by Mike Reeve. The concentrations of *Phaeodactylum* used in the lab were much higher than one would meet in nature generally.

CONOVER: With the possible exception of where *Artemia* lives.

WALNE: I've seen *Artemia* in various places and the water always seemed remarkably clear to me.

CONOVER: Probably because they are very good at filtering.

OPPENHEIMER: We made some rather interesting observations when we tried to determine the feeding behavior of shrimps by using electrophysiology techniques to measure the frequency of potentials in food ingestion. We found that under average laboratory conditions one can measure up to three or four volts of potential difference through the running sea water system to the ground, and the spurious potentials may easily confound the membrane measurement. Such potential might also influence situations where one wants to measure surface potentials or electrochemical aspects influencing the ingestion of what the animals select.

RAYMONT: The technique of microencapsulation is quite an interesting one, with regard to the question of tasting. A Bangor group have now done some quite successful experiments with feeding copepods and crustaceans like *Artemia* on these small capsules. The animals can apparently appreciate the different sizes of capsules; not only do they feed, but they grow on this food. This seems a case of automatic filtration, because I cannot believe that they *taste* the surface of the capsules.

CONOVER: I would want to know more about this work.

WALNE: I can add some more information; they are using nylon capsules, with haemoglobin as the content. The capsules are only successful when the animal is capable of cracking the nylon shell. *Artemia* has been brought to the breeding stage with this diet.

CONOVER: How does *Artemia* crack the shell?

WALNE: With its mouth parts, presumably.

CONOVER: I suppose *Artemia* is one of these organisms which might shove living cells right through the gut. Copepods generally crush anything they ingest unless the particles are very small.

PERSOONE: It might be that some of the capsules have either dissolved or cracked so that there is a certain amount of dissolved organics in the water. We don't know very much about the chemical sensitivity of the organism, and probably not only the food is important, but also the organic dissolved material around it.

PILLAY: It was my hope that after the presentation of these papers we will have enough time to discuss the basic problem. It has certainly been clearly shown that waste heat and the likes of it can be used for increasing primary production. It is being utilized to a certain extent in some countries at least in very simple systems of aquaculture, but certainly there are some major constraints preventing a larger scale of use, particularly when we are thinking of marine aquaculture. I think that in our future discussions here we should give sufficient attention to identifying these major constraints and to suggesting the means of overcoming these. The main function of a workshop of this nature would be to be able to make very specific suggestions for the solutions of some of these problems so that our present basic knowledge could be utilized more effectively to meet the food shortage of the world.

Large-Scale Culturing Systems

Salient Features of Coastal Waters
for Aquaculture

Torkild Carstens

River and Harbor Laboratory
University of Trondheim
Trondheim, Norway

INTRODUCTION

A useful starting point for the discussion of aquaculture potentials in any given region is the listing in Figure 1 of reasons why a species is absent (Krebs, 1972) or limited.

The first answer, inaccessibility, is not as trivial as it may seem at first glance. To the cultivator it immediately opens up exciting possibilities for imports, a subject which I trust will be seriously considered in any practical case.

The second diamond also points out a very interesting suboptimal situation which can be improved. Previous experience in many countries, as well as current research in this country, show selective breeding and hybridization to be extremely promising.

The third diamond sorts out various biological hazards in the particular ecosystem.

The last set of limits to the species distribution or abundance consists of physicochemical factors, and the discussion below is centered on these environmental factors.

For a given locality each of these parameters has its own regime. For many localities the natural regime has changed due to human activities, and today probably most of the coastlines of the world undergo gradual changes induced by man.

Although the type of aquaculture we discuss here assumes seaboard or hinter-

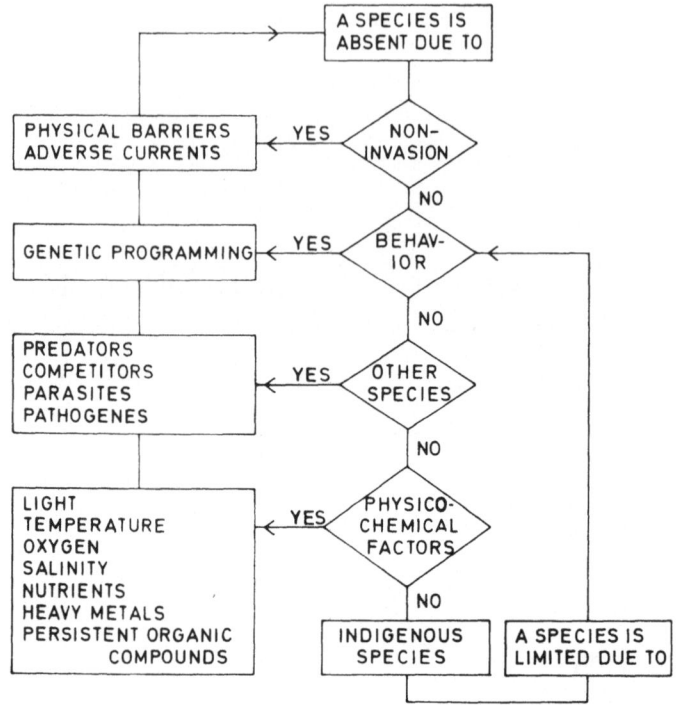

Fig. 1. Reasons for absence or limited presence of a species in an environment.

land activities that generate waste heat and nutrients, the fact remains that aquaculture is a rather exclusive use of the coastline. If the aquaculture is to thrive, there must be restrictions on other uses of the area, commercial as well as recreational. With this in mind, the discussion below will be limited to the less pressured inshore areas where aquaculture, in my opinion, is most likely to succeed: the Pacific coast north of, say, Seattle; the west coast of Scotland; and the Norwegian coast.

SHELTER AND IMPOUNDMENT

The Necessity of Shelter

The one universal requirement in all kinds of culture, including aquaculture, is protection from physical natural forces—violent winds, floods, and high waves. For this reason there is little chance ever of an oceanic mariculture, with a possible exception for fully submerged bottom operations. It is no coincidence that terrestrial freshwater aquaculture and inshore marine aquaculture are well developed and several thousand years old, while oceanic mariculture is still in its infancy. Clearly, what we have ahead of us in the foreseeable future is an expansion of a coastal mariculture, sheltered in lagoons, fjords, archipelagos—any place which is not directly exposed to ocean waves.

It is my feeling that the future structures of aquaculture systems and their operation will tolerate no more wave action than present-day harbors, that is, a maximum wave height of 0.5-1 m. This means that the fetch should not exceed 1 km, except in narrow sectors corresponding to the exposure through breakwater gaps in harbors.

While aquaculture requires a wave protection comparable to that of a marina, the flushing requirements are quite different. Harbor basins are notorious for their foul condition, brought about by reduced flushing and increased waste input compared with the surrounding waters.

The wastes of fish farms or other fertilized systems are probably no less than the waste input in well-managed harbors. Adequate flushing is essential to prevent septic bottom water, too high summer temperatures, or too low winter temperatures. We cannot, therefore, expect to succeed as cultivators simply by enclosing small basins with massive breakwaters, as is done in harbor construction. With a possible exception for Alaska and other places with large tidal amplitude, the necessary water renewal must be secured by forced flushing.

The conflicting requirements of shelter and flushing may be happily combined when the breakwater is a string of islands. The spacing of the islands should be small enough to admit only an acceptable amount of wave energy, yet large enough to provide sufficient flushing.

The Case against Impoundment

There are on record some rather amazing aquacultural successes with release of young salmon from ponds directly into the sea. Upon reaching sexual maturity the salmon, responding only to their own homing instinct, have returned to the pond. This mode of operation is aptly termed *ocean ranching*. It is biologically sound and economically attractive, but is full of legal tangles.

In Japan sea trout have been trained to obey acoustic food bells. While feeding in high concentration the fish can be harvested. Aquaculture research in this country also includes a project of this kind, with acoustic attraction or repulsion of free-swimming fish. The potentials of acoustic and other electronic manipulation of marine animals are exciting. Today we do not know whether such gear will develop into new hunting devices, or be more useful in connection with feeding as a cultivating tool.

A great advantage of unimpounded systems is that oxygen crises are ruled out for swimming species. The nets of impoundments cut down further the already reduced flushing inside the breakwaters. Expensive removal of trash accumulating on the nets is necessary to maintain healthy oxygen levels inside.

The Case for Impoundment

Conservation and productivity problems are perhaps less likely to favor impoundment in aquaculture than they are in agriculture, where fenced private property superseded common grounds centuries ago.

The dream of aquatic farmers is to set up an ecologically balanced farm which produces a desired set of species for harvesting. Among the infinite number of sets that may fill the niches in an open system, the farmer prefers one particular combination of economically attractive species. To minimize his toil, the farmer wants a closed and not an open range.

As in agriculture, there are large savings in single crop cultures, and so the marine farmer has a staple species which he harvests. The remaining species serve either as food for the top predator or as scavengers keeping the bottom clean.

Depending on the end product, the impoundment takes on various forms. For sessile populations no impoundment is necessary. The only structures required are rafts or other platforms away from the bottom with its predators and competitors.

Planktonic populations, on the other hand, require full vertical barriers if they are to be harvested under control. Devik's (1973) idea is to trap plankton in open-ended vertical cylinders so designed as to allow the necessary exchange of water through the open bottom.

Fish and moving crustaceans are normally confined by nets, screens, or bars with sufficiently small mesh size or bar spacing to prevent escape. These hindrances could be placed in the breakwater openings, allowing the animals to move freely within the basin. Current practice seems to favor pens with high fish concentration placed in basins that are sheltered against waves, but have unrestricted flushing gaps.

Perhaps the best argument for impoundment is heating. By confining the animals, they can be kept at optimum temperature with a minimum of heat input.

INSHORE WATERS

The sheltered inshore waters on which we have focused our attention are bounded by oceans on one side and by continents on the other. From either side a set of boundary conditions is imposed on the inshore water masses.

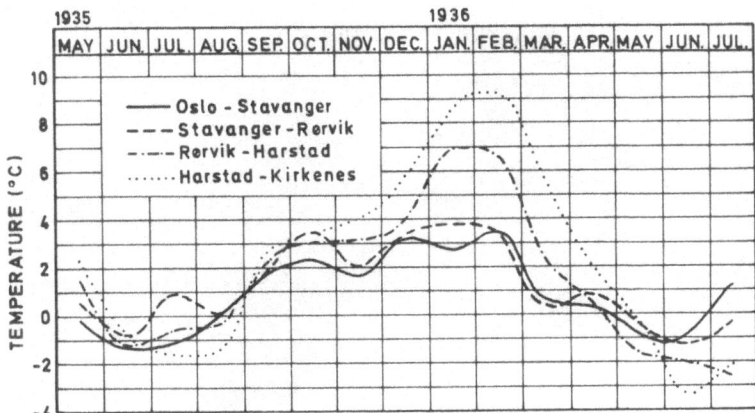

Fig. 2. Monthly mean difference between sea and air temperature (May, 1935, to June, 1936) on different sections of the coast.

Temperature. The continental climate differs from the climate of adjacent oceans for a number of reasons related, in the final analysis, to physical properties such as reflection and absorption of sunlight, heat conductivity, and specific heat.

There will always exist temperature differences between land and sea, and these differences will set up fluxes of heat and mass in a direction normal to the coast. In the water masses such transports reveal themselves in gradients of temperature and salinity; however, a more prominent source of salinity gradients is the freshwater runoff.

Figure 2 shows how the monthly temperature difference between sea and air varied in 1935–36 along the Norwegian coast. The southern section is more influenced by land than the northern section and shows fewer deviations, which means a greater temperature range in the sea.

Some observations by Hollister (1971) from British Columbia show the same trend. In Figure 3 are plotted the temperature and salinity range of daily records in 1967 for stations around Vancouver Island. Again, the exposed stations at either end of the island have shorter ranges than the more protected stations along the inside passage.

Some interesting details in the annual cycles at Cape Mudge are presented in Figures 4 and 5, showing 7-day normally-weighted running means of temperature and salinity, respectively. The fluctuations are greatest during the summer, when the temperature has its maximum and the salinity its minimum.

This increase in variance with increase in the seasonal mean compounds the environmental stress by imposing simultaneously extremes of temperature/salinity and of temperature/salinity shocks. The observed reduction in number of species with decreasing exposure may well be a result of such oscillations in climatic factors.

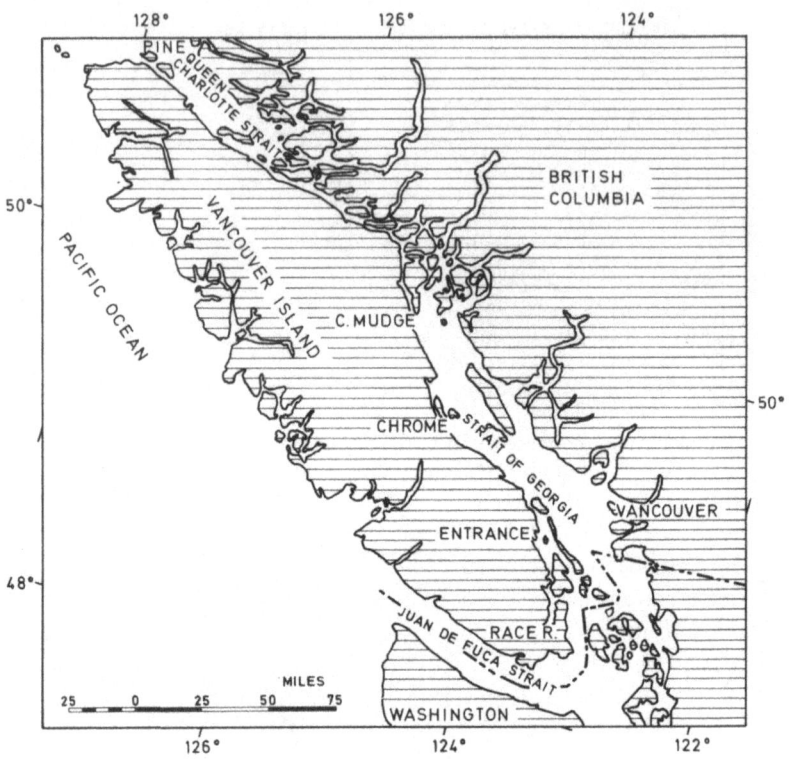

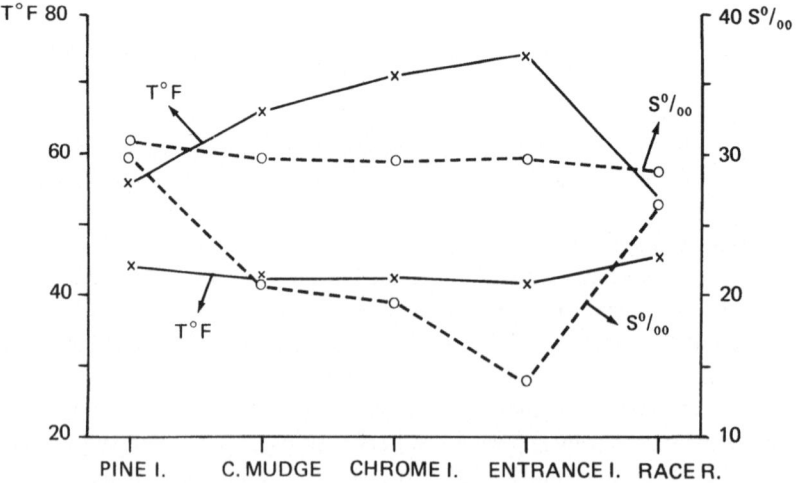

Fig. 3. Temperature and salinity range around Vancouver Island.

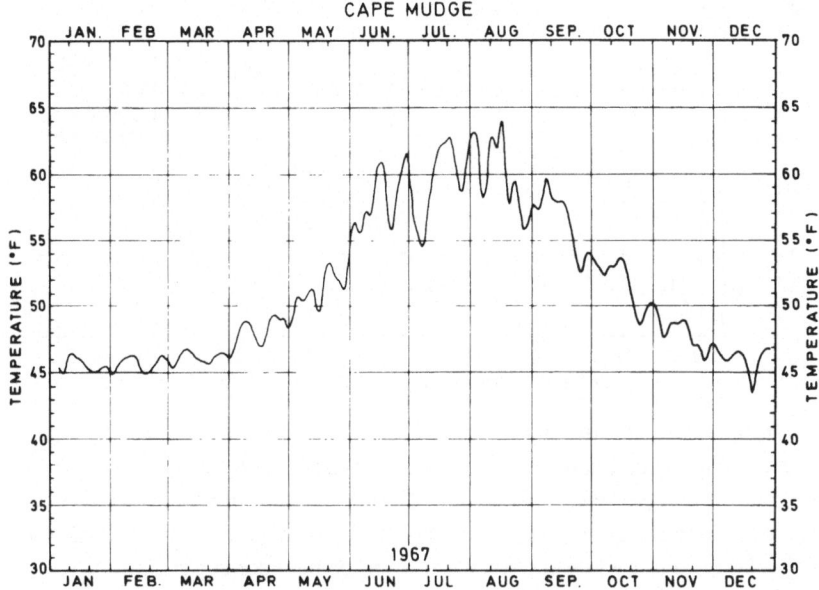

Fig. 4. Annual temperature cycle for an inshore station.

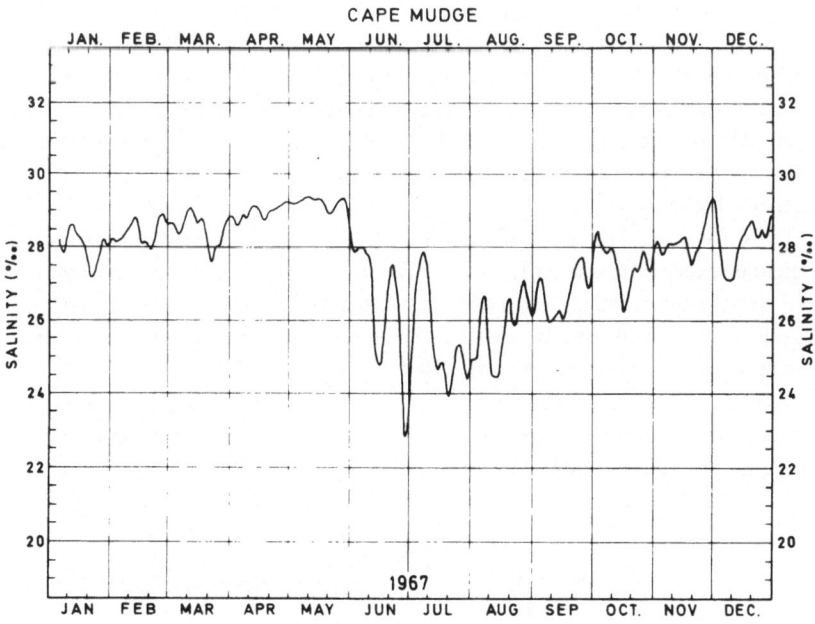

Fig. 5. Annual salinity cycle for an inshore station.

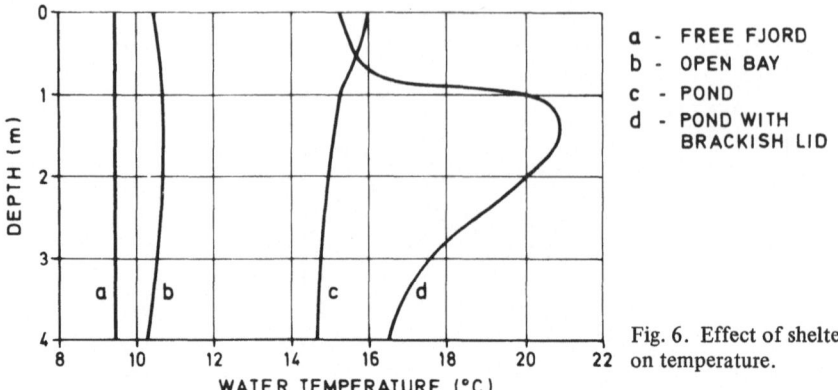

Fig. 6. Effect of shelter on temperature.

Greenhouses. A still more striking effect of the land masses on adjacent waters is shown in Figure 6. Water temperatures for localities with decreasing exposure were observed the same day. Clearly, coastal ponds may get very warm during hot spells in the summer, especially those with a lid of brackish water. Such a lid produces a greenhouse effect by trapping incoming radiation.

Gaarder and Bjerkan (1934) describe a century-old oyster culture in this country based on the subtropical climate obtained in layered coastal ponds. Although commercially unimportant at present, these breeding ponds for sub-tropical species in high latitudes (65°N) are extremely interesting as demonstrations of what is actually feasible.

Freezing. As anybody can readily observe, the more sheltered a body of water is, the more likely it is to develop a seasonal ice cover. More often than not the cause of the ice formation is a local freshwater source which stabilizes the water column. Thus there may be plenty of heat within half a meter from the water surface, but the density gradient makes this heat unavailable without mechanical stirring.

In Norway we are used to thinking of coastal ice problems as trivial and negligible except occasionally in the Oslofjord. However, the ice covers on fjord arms, smaller bays, and coves cut down on the limited acreage naturally suitable for aquaculture. In eastern Canada the ice regime may be a serious deterrent to aquaculture in an otherwise attractive area.

Salinity. The coastal source distribution of freshwater depends not only on the local precipitation regime, but also on the hinterland topography and the consequent drainage systems.

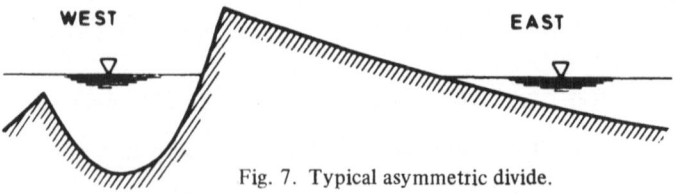

Fig. 7. Typical asymmetric divide.

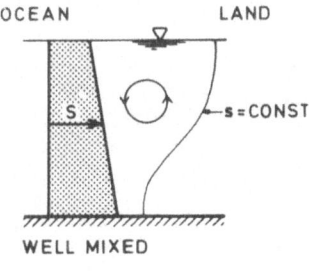

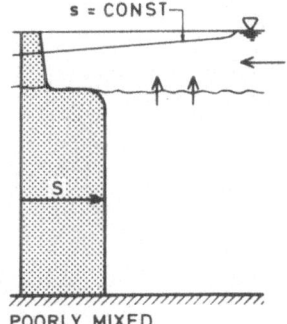

Fig. 8. The two estuary extremes.

POORLY MIXED

Typically, the west coasts of Canada and Norway have steep grades from relatively high mountains (Figure 7). The mountain slopes draw heavy precipitation when the wet air masses from the ocean are forced upward on their course from west to east (orographic precipitation). Unit runoff is therefore high per square kilometer, but not necessarily per kilometer coastline because the catchment area is relatively small.

With the divide close to the west coast, most of the land drains in other directions, and so the average runoff per km coastline is more evenly distributed than the areal runoff. The actual runoff occurs concentrated through river mouths. Rivers in turn follow topographic grades and therefore end up in fjords and bays rather than on open coasts.

Estuaries. Estuaries receive inputs of freshwater which are gradually mixed, during transit, with the underlying water masses. If the tidal currents are strong, freshwater is mixed throughout the entire water column by turbulent vertical diffusion. The isohalines are strongly tilted, and the pycnocline is weak in such well-mixed estuaries. If tidal currents are generally weak, mixing is primarily by entrainment of seawater upward into the brackish overcurrent, with little loss of freshwater downward from the surface layers.

Figure 8 illustrates the two extremes among estuaries. The poorly mixed, layered fjord is predominant on the coasts we focus attention on here.

The isohalines in estuaries move in and out with the seasons in response to the freshwater runoff cycle. They are also strongly influenced by meteorological factors such as winds and changes in barometric pressure. For these reasons salinity fluctuations are as commonplace in inshore waters as temperature fluctuations. Again, we find the highest variability where large land masses surround the water body, as in narrow fjords.

Figure 9 shows the annual salinity range in Hardangerfjorden in 1966. The source salinity outside the mouth varied 4%, while the range at the head of the fjord was 34% for the water surface. The reduction in range with increasing depth is also seen.

The osmotic problems imposed on the organisms by salinity fluctuations of

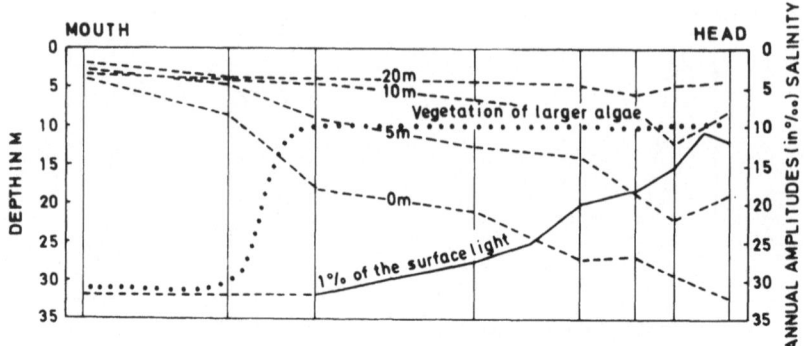

Fig. 9. Salinity range in the Hardangerfjord (dashed lines).

this magnitude must be rather trying, especially since they may occur quickly. In all probability, we have here another important environmental factor that sets inshore limits to a species distribution.

COASTAL CURRENTS

Upon leaving the estuary, the low-salinity effluent is not immediately mixed with seawater. The process takes some time, and during this time the effluent travels some distance. Normally this travel is a short distance offshore and a considerable distance longshore. Typically, the runoff joins a coastal current for several hundred kilometers, forced up against the shelf and shores by the rotation of the earth. Vortices are continually peeled off on the ocean side, while at the same time ocean water is entrained in the coastal current. Eventually, perhaps after several thousand kilometers, the current peters out offshore, diluted and powerless.

Isohalines and isotherms normally run parallel to the bottom contours, forming gradients normal to the coast as in the cross section between Norway and Denmark shown in Figure 10.

Figures 11 and 12 show long-term average temperature and salinity ranges at 4-m depth at stations in the Norwegian coastal current plotted in Figure 13 (after Braaten and Sætre, 1973). Rivers and outflows from estuaries perturb this picture with their plumes of low salinity and anomalous temperature; however, several mechanisms prevent these "rivers" from forming straight jets in continuation of their drainage channels or fjord axis.

First, the simple hydrodynamic drag of the coastal current bends the river jet into the current. Second, the water entrained from the current into the jet preserves its longshore momentum, which gradually reduces the angle between the jet and the current. Third, in inshore waters with weak ambient currents, the socalled Coanda effect, or the tendency of jets to attach themselves to the nearest boundary, may gain importance. As a result of entrainment into the jet, a hydrodynamic pressure force develops across the jet, forcing it toward the boundary.

In the absence of ocean currents, the Coriolis force, which deflects any moving mass on the northern hemisphere to the right, would eventually turn east cost rivers to the south and west coast rivers to the north. Coastal currents usually have directions coinciding with such Coriolis-induced (cyclonic) motion; however, there are exceptions. The large-scale wind-induced circulations of the oceans also feature anticyclonic motions—for instance, the southbound California current. On a much smaller scale the momentum of river or tidal jets may propel anticyclonic circulations in inshore waters when deflected by the opposite shore.

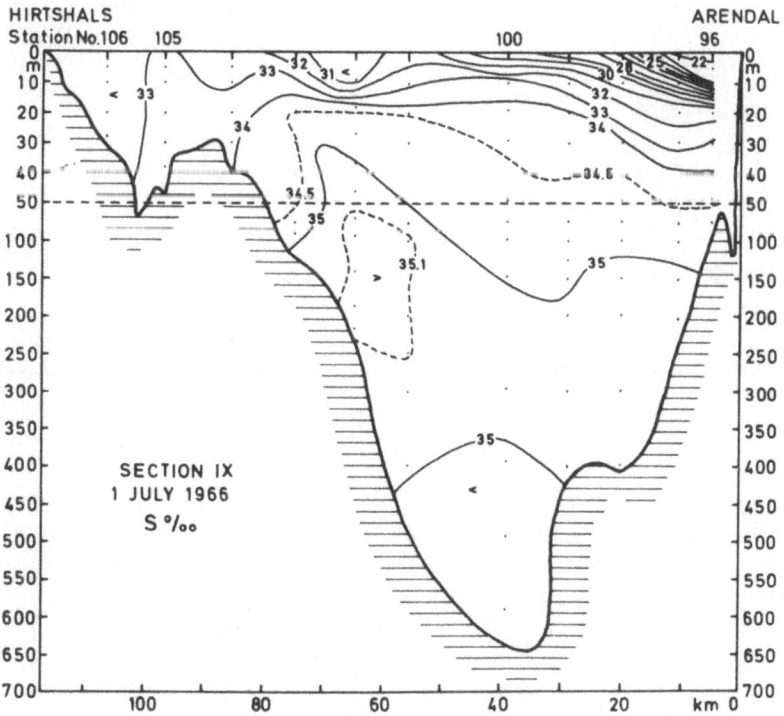

Fig. 10. Isohalines in Skagerrak.

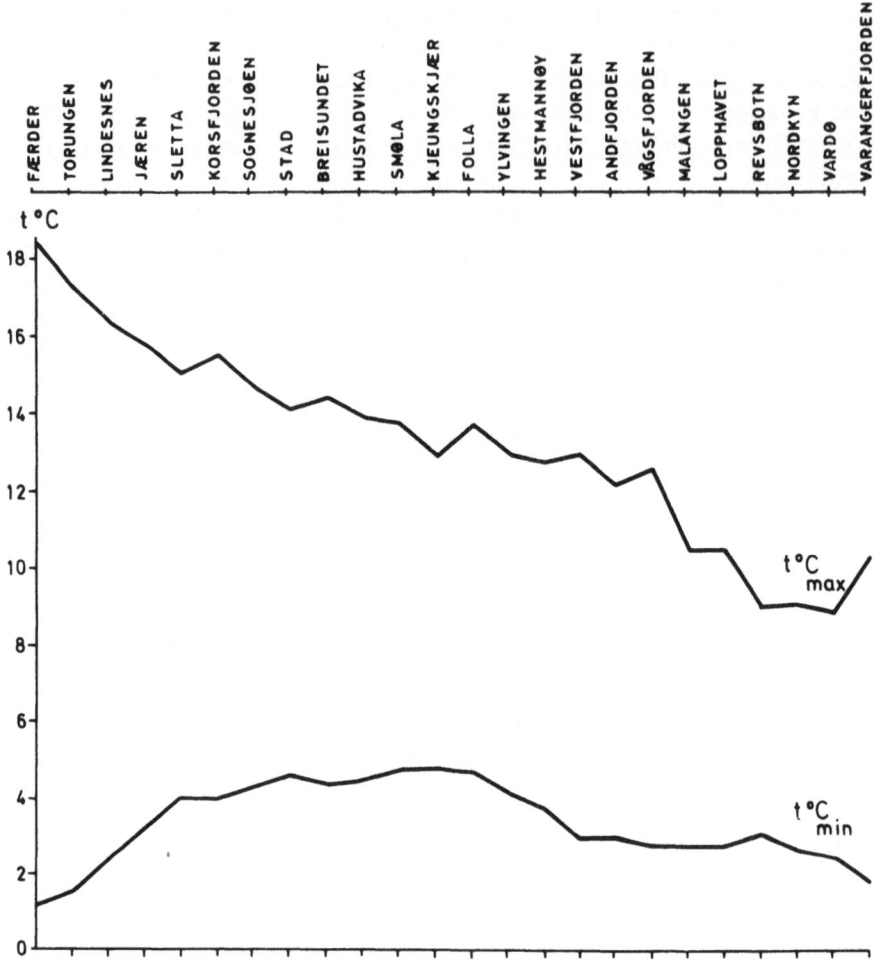

Fig. 11. Annual temperature extremes in the Norwegian coastal current.

Usually all lateral forces work together, producing a strong bending of the river jet. The largest humps on the isohalines are formed when the initial jet momentum is directed against the coastal current. In any case, lateral equilibrium is nonexistent until the river plume is forced into a coastal current or up against the coast itself.

The longshore gradients are weak, with a gradual approach towards oceanic values for t_{min} and S_{max}. The fluctuations in t_{max} and S_{min} are due to local runoff from fjords.

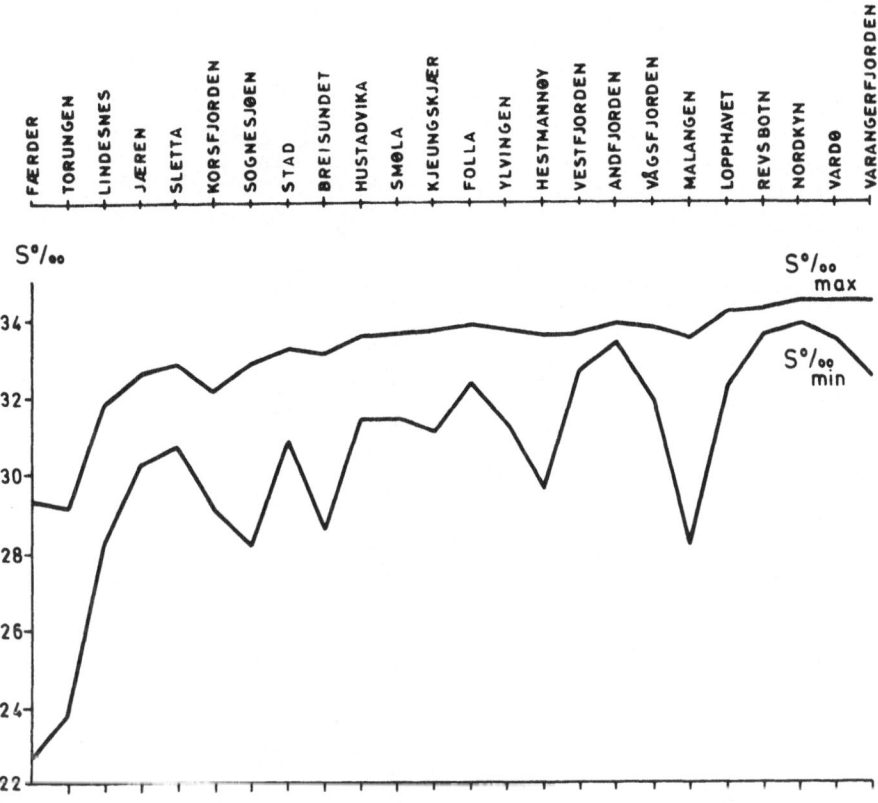

Fig. 12. Annual salinity extremes in the Norwegian coastal current.

The importance to us of these coastal currents is that they displace oceanic water away from the surface layer near the coasts. If, for instance, the coastal current becomes seriously contaminated, we shall have to go a fair distance offshore, on a shallow shelf, to find unpolluted ocean water. On a deep shelf the coastal current may be overriding ocean water. Figure 14 illustrates these two cases, and Figure 14 and the map in Figure 15 show the Norwegian coastal current.

The pollution hazard was demonstrated by the detection in 1970 of chlorinated hydrocarbons in water and animal samples in the Norwegian coastal current (Figure 16). The source of such persistent organic compounds may be thousands of kilometers away, and yet they may contaminate the entire coast if they are released in the coastal current. By contrast, urban sewage consisting of biodegradable components may have serious local effects, but no significant effects far from the outfall.

Torkild Carstens

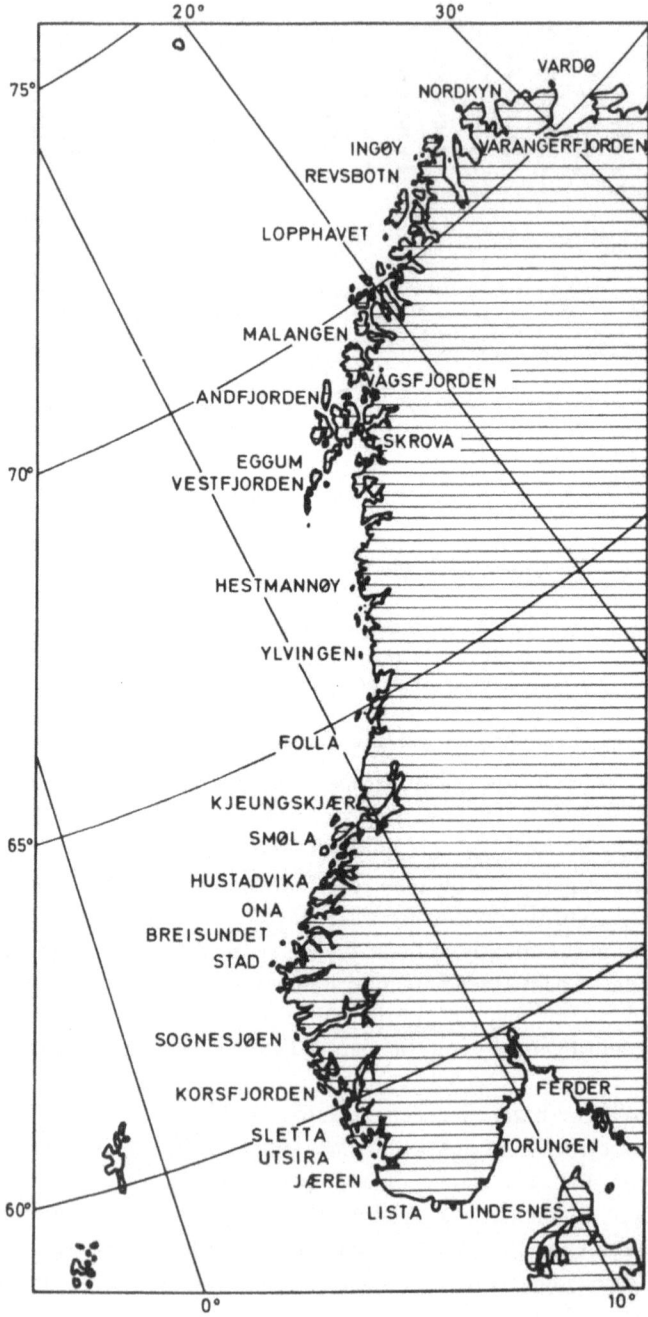

Fig. 13. Map of stations.

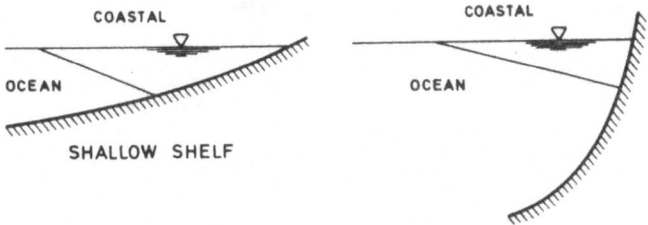

Fig. 14. Coastal current displacing oceanic water.   DEEP SHELF

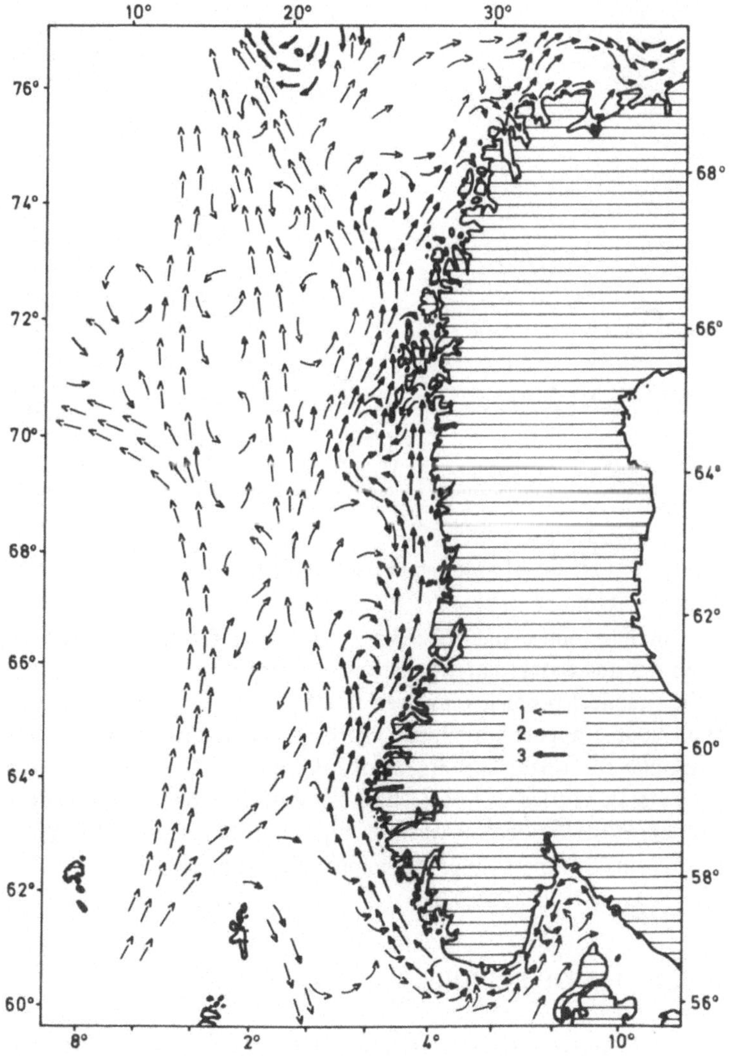

Fig. 15. Distribution of water masses and current. (1) Atlantic water; (2) coastal water;
(3) polar water.

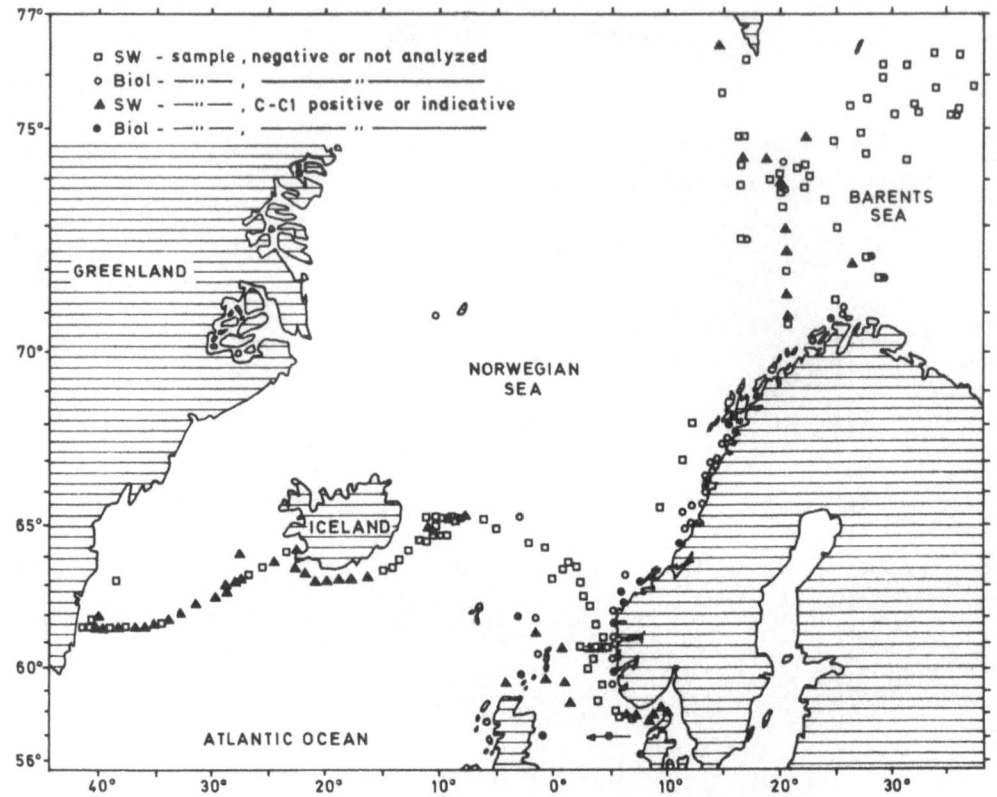

Fig. 16. Chlorinated hydrocarbons, summer, 1970. After Jensen *et al.*, 1970.

In principle, one could conceive a cultivation plant as a vessel floating in sheltered coastal water, but supplied with ocean water for its operation. However, the practical problems involved in drawing large quantities of water from a distant source appear prohibitive. A deep source directly below the vessel is much simpler to utilize. This argument rules out cultures requiring ocean water in shallow seas, but leaves open a possibility of such operations on deep coasts. The thickness of the coastal water is typically 50-100 m but it goes without saying that this thickness varies, geographically and seasonally.

PRODUCTIVITY OF INSHORE WATERS

Any productivity model sorts out light, nutrients, and oxygen as the parameters of first importance. Light is the indispensable energy source, and nutrients and oxygen must be replenished if any member of the ecosystem is to be harvested.

We can supply nutrients, and oxygen levels are readily controllable by mechanical means. When it comes to light, there is nothing we can do at present to enhance its intensity in large-scale systems. However, high productivity is not distributed globally according to the input of solar radiation. A comparison of the insolation (Figure 17) with the phytoplankton productivity (Figure 18) brings home this point.

The most productive zones of the oceans are those areas that have a dependable supply of nutrient-rich water. We can safely extrapolate this knowledge to inshore waters and limit this lowest order survey of productivity to the nutrient situation.

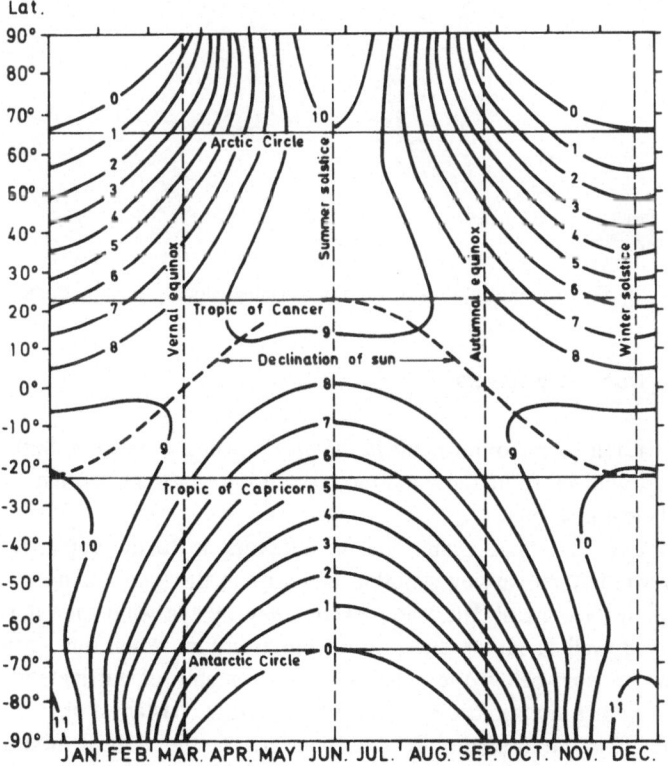

Fig. 17. Solar radiation on a horizontal surface outside the earth's shadow in hundreds of $gcal/cm^2/day$ (from von Arx, 1962).

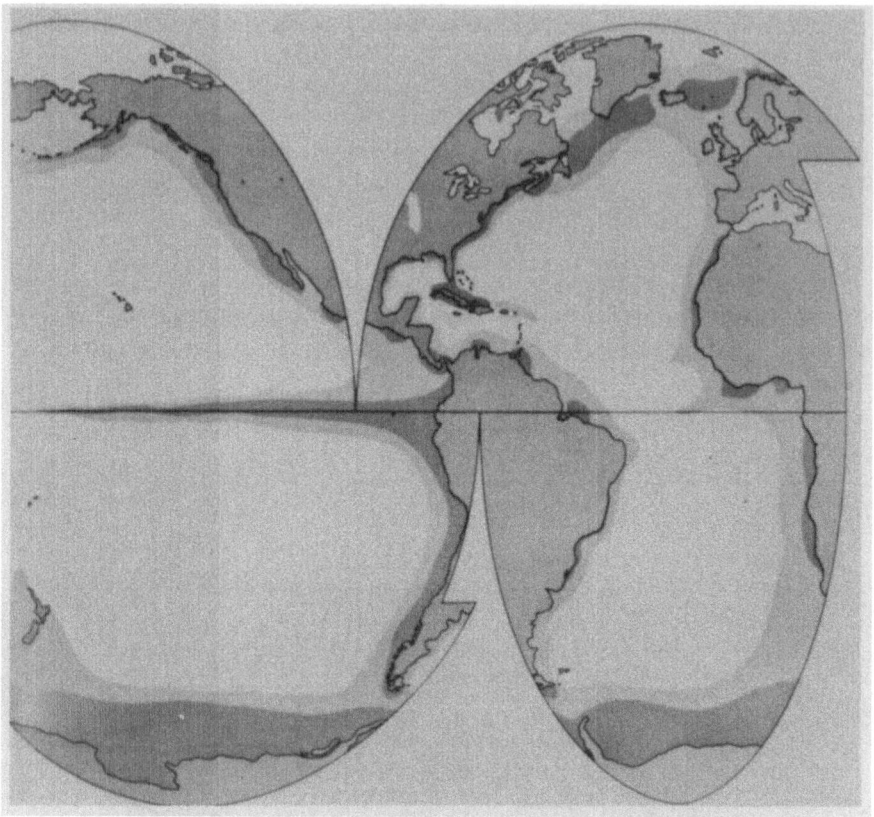

Fig. 18. Phytoplankton productivity. Dark shading indicates high productivity.

Nutrients from Deep Layers

The natural source of nutrients in the open ocean is deepwater. Sinking dead plants and animals from the euphotic layer disintegrate and enrich the water column below this layer.

Upwelling. Lifting of water rich in nutrients up to the euphotic layer is done by wind stress causing simultaneous drift in the same offshore direction over large surface areas. The water swept away by the wind then must be replaced by upwelling rather than by horizontal inflow. When this vertical flow carries water from below the euphotic zone, a supply of nutrients is secured. Onshore summer monsoons do not favor upwelling in coastal waters.

Mixing. The atmosphere also is responsible for the vertical mixing, which is most active during the cold season. The decrease in runoff weakens the vertical salinity gradient, allowing the seasonal increase in wind-generated turbulence to overturn the water column down to perhaps a few hundred meters. The initial cooling, when the water temperature is reduced from, say, 15 to 5°C, may also promote vertical exchange by thermal convection. Seasonal mixing is the chief mechanism that restores nutrient levels in inshore waters after depletion during the period with a strong pycnocline.

Compared with the offshore shelf areas, inshore waters have restricted access to deepwater nutrients. Most important is the decoupling of the upper and lower layers due to vertical stability, which imposes a shorter vertical range on upwelling and wind mixing.

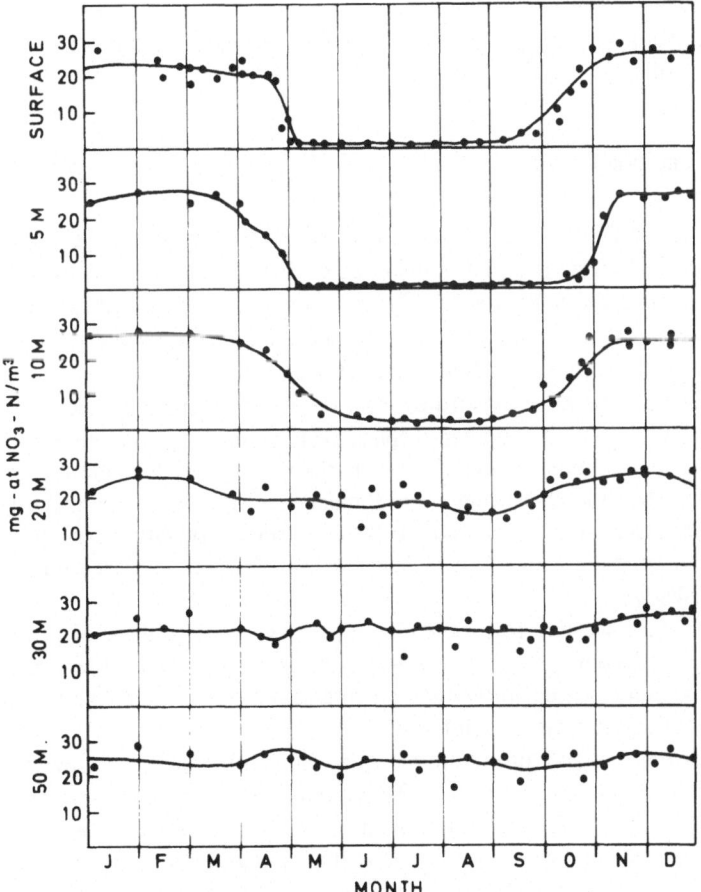

Fig. 19. Seasonal cycle of nitrate in Auke Bay, Alaska.

Nutrients from Land

Figure 19 shows the seasonal cycle of nitrate in Auke Bay, Alaska, and Figure 20 shows the associated phytoplankton blooms as reported by Curl (1972). Figure 21 is a location map.

There is a conspicuous lack of nitrate in the euphotic layer during the period of high stability. Nevertheless, several blooms are still observed and explained by local events such as occasional wind mixing, streamflow, rainfall, and *in situ* regeneration.

The pattern is familiar from other inshore areas. A predictable spring bloom after the onset of stability is followed by a series of more random events throughout the summer. Then in the fall a major bloom occurs as a result of reduced stability and renewed wind mixing.

Sakshaug (1970) found that riverborne nutrients are important in Trondheimsfjord.

Other Limiting Factors

An example of the ecological impact of natural environmental stress is given in Table I, showing the total number of species in the four parts of the Hardangerfjord outlined in Figure 22. The numbers drop off toward the head of the fjord. One of the factors contributing to increasing stress with distance from the mouth of the fjord is the salinity range (Figure 9).

Stress is caused by departures from the mean parameter value. In case of a permanent change, the number of species may increase or decrease. For instance, in polar and subpolar waters, a rise in the mean water temperature will cause more immigration than emigration. Similarly, a lowering of the mean temperature will attract fewer species than those repelled. Oscillations around a steady mean are stressing and result in fewer species than the expected number without oscillations.

The price we pay for shelter is an increase in variability, of temperature, salinity, and composition of the water masses. By reducing the variability of the waves and consequently their mixing action, we increase the variability of just about any parameter of interest.

Indentations and promontories offer numerous points of flow separation and areas of closed circulation. These traps, in fjords and bays and behind headlands, have individual flushing times that are sometimes poorly related to any general driving mechanism such as the coastal current, the monsoon winds or the freshwater runoff. To understand these local flow configurations, it is necessary to look at the current and wind fields in great detail.

It is precisely in such sheltered inshore localities, with their complex flushing regime, that future maricultural facilities are likely to be built, but before

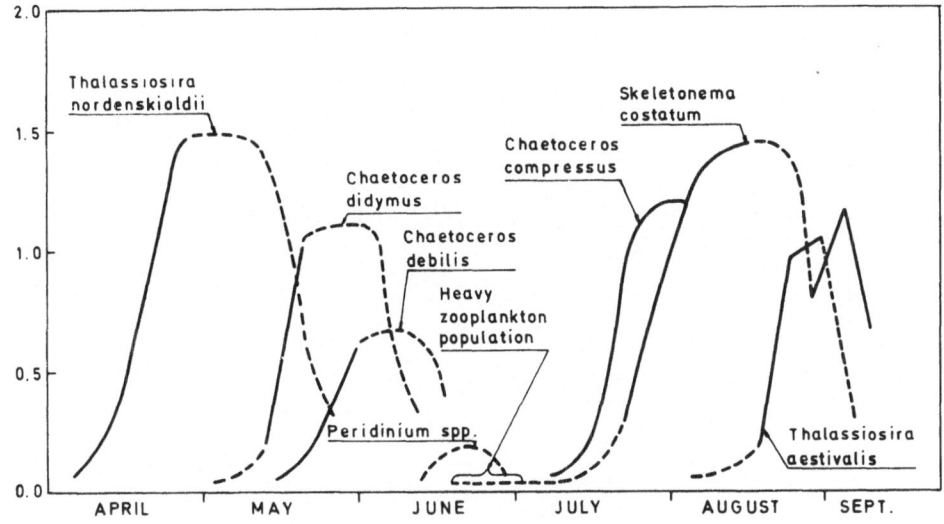

Fig. 20. Order of succession of phytoplankton in Auke Bay, 1967.

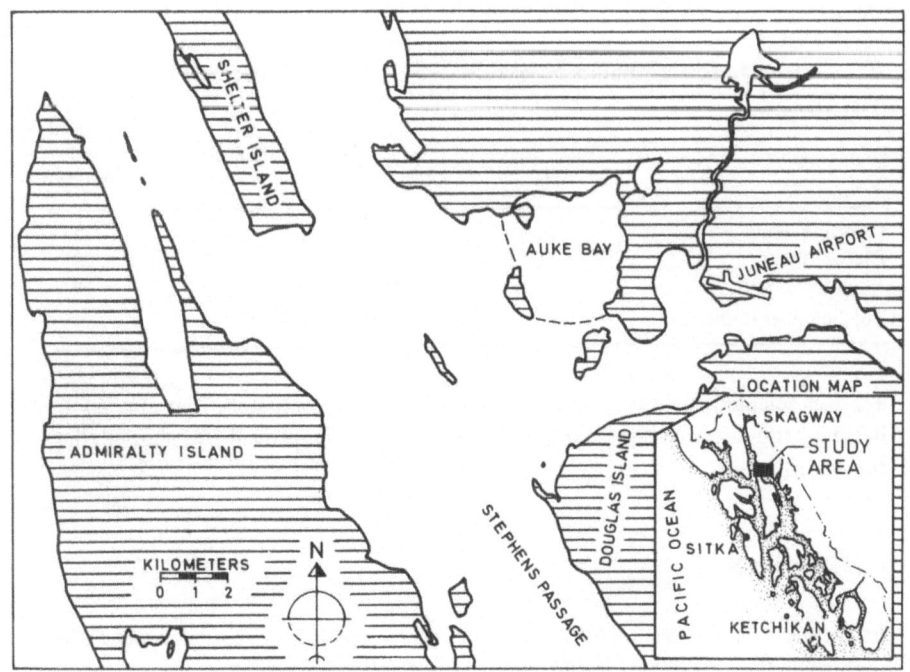

Fig. 21. Location of Auke Bay, Alaska.

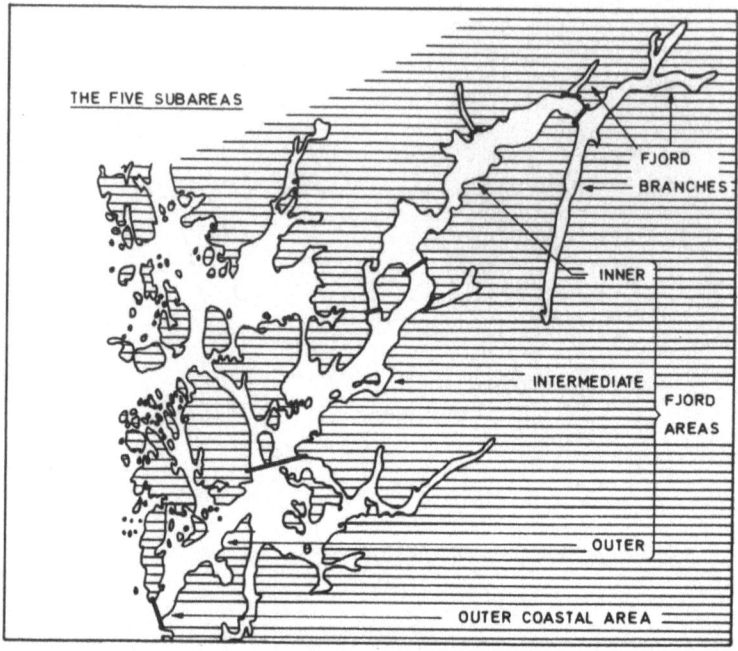

Fig. 22. Hardangerfjord and its subdivisions

Table I. Total Number of Species Recorded in the Four Parts of the Fjord

	Outer fjord area	Intermediate fjord area	Inner fjord area	Sörfjord and Eidfjord
Chlorophyceae	21	27	27	11
Phaeophyceae	70	49	48	29
Rhodophyceae	75	53	36	19
Total number in the area	166	129	111	59

a site is selected, the environmental stress should be evaluated. If it appears to be high, the site may be abandoned, or it may be improved by manipulating the environment.

REFERENCES

Braaten, B. R., and R. Saetre. 1973. Breeding of salmon in Norwegian coastal waters. *Fisken og havet B:* 2. Institute of Marine Research, Bergen (In Norwegian).

Curl, H. J. 1972. An ecosystem study in the inside passage of southeastern Alaska. 2nd ann. conf. on estuaries, Oregon State University.

Devik, O. 1973. Bruk av flytende veksthussystem som modell for estuarmiljø. Niende nordiske symposium om vannforskning, Trondheim (In Norwegian).

Gaarder, T., and P. Bjerkan. 1934. Østers og østerskultur i Norge. AS John Griegs Boktrykkeri, Bergen (In Norwegian).

Hollister, H. J. 1971. Observations of seawater temperature and salinity at British Columbia shore stations 1967. Fish. Res. Board of Canada, M.R.S. No. 1133.

Jensen, S., A. Jernelov, R. Lange, and K. H. Palmork. 1970. Chlorinated byproducts from vinyl chloride production. FAO Tech. Conf. Marine Pollution, Rome, MP/70/E-88.

Jorde, I., and N. Klavestad. 1963. The natural history of the Hardangerfjord. 4. The benthonic algal vegetation. Sarsia 9.

Krebs, C. J. 1972. Ecology, Harper & Row. P. 16.

Sakshaug, E. 1972. Phytoplankton investigations in Trondheimsfjord, 1963–1966. Contribution No. 153, Biological Station, Trondheim.

Marine Greenhouse Systems: The Entrainment of Large Water Masses

Ole Devik

Christian Michelsen Institute
Bergen, Norway

Enclosures made of floating plastic curtains were tested for water exchange to the ambient through the open bottom. The dimensions of the enclosures were 6 m in diameter and 12 m deep.

The water column was stabilized by a density gradient, and measurements of salinity and temperature indicated that the elimination of the horizontal exchange in the upper layers will lower the rate of the vertical exchange of the enclosed water sufficiently to allow complete utilization of added nutrient before they are dissipated.

Repeated weekly fertilization with nitrate (about 100 mgat/m³) and phosphate (3.5 mgat/m³) during three weeks gave a plankton bloom that depleted the nitrate within one week of the final addition. Values of ammonia after one week of fertilization reached a relatively constant level, while the level of inorganic phosphate remained relatively high during the whole period.

INTRODUCTION

The waters of the coastal areas of Norway can be characterized as sheltered, with relatively steep shores, so that the major part of the waters have a depth larger than 10 m. Apart from the landlocked fjords, the water moves relatively freely, and, at least in the upper layers, the horizontal exchange is relatively rapid.

It is these layers that are the productive ones from the point of view of aquaculture. The exchange rates are, however, generally too high to permit a localized increase of nutrients by means of fertilization. Moreover, where such conditions exist, intensive fertilization will often lead to undesirable changes in the bottom layers. Here, exchanges will occur at infrequent intervals, and anoxic conditions will occur regularly.

The direct utilization of fertilized waters requires areas of a magnitude that make a complete enclosure difficult, particularly when one objective is to minimize the undesirable changes in the substrata. The evolution of large-scale aquaculture in sheltered, deepwater areas may be said to depend upon development of arrangements to contain large volumes of water and methods for regulating their productivity.

OBJECTIVES OF THE ARRANGEMENTS

We can cite several objectives for a suitable arrangement: Lowering the exchange rates of water within the arrangement to give holdup times of added water for periods of five days and upward, control of the input of phytoplankton nutrients; control over the species composition in the system, both of primary procedures and of grazers and predators; rugged simplicity, to give convenience in the handling of the enclosures and durability against adverse weather conditions.

In addition to these general objectives, to be suitable for aquaculture the arrangements should also give adequate return of the invested efforts in terms of products obtained and savings effected for measures that are demanded for the protection of the environment; and develop culturing techniques that do not need complicated instrumentation or control procedures that might be called for in the development of such techniques.

This chapter will describe experiments with a marine greenhouse system based upon the use of closed plastic curtains, where the water column is stabilized by a density gradient. The main objectives of the experiments have been to see whether an elimination of the horizontal exchange from the surface down will lower the vertical transport sufficiently, and to make preliminary investigations with regard to the persistence of fertilizer effects in terms of recycling.

Schematically the system can be described as in Figure 1. The diagram indicates the main flow paths of the various components of the system.

Similar arrangements have been used by MacAllister *et al.* (1961) in the form of closed plastic spheres with their buoyancy controlled by regulating an air volume enclosed in the sphere. Goldman (1962) has used long plastic bags, open at the bottom, for experiments with semienclosed systems in freshwater, which limits the possibilities for the stabilization of the water columns.

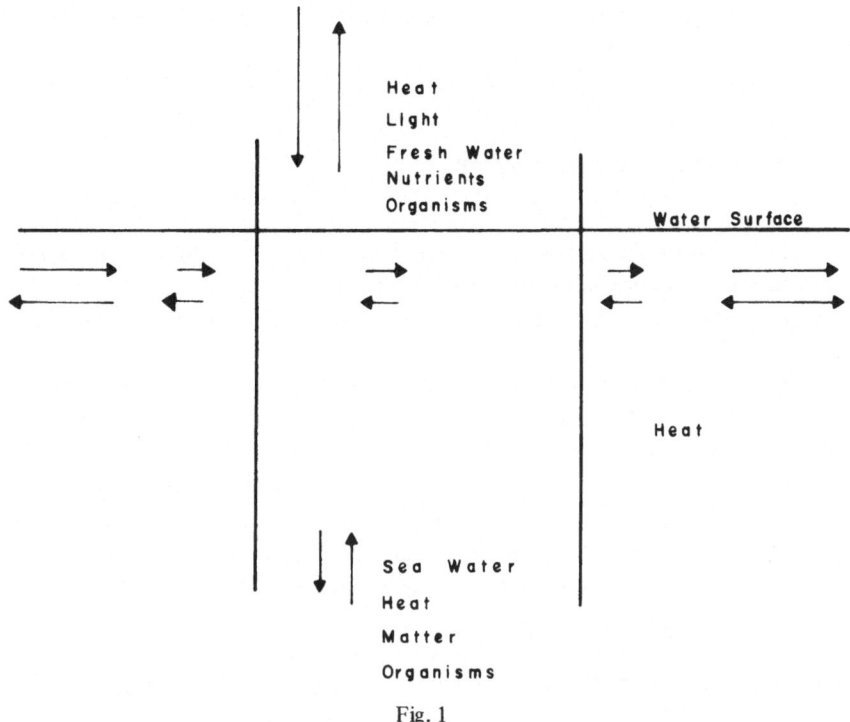

Fig. 1

In the systems outlined in Figure 1, the flow of heat to and from the surroundings will occur in all three coordinate directions, while the overall exchange of matter between the system and the ambient will only occur in the
vertical directions.

The horizontal transport of heat will be coupled with the vertical transport
of matter. Through the lower surface, which is defined by the lower edge of the
curtain, there will be a net diffusive flux of the various components of the
seawater.

Through the upper surface, there will be an input of light, heat, and freshwater as precipitation or by pipeline, as well as an input of nutrient and organisms according to schedule.

In the experiments we wanted to evaluate the stability of the water column
inside the enclosure, the magnitude of the pycnocline needed for the stabilization, the principal factors that determine the stability, what the probable ratio
cross section: depth of the system and the possible dimensions will be, how much
rapid exchanges in the ambient water will make themselves felt inside (periods
with a marked surface cooling are of particular interest), what the possibilities are
for the control of unwanted grazers, how far we can control the species spectrum

of the primary producers, and how adverse weather will affect the structural elements of the arrangement.

EXPERIMENTAL ARRANGEMENT

The enclosures should withstand wind up to gale strength, and effectively eliminate the entrance from outside of waves with heights up to 1.0-1.5 m. This wave height will, for winds of gale strength, correspond to a fetch of about 3 km or less.

The site was chosen so that the exchange to the ocean was relatively free. It was thought that this would give more rapid exchanges than one would encounter in the fjords themselves.

Figure 2 indicates the location of the experiments. The place is near Bergen, on the western side of the island of Sotra, about 7 km from the open sea. At the site there is a sloping bottom, the depth ranging from 15 to 20 m. The bay forms part of a larger basin, with a surface area of about 5 km^2, connected to the ocean by sills of 15-30 m.

The tide has an amplitude of 1.0 to 1.5 m. In the outlet just north of the site, measurements showed that the current changed direction in phase with the tides, with peak velocities in each direction of 7-10 cm/sec. No current measurements were done on the site itself, but from general observations it seems that there is a general water exchange following the tide, and that currents usually have had a velocity up to 2.5 cm/sec.

The experimental arrangement is shown in Figure 3. The enclosure itself was made of split-fiber nylon material, reinforced in the upper and lower parts by tarpaulin cloth. The vertical dimension of the curtain was 12 m, its length was 20 m. It was closed by a slip-tubing arrangement that ran along the vertical roping of the curtain. Weights of about 15 kg/m secured the vertical position of the curtain, which had a neutral buoyancy because foam sheeting was attached to the upper 3 m to compensate for the load.

The curtain was surrounded by a collar made of polyethylene tubing carrying davits which secured a freeboard of 25 to 30 cm of the curtain. The buoyancy of the collar was about 10 kg/m, to provide some margin for any additional load from fouling (by mussels, for example).

Freshwater and mineral nutrients were supplied by a plastic hose taking water from a small stream near by, so that additional water at a rate of 0.25 m^3/m^2/day could be obtained.

The system was operated during April-October 1972. After an initial testing period of two months, fertilization experiments were started. The composition of the nutrient mixture was made up according to Guilliard and Ryther (1962),

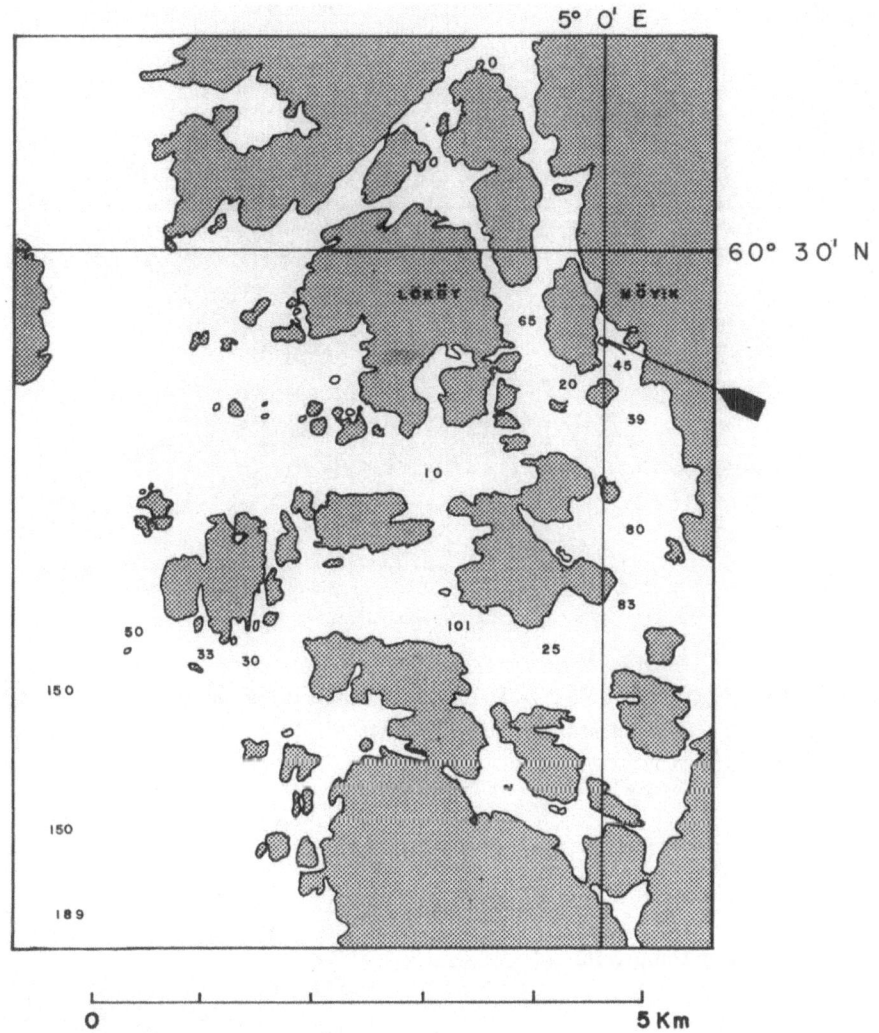

Fig. 2. Location of the experimental site M = 1 : 50 000.

and the amount adjusted so that the level of NO_3-N and PO_4-P during the fertilization period should be at an average roughly eight times the level of the ambient. The fertilizer was added as a concentrated solution in three portions, one week apart, to the surface layer.

Salinity and temperature was measured daily to 0.005% and 0.1°C, respectively, with a temperature-compensated conductivity cell and a thermistor-measuring bridge. Changes in the nutrient conditions were followed by analyses

Fig. 3. The enclosure.

of NO_3-N, NH_3-N, and PO_4-P. A rough impression of the primary production was obtained by measuring the reflectance of Millipore filters obtained from samples drawn at various depths.

RESULTS

Salinity and Temperature

The changes in salinity and density are summarized in Figures 4, 5, and 6 for both the system itself and the ambient. Figure 7 indicates how the average salinity of the 0–3 m and the 9–12 m layers vary inside the system, compared to the salinity at 12 m.

The measurements of the ambient indicate water movements with a period coinciding with that of the tides, on which are superimposed periods of lower frequencies. The stability of the ambient water was high during spring and summer, while during the fall and the winter months the mixing in the upper layers was rapid.

The diagrams indicate for the period of July a relatively high degree of uncoupling between the inside and the outside. During the period of August and

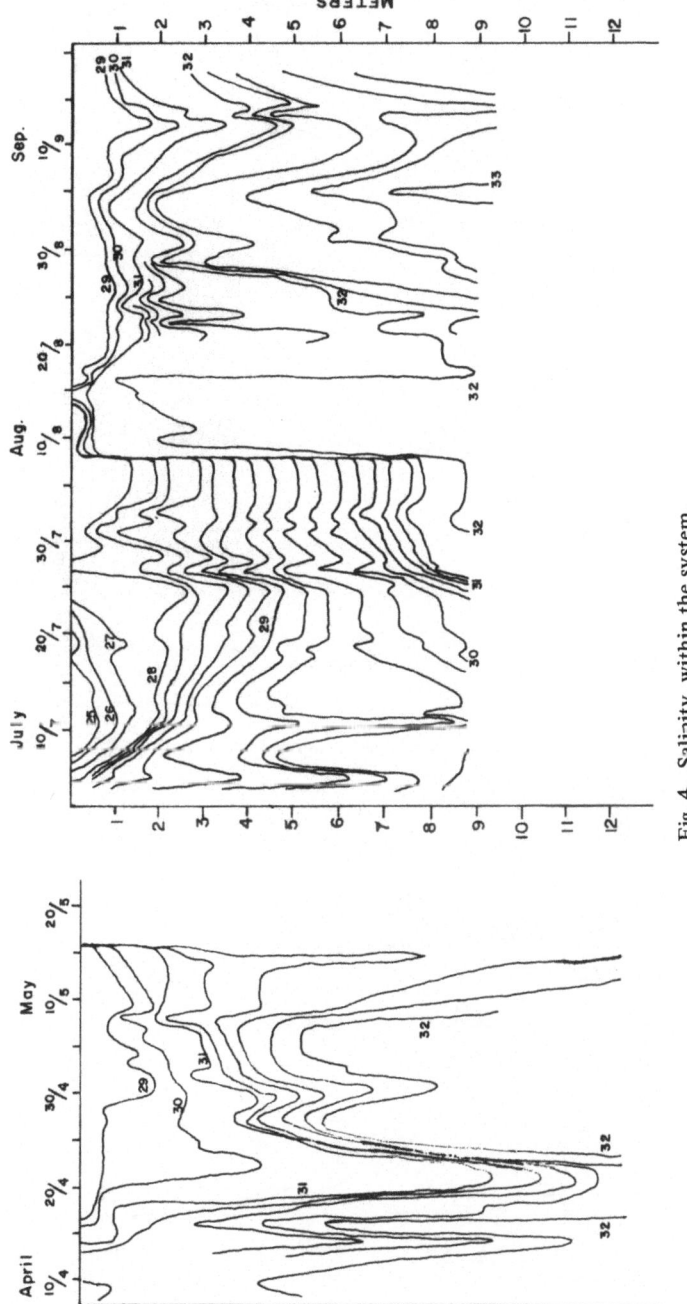

Fig. 4. Salinity, within the system.

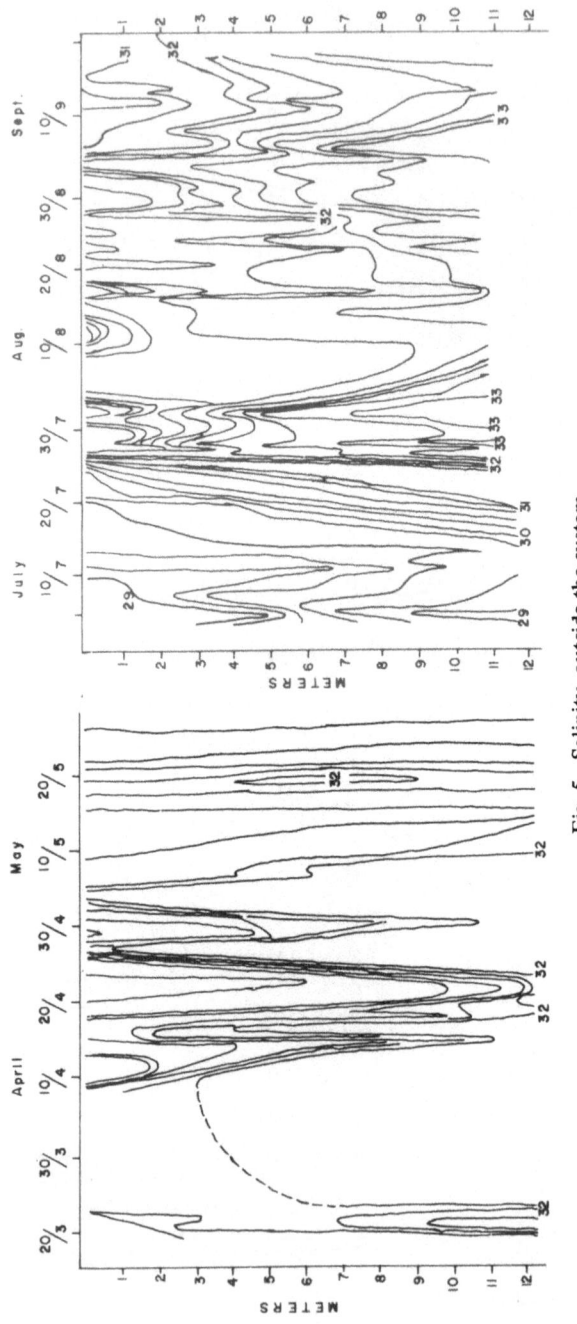

Fig. 5. Salinity, outside the system.

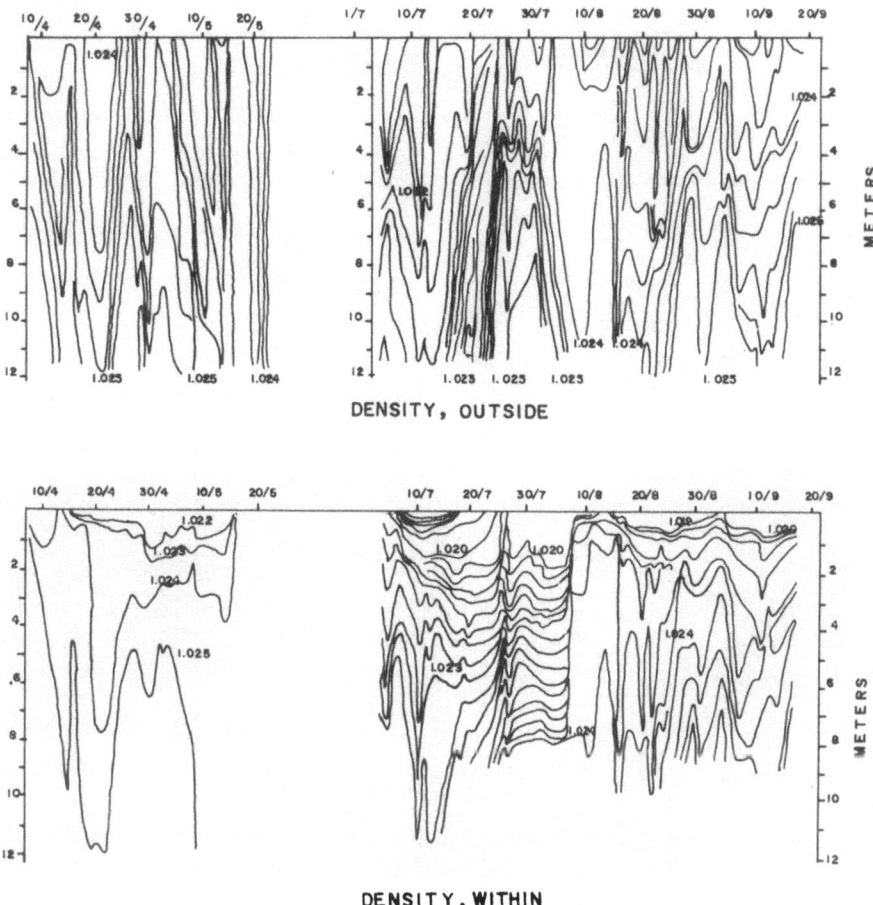

Fig. 6. Density, within the system and outside.

September the damping of the changes from the outside is rather less, and mainly restricted to the upper third of the system.

The Effect of Fertilizing

A standard nutrient mixture (Guilliard and Ryther, 1962) was introduced in the upper layers on July 7, 14, and 20. In Figure 8 the total amount added of NO_3-N and PO_4-P are indicated, to compare with the total amount of NO_3-N, NH_4-N, and PO_4-P determined by analyses.

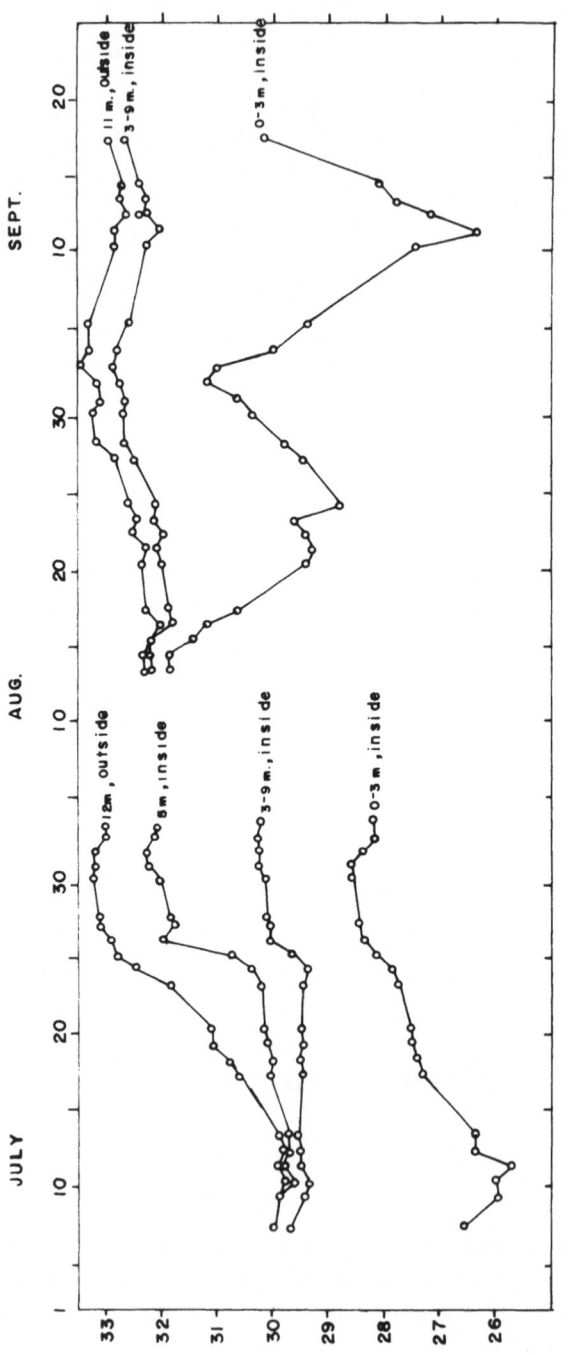

Fig. 7. Average salinity of the system.

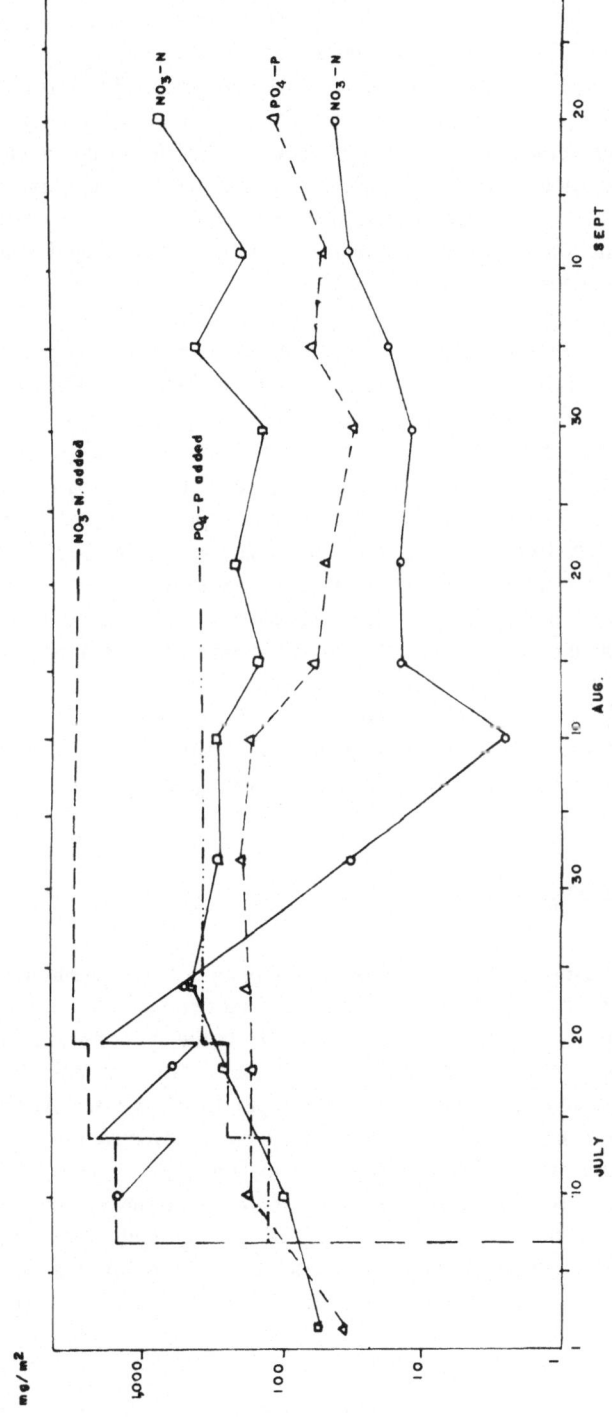

Fig. 8. Total amount of NO_3–N and PO_4–P input to the greenhouse system and the subsequent change of concentration.

The analyses indicate a rapid consumption of NO_3-N. Ten days after the last addition the level has dropped to about 4% of the original, corresponding to a half-life of about 2.2 days, or a time constant of about 3.1 days.

Figures 9, 10, and 11 demonstrate in more detail the distribution of the nutrients inside, compared to the values outside. The persistent high level of PO_4-P is interesting when compared to the sharp decline in NO_3-N during the same periods. One explanation could be that NO_3-N is assimilated at a higher rate, so that PO_4-P is in excess and may be regarded as a semiconservative component. Another explanation might be that grazing organisms create a rapid turnover of PO_4-P, with a concomitant increase in NH_3-N.

To obtain a rough idea of the phytoplankton density, 250-ml samples from various depths were filtered through Millipore filters and their reflectance was measured and compared to the reflectance values of samples from the outside. Figure 12 shows that the outside waters show little growth during the period. The values of the inside samples from the upper layers indicate that the phytoplankton growth is concentrated in these layers, and that there is a fairly sharp decrease in the abundance after the fertilization is stopped (Figure 13).

Pooled samples from the inside and the outside were preserved with formalin, for later determination of the abundance of various species of the phytoplankton. Table I summarizes the results. The main difference between the inside and the outside seems to be the preponderance of *Heterocapsa triquetra* inside the system, and that parallel to this there is a somewhat larger population of *Coccolithus huxleyii* within the system. The dominance of these species seems to disappear during the last experimental period (8/28), at which time there seem to be only small differences in the species spectrum.

SUMMARY OF THE RESULTS

For the evaluation of the possibilities of the open-bottom greenhouse system to sustain a continuous culture of relatively constant composition, it is essential to obtain an idea of the exchange rates of water from the upper productive layers with water from the lower layers. The continuous culture system will be one where nutrients and inoculum serve as the input to the surface layers, while unwanted phytoplankton will enter from below by an advective mechanism, together with grazers which will be more or less freely moving.

The exchange is conveniently expressed in terms of the vertical eddy diffusivity, which will be a function primarily of the density gradient of the system. By definition, the eddy diffusivity is the ratio between the flux of a component due to the random transport processes across a boundary and the density gradi-

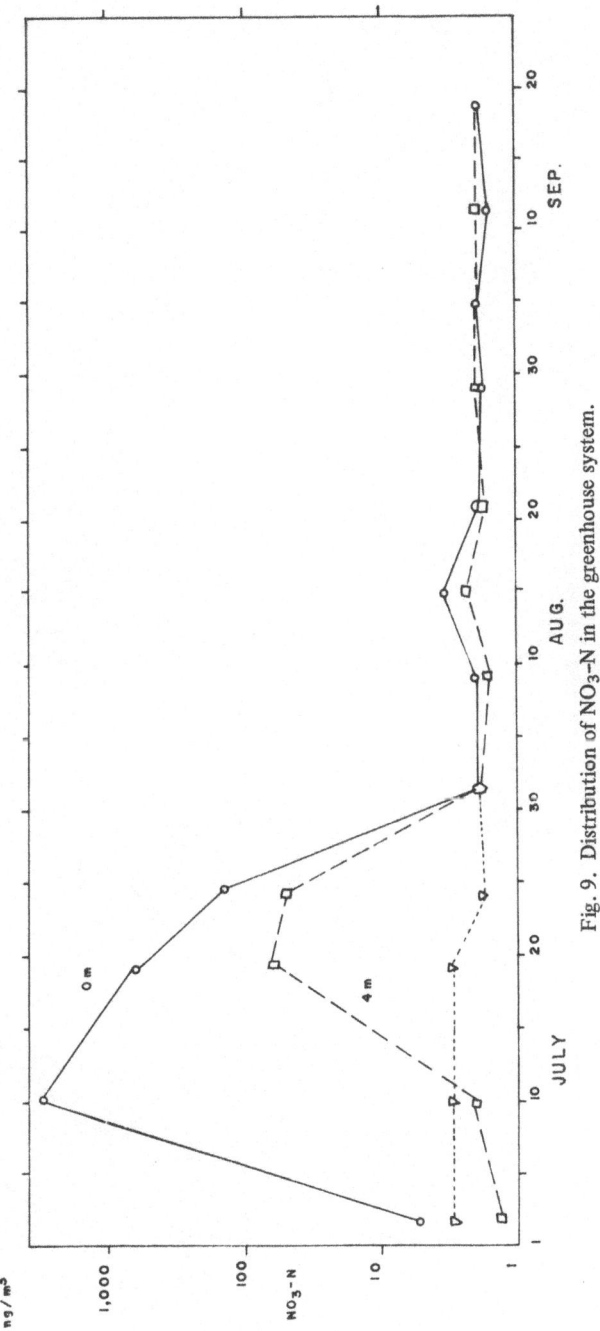

Fig. 9. Distribution of NO₃–N in the greenhouse system.

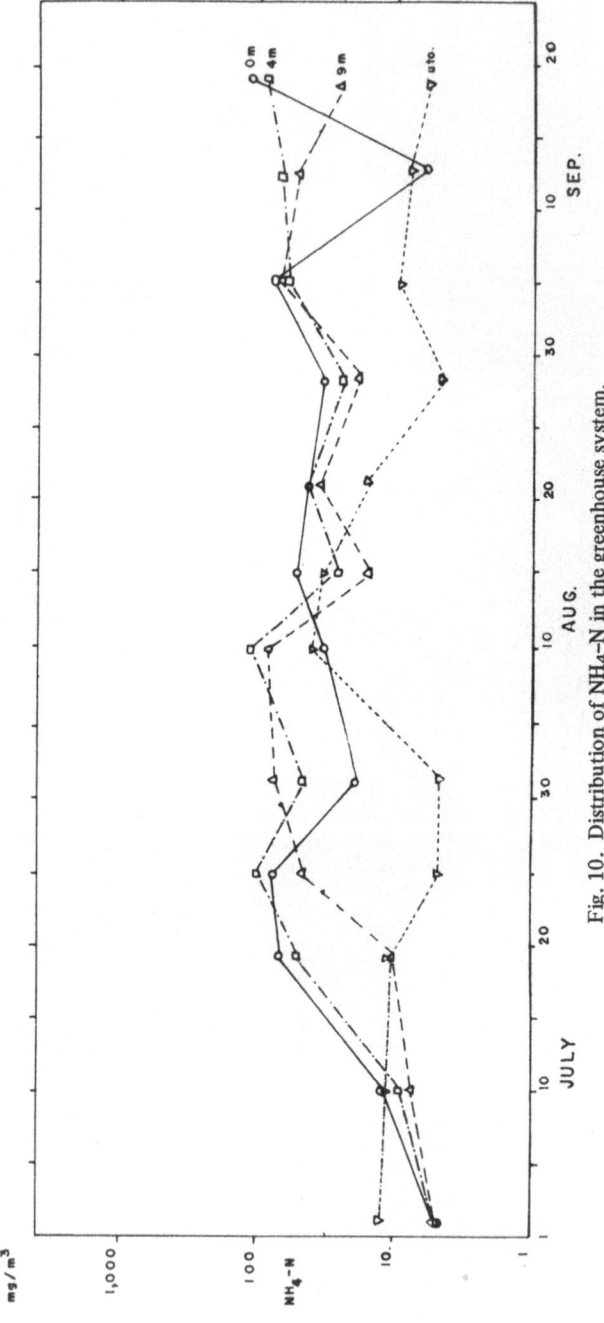

Fig. 10. Distribution of NH₄-N in the greenhouse system.

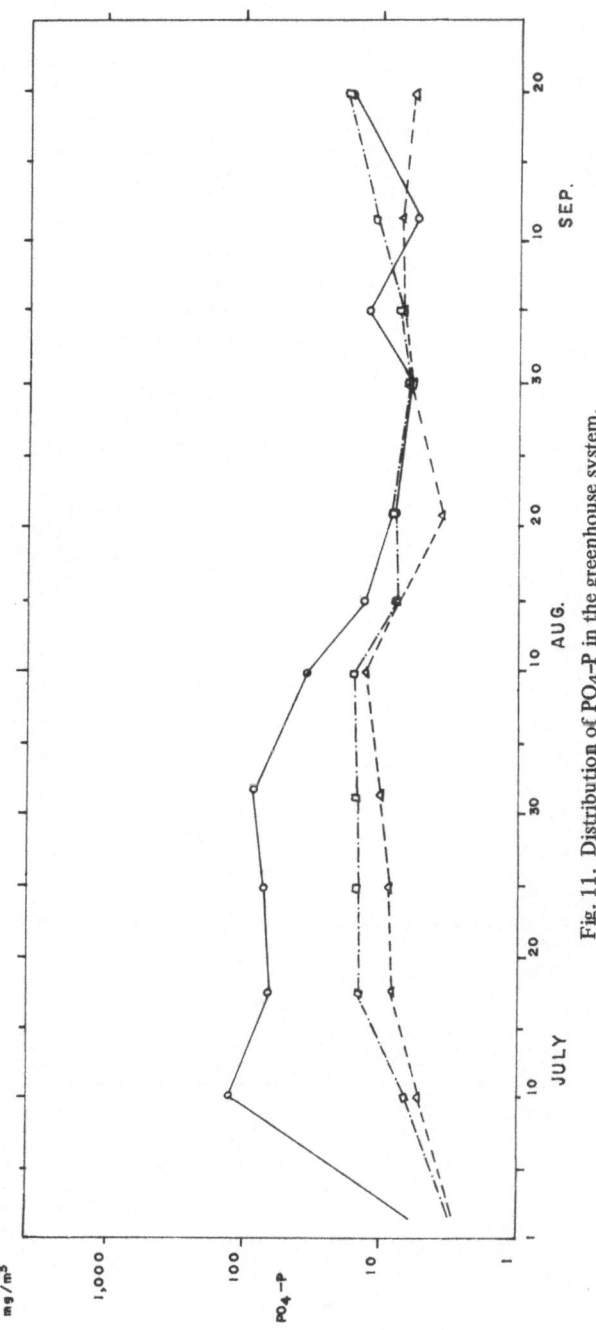

Fig. 11. Distribution of PO₄–P in the greenhouse system.

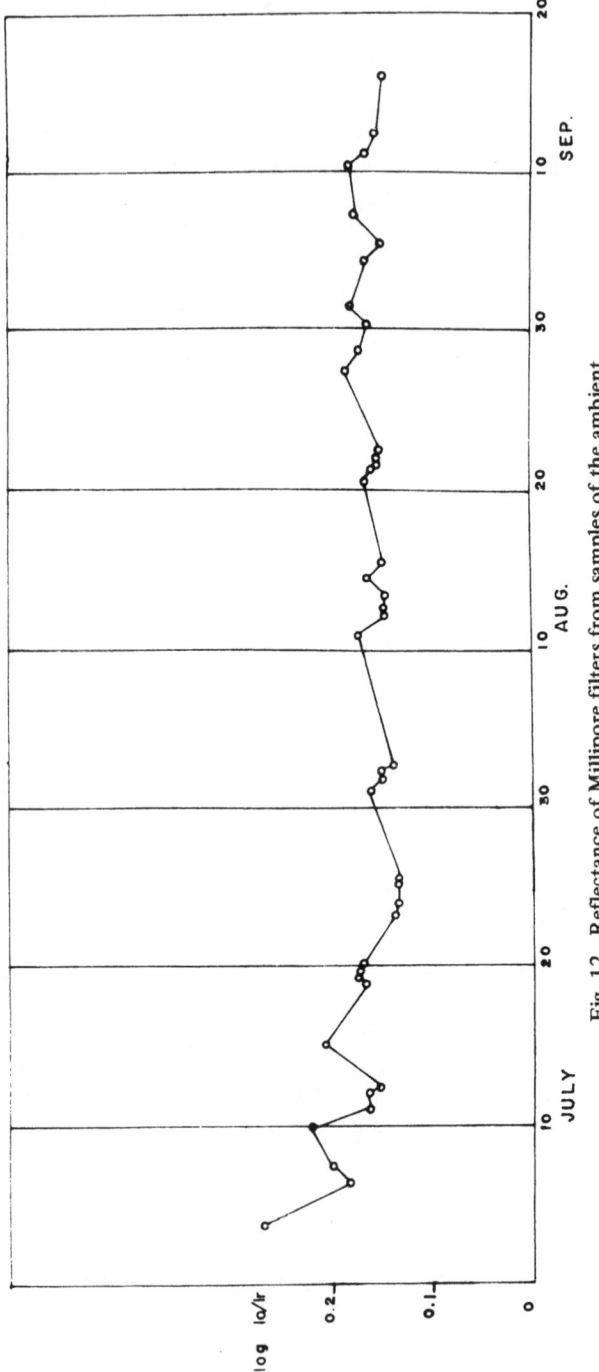

Fig. 12. Reflectance of Millipore filters from samples of the ambient.

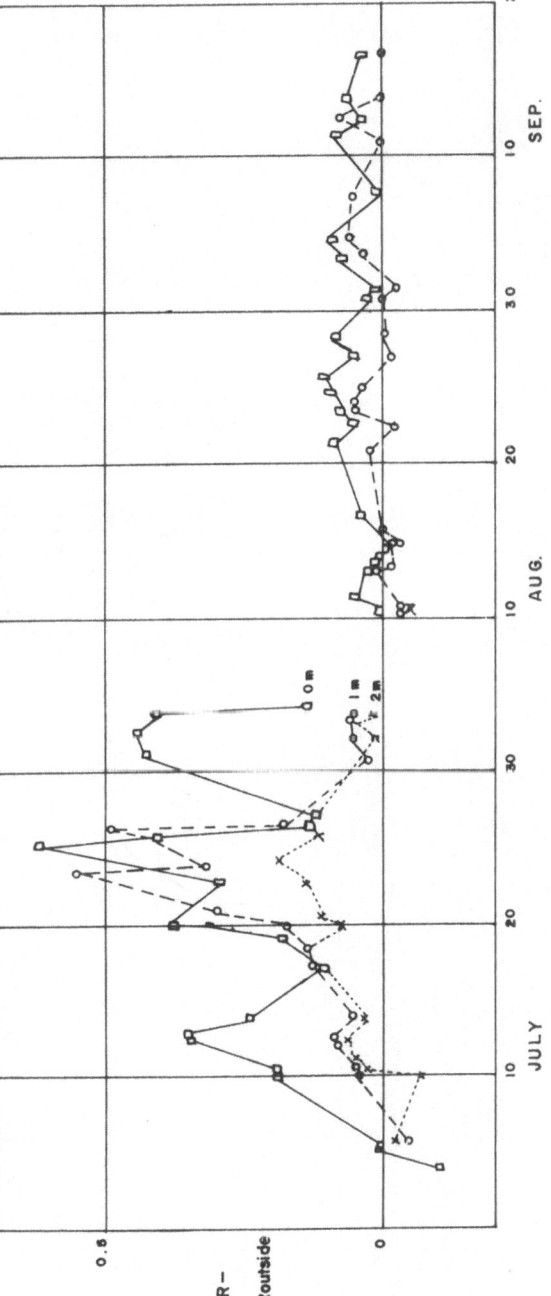

Fig. 13.   Reflectance difference between Millipore filters of pooled samples from the interior and samples from the outside.

Table I. Species Composition of Plankton in the Greenhouse System, Compared to the Outside, July–August, 1972
Organisms per milliliter

	7/24		7/25		8/2		8/3		8/28	
	Exterior	Interior	Exterior	Interior	Exterior	Interior	Exterior	Interior	Exterior	Interior
Diatoms										
Chaeteceros spp	—	—	—	—	1.6	—	1.2	—	—	—
Eucampia zoodiacus	—	—	—	—	—	—	0.4	—	—	—
Nitzschia "delicatissima" type	—	0.4	0.8	—	32	—	33	2.4	0.4	3.2
Nitzschia closterium	—	—	—	6	2	1.6	2	3.2	0.8	—
Skeletonema costatum	—	—	—	0.8	100	—	88	0.8	—	—
Thalassionema nitzschioides	—	—	—	—	—	—	0.8	—	—	—
Thalassiosira sp	—	—	—	—	—	—	—	—	0.2	—
Rhizosolenia alata	—	—	—	—	—	—	—	—	2.4	—
Undet. pennat.	—	—	0.4	—	—	0.8	0.8	—	—	—
Dinoflagellates										
Ceratium horridum	—	0.8	—	0.2	—	—	—	—	—	—
Heterocapsa triquetra	0.8	1920	1.2	3140	2.8	1472	1.6	2232	—	—
Gymnodinium spp	—	—	1.2	—	1.6	—	0.8	—	1.6	—
Peridinium pellucidum	—	—	—	—	0.4	—	—	—	—	—
Other groups										
Coccolithus huxleyii	80	280	64	440	120	780	146	800	100	120
Dictyocha speculum	—	0.8	—	1.6	—	2.8	—	1.2	—	0.8
Leucocryptus marina	—	1.2	0.8	0.4	0.4	—	—	—	—	0.8
Undeterm. flagellates	0.8	—	1.6	—	1.2	—	—	—	0.8	1.2
Ciliates	—	0.8	—	—	—	—	—	—	—	—

ent of the same component across the same boundary, expressed in consistent units.

We are concerned with the exchange of seawater of the 0–3 m level, say, with the seawater at 12 m, assuming a random process. For any section, we can express the rate of increase in mass per unit volume

$$\frac{\partial m}{\partial t} = \frac{\partial}{\partial t}\left(D_z \frac{\partial m}{\partial z}\right) \tag{1}$$

where m is the mass in kg/m^3, t the time expressed in days, D_z the vertical eddy diffusivity in m^2/day, and z the distance measured from the surface and positive downward.

For the unit volume (1 m^3)

$$1 = \frac{V}{\rho} \cdot V_V + \frac{S}{\rho} \cdot V_S = \frac{m}{\rho} \tag{2}$$

where V is the concentration of water, in $^0\!/_{00}$, V_V the specific partial volume of water, ρ the density of seawater in kg/m^3 at the salinity S, expressed as $^0\!/_{00}$, and V_S the specific partial volume of salt.

By definition,

$$V + S = 1000 \tag{3}$$

By combination of equations (2) and (3) and taking the partial derivatives, (1) can be transformed to

$$\frac{\partial S}{\partial t} = \frac{\partial}{\partial z}\left(D_z \frac{\partial S}{\partial z}\right) \tag{4}$$

To the first approximation, the vertical eddy diffusitivity can be assumed to be inversely related to the density gradient across the boundary considered

$$D_z = \frac{D_0}{(d\rho/dz)} \tag{5}$$

where the density will be a function of salinity and temperature.

From these considerations we may express the stability conditions for situations where there is an increase in the temperature with increasing depth, as well as an increase in salinity.

As a first step, however, it will be of interest to calculate the values of the vertical eddy diffusivity for the actual conditions encountered during the experiment. For the experimental period of July this is relatively simple, because the fresh water supply was shut off during that period and there was practically no rain, and because the salinity in the middle layers was practically constant, with the exception of two days in the middle of the period, when there was a sharp increase (Figure 7).

During the period July 12 to August 4 the average salinity gradient increased from 0.03 to 0.3 $\%_{oo}$ per m, due to a steady increase in salinity at the 12-m level. The flux of salt through the 3-m level was at an average 0.42 kg/m²/day, and the eddy diffusivity can be calculated to about 1.3 m²/day, as an average.

During the subsequent period, the freshwater supply varied between 0.06 to 0.24 m³/m²/day, and salinity was not constant at any one level. Fluxes were calculated for the sections 0–3 m and 3–12 m from the average change in salinity per day, taking into account the downward flux of seawater equivalent to the flux of the freshwater added. Eddy diffusivities were calculated by means of the average salinity gradient of the section during the same period.

The values of the eddy diffusivity through the 3-m level varied between 1 m²/day and 20 m²/day, with an average value of 6 m²/day. At the 12-m level, this simplified calculation gave values ranging from 11 m²/day to 130 m²/day, with an average of 39 m²/day.

There is a tendency for the high values of the eddy diffusivity to coincide with conditions of turbulence and rapid mixing in the outside waters. The values of the eddy diffusivity allow for an estimate of the time range for a 50% exchange of the water in the enclosure with water from the outside. Assuming a constant eddy diffusivity over the system, and a supply of outside sea water at the 12-m level of constant properties, we can approximate the system to a one-dimensional semiinfinite medium, with the diffusing substance entering at one interface—here the 12-m level. The boundary conditions will be that there is no flux across the upper boundary, and that the initial concentration of diffusing substance is zero for all distances within the boundaries. In such a case the time for a 50% exchange of the water in the system with that outside may be expressed (Crank, 1958) by the parameter value of

$$\frac{Dt}{z^2} \sim 0.20$$

For a depth of 12 m we can estimate the 50% mixing time for several values of the eddy diffusivity (Table II). At the time of 50% mixing, the upper three

Table II. Estimates of the 50% Mixing Time for a System
of Thickness 12 m, Assuming Different Values
of the Eddy Diffusivity

D_z, m²/day	t, days
1	29
5	6
20	1.4
50	0.6

meters will at an average have exchanged 25% of the original water against water from the outside, with a delay as indicated by the elapsed time.

We may expect that the various water masses entering the main basin will have varying properties, and be separated by more or less sharp fronts or boundaries. Instead of a constant composition of the water at the lower boundary, the changes in composition may be described better by a periodically varied composition, or as a series of pulses. The most important periodic influence are the tides, together with the slower changes with periods between 3 and 8 days. It is of most interest to evaluate what exchange rate to expect from the tidal variations, which have a period of about 12 hours–0.5 days.

A maximum occurring with a spacing of $\tau_0 = 0.5$ days can be said to propagate upward from the lower boundary at a rate of

$$W = 2 \left(\frac{\pi Dz}{\tau_0} \right)^{1/2} \tag{6}$$

and the delay at a distance z from the boundary will be

$$\tau_{\text{delay}} = \frac{z}{2(\pi D_z/\tau_0)^{1/2}} \tag{7}$$

If the property of interest fluctuates around a mean of magnitude C_0, the amplitude of the mean will at this time be diminished by a factor $\exp(-2\pi\tau_{\text{delay}}/\tau_0)$, i.e., $C = C_0 \exp(-2\pi\tau_{\text{delay}}/\tau_0)$.

For the range of the eddy diffusivities that were found previously, we can then calculate the velocity of propagation, the time of delay at the 9-m level above the lower boundary, and the attenuation of the amplitude. Calculated values are indicated in Table III.

For comparison, exchange between the upper 3-m and the 12-m levels have been calculated from the experimental data, and for the period August–September found to vary between 1 and 20% per 24 hours. These ratios seem to be more in accord with the larger value of D.

Table III. The Propagation Velocity of Periodic Maxima
of 12 hr Period into the Greenhouse System, the Delay
at 9 m above the Lower Boundary, and the Attenuation
of the Amplitude

D_z, m²/day	w, m/day	Delay at 9 m, days	Attenuation, % of original maximum
1	5	1.8	10^{-3}
5	11.2	0.8	5.10^{-3}
20	22.4	0.4	0.6
50	35	0.25	4.1

For any component that is being transported by advection, we would then expect that on the average the component will be diluted 1 to 20% of the original concentration once it reaches the upper 3 m, which will constitute the production zone, and have a delay of some 3 to 40 hours, depending on the stability of the water column. It is intended to work out a model of the transport processes along these lines to evaluate the growth of wanted and unwanted species of primary producers, with varying input of nutrients and inoculum.

From the three experimental runs it also appears that at a density gradient of about 0.3 kg/m^3/m the system is sufficiently stable toward sharp temperature fluctuations, as indicated by the situation in the system during days 24 to 27 of July. Figure 14 illustrates this.

On the basis of the salinity measurements, the degree of mixing of water from the lower boundary is calculated for these days, and also for the following 10 days. The salinitiy of water at the lower boundary was sufficiently constant to allow for such an approximation, and it is apparent that the outside water has penetrated only to a limited extent into the system. On the other hand, the temperature fluctuations penetrate more easily, as indicated in Figure 15, but the changes manifest themselves more as an internal mixing.

As regards the possible dimensions of a large-scale system, it is reasonable to evaluate this in terms of the ratio between the vertical and the horizontal diffusivity.

Values of the horizontal eddy diffusivity of $4 \cdot 10^3$ cm^2/sec (35,000 m^2/day) are quoted for inshore waters by Okubo (1971). This leads to a ratio between

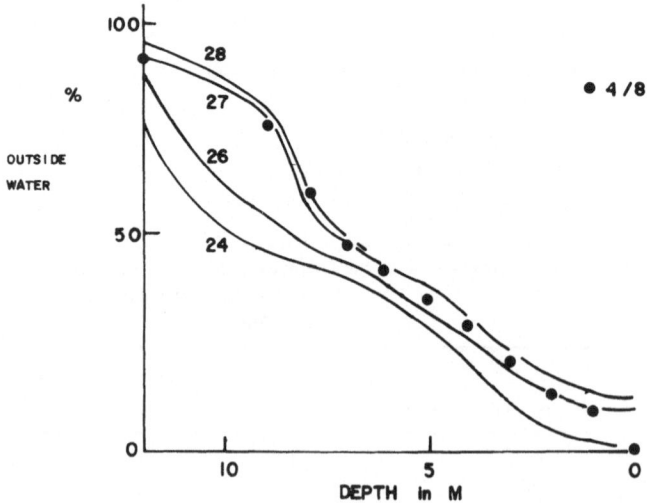

Fig. 14. Exchange of water 7/24–8/4; salinity, surface 7/24:27.6%; salinity, 12 m, 7/24:33.5%.

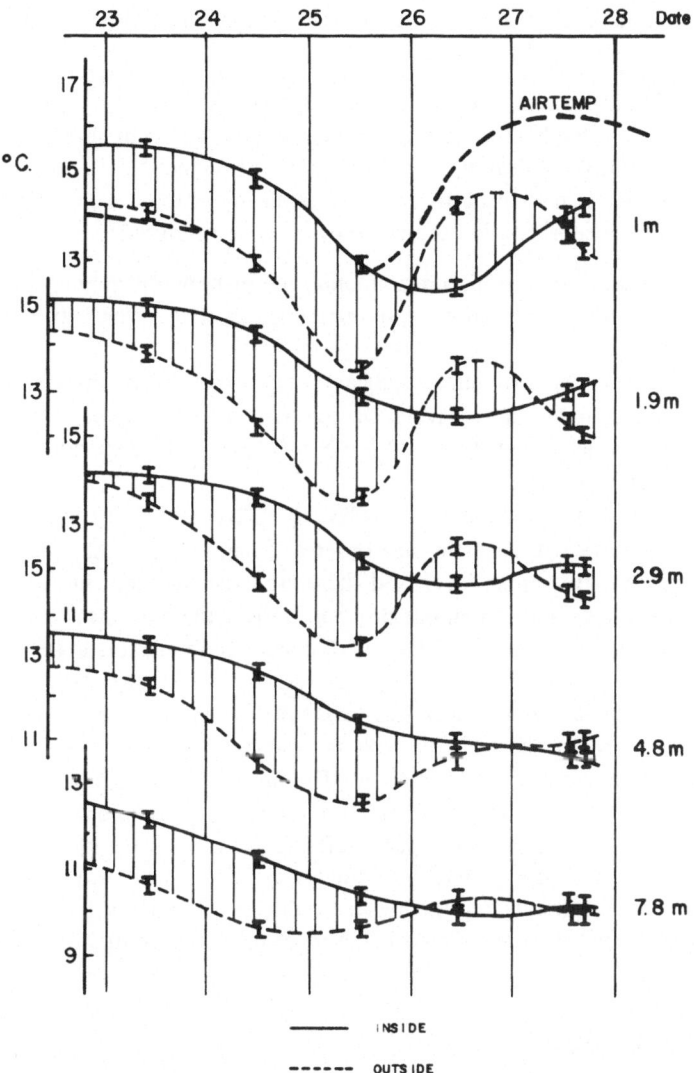

Fig. 15. Temperature changes observed at different levels in the interior and outside.

the horizontal and vertical eddy diffusivities of $3 \cdot 10^4$ to $2 \cdot 10^3$, according to the time of the year. We can then assume that a width : depth ratio of between 50:1 and 100:1 might be feasible, provided that the curtains are sufficiently deep to create a layer that will also attenuate the more extreme changes, short of a large decrease in density which destroys the stability entirely.

The indicated ratio between width and depth should make it feasible to obtain enclosures of areas up to the order of 100,000 m². This raises additional questions that have to be answered:

1. Will added nutrient spread itself evenly over the area?
2. What are the heat transfer properties of such a system, and what are the possibilities of heating the volume from the outside?
3. How far is it realistic to increase the eutrophication in the upper layer, with due consideration to the use of the food produced?

The measurements of nitrate, ammonia, and orthophosphate may be used to obtain preliminary information on the possible recycling of the nutrients within the volume.

The measurements of nitrate indicate a rapid rate of assimilation, with the growth mainly in the upper layers. On the basis of the analyses of nitrate one can calculate a net assimilation of 4.5 g $N/m^2/20$ days, corresponding to a fixation of about 1 g $C/m^2/day$.

By contrast, the ammonia is more evenly distributed in the whole volume, the level of nutrient is generally one order of magnitude less than the initial nitrogen values of the upper layer, and there are large variations in the values of the various levels. These variations probably reflect the movements of the grazing population, indicating a domination of grazers and predators in the recirculation of nitrogen.

The phosphate values show less variability, even if these values also are relatively evenly distributed in the volume, and the values resemble superficially those of a semiconservative component. The most probable explanation is that there is an abundance of phosphate in comparison to ammonia, so that the regeneration of phosphate only partly affects the level of the phosphate pool.

The analyses of nitrate, ammonia, and phosphate indicate then a rapid assimilation of these nutrients into the higher trophic levels, and that the excretion of the nutrients from this population is the most important route of recirculation.

ACKNOWLEDGMENT

The author wishes to acknowledge the financial support of the Royal Norwegian Council for Scientific and Industrial Research.

REFERENCES

Carslaw, H. S., and J. C. Jaeger. 1959. *Conduction of heat in solids*, 2nd ed. Oxford University Press, Oxford.

Crank, J. *The mathematics of diffusion.* 1958. Clarendon Press, Oxford.

Goldman, C. R. 1962. *Limnol. Oceanogr.* 7: 99.

Guilliard, R. R. L., and J. H. Ryther. 1962. *Can J. Microbiol.* 8: 229.

Macallister, C. D., T. R. Parsons, K. Stephens, and J. D. H. Strickland. 1961. *Limnol. Oceanogr.* 6: 237.

Okubo, A. 1971. Pages 141 ff. *in*: Donald W. Hood, ed. *Impingement of man on the oceans.* Wiley-Interscience, New York.

DISCUSSION

CONOVER: The chlorophyll in your system seemed to be confined to the surface, which certainly seems to indicate great stability. It might be too stable. Would light be limiting the production below 1 meter or was grazing controlling the production?

DEVIK: In the experiments we deliberately placed the nutrient in the upper meter, and because of this we cannot separate the effect of light and of the nutrient on production. From the decay rate of nitrate nitrogen we may calculate a production rate of about 1 gram carbon per square meter per day, which are production values that are already reached under natural circumstances in the month of March at these latitudes. The experiments were done during July, the weather was very nice, and accordingly the light intensity would be two to three times as high as in the beginning of March.

PERSOONE: Would it not be interesting to try to follow the nitrate gradient from the surface to the bottom? This might be helpful to correlate the plankton production with the nitrogen content of the water.

DEVIK: In the problem of analysis we had to economize with the chemical analysis. We mapped the water movements by temperature and salinity measurements and the chemical analyses were done to get an overall impression of the changes. In later experiments we certainly will improve on this point.

PERSOONE: Is there a chance that the productivity could have changed very much during the experiment due to migration of other types of plankton species into the fertilized system? In other words, would the natural change of phytoplankton in the environment influence the species composition within the system?

DEVIK: The species composition was determined on samples taken two weeks after fertilization and onward. The values are included in the paper, but were not referred to in the lecture. From the table it is seen that inside the dominating species was *Heterocapsa triquetra* in the main experimental period. The main experimental period may be said to have lasted until August 4, when the stabilizing top layer disappeared and a new one had to be built up. Moreover, on the inside there is indicated occurrence of *Nitschia closterium* which occurs scarcely on the outside, while species like *Skeletonema* has a higher frequency on the outside than on the inside.

These data, to my mind, indicate that we have effected quite a good degree of isolation and this may form a basis for building up a continuous culture system. In such a continuous culture system it might be possible to impose from the top the organisms we want to cultivate, without too much interference from infections from the outside. It would be of particular interest to define conditions so that the imposed organisms at any time are able to squeeze out competitors entering from below. The next question would then be the utilization of such productivity, for example, in terms of bivalves, when we think of ordinary planktonic organisms. Of the potential users would be harvestable species, which then could be utilized in, for example, animal husbandry.

PERSOONE: What is the natural phytoplankton in this area? From the table the flora seems to be dominated by diatoms with some dinoflagellates.

DEVIK: I would like to turn that question over to my colleagues from Espegrend.

TAASEN: Usually *Skeletonema* is the most common species in the phytoplankton in this area, but in this case it seems to be very rare, other species, such as *Chaetoceros* spp., usually should have been more numerous too. But as these results are from July, we might, from time to time, expect such results with very low numbers of diatoms and dinoflagellates.

PERSOONE: The scarcity of the plankton is remarkable.

DEVIK: Yes, it is a barren area.

CHRISTENSEN: Could this system of marine greenhouses be applied to heated water from power plants? As I see it, the primary objective of this system is to lower the exchange rates of water from the bottom to the surface, and this is of course contrary to using the heated water, because with an exchange rate of five days or more the added water would be cooled by wind and radiation.

DEVIK: In the case of a hot water effluent, a possible arrangement would be to let the hot water effluent heat the inside through the plastic curtains. We have made some preliminary calculations on the heat exchange of such an arrangement. If we have a depth to width ratio of about 1:8, it seems possible that the heat transferred through the walls will compensate for the heat loss through the surface, so that we will obtain a net increase of temperature within the system itself. So far we have no idea how much such heat transport will affect the turbulence within the system. Our plan is to study these problems by suitable modeling.

RAYMONT: Judging from the limited density of diatoms in the system you have obviously restricted nutrient circulation effectively. Do you think there is a question of a lack of silicon inside?

DEVIK: Yes, I think so. For the fertilization we used a standard medium of Ryther and Gilliard, which contains a certain amount of silica. I believe the available silicon has been relatively rapidly depleted, but I have no analysis to corroborate this.

FORSBERG: You had no mechanical problem with your plastic enclosure. In the middle sixties, if I remember right, the Windemere group tried to divide a small part of a lake by means of a plastic curtain, but the experiment was unsuccessful because the curtain broke. What would you consider the upper limits of the diameter of your enclosure?

DEVIK: With a proper design of the floating supports to eliminate the wear on the plastic, I don't think the strength of the plastic is a limiting factor. In this particular application we used split-fiber type nylon of a heavy gauge, to lessen the risk of mechanical failure. Plastic of this type retained a considerable amount of strength even after one year in the sea. For shorter periods I think it would be feasible to use a lighter dimension, maybe down to a thickness corresponding to 30 grams per square meter. In this context experiences with oil booms would be valuable.

CARSTENS: At present none of the oil boom constructions are good. The main short-coming is that they do not stand up to the constant wear of the sea movements. What is your experience with regard to currents in the sea?

DEVIK: We have only limited experience. The site was chosen so that the currents would be small. We have no direct measurement for current, but from the drag on the mooring we could estimate a current of about 2 to 3 centimeters per second, at an average. It seems that the current was predominant from the surface down to about 3 meters, and it apparently was not coupled with the wind along the surface. So at a guess one could say that the system should be able to withstand currents of up to about 5 cm per second.

OPPENHEIMER: The only failure Strickland had with his bag experiments in the Pacific was due to the buoyancy of the upper part and the friction of the walls. It seems he was using plastic tubes in lengths of 100 feet or so, and 10 to 15 feet in diameter and the friction on the plastic as it moved through the water with the heave of the waves on the surface pulled the plastic apart.

DEVIK: We got around that by having a very heavy tarpaulin and top.

OPPENHEIMER: And less waves. His experiments were done out in the Pacific Ocean.

DEVIK: Yes, but a tarpaulin would even out this stress from the waves, the localized stresses would be taken up by a larger part of the solid tarpaulin, and besides there is no pull from the weight of the water column inside the system.

WALNE: You indicate a fixation rate of about 1 gram carbon per square meter per day, and you suggest bivalves. For the sake of discussion, we can think roughly of this 1 gram carbon per square meter per day sinking past a layer of bivalves. One gram of carbon or about two grams of dry matter is a very small ration to get a maximum growth of juvenile bivalves. For juvenile bivalves we feed about 40% of the dry body weight per day, falling to about 10% for a fairly large bivalve. In any case this will mean a fairly small standing crop of animals per square meter.

DEVIK: That is right. A productivity of one gram carbon per day per square meter will hardly be enough to sustain growth of an interesting amount of bivalves; to make such a system interesting the productivity must probably increase three to five times. This may in theory be accomplished by increasing the rate of fertilization, up to the level where the light becomes limiting. There is still quite a way to go before we reach that level, but then we have very little information on how a combined system will behave.

PERSOONE: What about the possible grazing? There was no mention about the zooplankton composition outside as compared to the inside of the system. Do you have any data on the zooplankton, and what would be the possible effects of grazing on the productivity?

DEVIK: To take the estimate of the gross productivity first: My estimate is based upon a proportion of nitrogen fixation to carbon fixation of 1:4, where the nitrogen fixation rate was calculated from the decline of the nitrate concentration. In this calculation I have not taken into account any recirculation of ammonia. The calculated rate, however, is comparable to the rates you find in fertile areas along the coast of Norway.

With regard to the feeding of the bivalves in such a volume, the optimal feeding of the bivalves is in a moderate current of water. In a system like the one we are discussing, such currents will be absent, but instead there might be a turbulence which might serve the same purpose. Only further experiments can indicate whether this turbulence will be sufficient to secure enough food for the bivalves.

Regarding the occurrence of zooplankton, we have no quantitiative data. But simply by inspection it was apparent there was much more animal life inside the system than outside.

CONOVER: I am trying to get an idea of the general structure of the system. In a sense it seems to me that you only need the upper part. You are not really taking advantage of the fact that there is a water column below. Would it be possible to bring this lower column of water with its content of nutrient into the overall culture scheme, that is, to induce upwelling of this water column to see whether you can increase the production without the necessity for added nutrients?

DEVIK: This might bring in conflicting purposes. In the present setup we probably need the lower layer as a buffer zone, but whether the buffer zone has to have a depth as large as at present, I am not certain. I believe, however, that in the summertime we can get by with less depth than at other times. We need, however, to calculate these things more carefully before making any definite statements.

NESTAAS: I wonder if Devik would like to expound a little on possible uses of these aquatic greenhouses, particularly in connection with waste heat water and nutrient-loaded effluent of some types.

DEVIK: Any remarks have to be speculative. One might take as a starting objective that it should be feasible to obtain a nutrient load or a potential productivity at the same level as you obtain in eutrophied areas in our part of the world. Then you have the same degree of primary productivity per unit area as we have in other systems. As one example we might point out the landlocked fjords or polls where there is a fairly good production rate of oysters, both small and large. This indicates that the greenhouse system might be utilized for some bivalve cultivation. In such systems we would want a certain increase of the temperature, but not too much. On the other hand, the influx of nutrient should probably be higher than what corresponds to the level in the cooling water. This then means that it would be convenient to separate the influx of nutrient from the maintenance of the heat. In practice one could achieve this by letting the system float in a stream of the hot water effluent, and add the nutrient solution—either sewage effluent or a made-up solution of fertilizer could be added directly to the system.

It is premature to discuss definite dimensions of such a system, the primary objective would be to see whether a moderate-sized system will function. In essence the system is a sheltered one which is rich in nutrients. The question is what kind of life will it sustain. We can probably obtain part of the output in the form of bivalves. It is difficult to make any definite forecast with regard to the amount of recirculation of nutrient from the excretions from the herbivores. One may also entertain the question whether such a system might be used as a shelter for the rearing or the gradual release of fish fry. Admittedly, this will mean stretching to the limit the possibilities of controlling the environment, and it will really be a matter by experiment to see how far we can afford to lose the predictability of such a system at the cost of simpler construction and operation.

RAYMONT: The productivity of the curtain enclosure seems to be on the low side for the amount of nutrient you used. What were the light conditions like during the experiment? Apart from the light during the middle of the day, would the light be hindered in reaching the interior by the sides of the curtain? In other words, were the sides of the curtain transparent? Or did you get some settlement, perhaps, of benthic algae on the side of the curtain? I know some workers who have tried this sort of experiment have found a very considerable reduction in light resulting in low productivity.

DEVIK: The sides of the curtain were made of green material, opaque to light. There were no measurements of light intensity made in the enclosure. The light intensity estimated from the meteorological data indicates during the experiment an average of 400 calories per square meter per day, referred to a horizontal surface. Of this amount, about 5% was in the ultraviolet range (290–390 nm). The daily total of sunshine was on an average 7 hours per day, or 44% of the maximum possible, corresponding to a mean daily length of about 16 hours. The sighting depth would probably correspond to about 1 meter if measured by a Zecchi disk, which should indicate a dense growth.

The estimate of productivity is, as mentioned before, based upon the decline of nitrate, and this figure will then not give an indication of the productivity from the recirculation of nitrogen in the form of ammonia. This recirculation might be quite high, because the density of zooplankton and larger animals was strikingly high during the whole experiment. We have, however, no data on the fauna. A certain amount of benthic algae settled on the side, and on the inside of the curtain some colonies of mussels had established themselves during the experiment.

SOEDER: Would you think that for the storage of solar energy a shallower coastal system would be more suitable? Because if your system is open toward the bottom and you are working in deep water, can you actually use the hothouse effect as pointed out by Mr. Carstens?

In shallow systems you can certainly store a lot of energy, as experienced in the Dead Sea, for example. There, you can cover the bottom with a black plastic material and have a cover of freshwater blocking the heat transfer to the atmosphere. Thus, you can reach temperatures up to $70^{\circ}C$, which is quite enough for thermal pumps, but I doubt whether such effects can be used in deep waters.

DEVIK: In this system we had an average temperature increase in the upper two meters of about 0.8 of a centigrade over the summer period. In the system as a whole there was a temperature increase over the ambient coincident with the temperature increase that can be calculated from the energy balance across the surface during summer time. This indicates that it should be possible to store heat corresponding to the solar heat of one to two days, in a relatively shallow layer underneath the surface. This heat will be dissipated in the course of one to two days if the weather turns cloudy.

Regarding the effect which Carstens described, which we observe in a closed system with a density gradient, this effect is best described by considering the system landlocked. Landlocked in this context means that the temperature fluctuations of the atmosphere will be relayed to the interior of the system via the interphase air/water, and via the interphase land/water. The thermal resistivity of the land will be considerably higher than of the water, and below a distance of 3 to 4 meters from the surface the seasonal temperature will virtually be evened out. At these levels, the average temperature of the ground will correspond to the yearly average, and as the average temperature of Norway in the southern part is around $7^{\circ}C$, this means that the wall temperatures of the landlocked fjord will have a corresponding average. As the seasonal increase depends upon the incidental solar energy, and the maximum retainable temperature will depend upon the lowest winter temperature, we obtain temperature effects like the ones described by Carstens.

In other words, the average temperature increase in a greenhouse system like the one described here will, to a large extent, depend upon the temperatures of the walls and the bottom. With the thickness of 12 meters as indicated, temperature fluctuations with periods less than, say, 36 hours will effectively be filtered out, while periodic changes with longer periods will induce larger changes with relatively little damping.

Dialysis Cultures in Integrated Aquaculture

Arne Jensen

Institute of Marine Biochemistry
University of Trondheim
N-7034 Trondheim, Norway

INTRODUCTION

On first inspection there does not seem to be any connection between dialysis culture techniques and integrated aquaculture. A closer look will, however, reveal several problems in aquaculture which may be successfully tackled by the use of dialysis techniques. Obvious examples are: bioassay for the monitoring of growth conditions for the primary producer; provision of algal inocula in large-scale cultural systems; and large-scale cultures for specific products or under special conditions. These aspects of the application of dialysis cultures will be discussed. We shall, however, have to start with the dialysis culture system itself and review some of its characteristic features.

DIALYSIS CULTURE

The technique itself is quite old and goes back to Metchnikoff, Roux, and Salimbeni (1896). In principle, the microorganisms are kept in a cage which retains the algae, bacteria, or what they may be and allows ready transport of nutrients and growth factors into, and metabolic products out of, the cage. The cage shall not form any barrier to temperature and low molecular weight chemicals. Light can also reach the microorganisms, provided the walls are transparent.

This means that most environmental factors are identical or nearly identical on both sides of the cage walls. This, therefore, seems to be as close as one can come to growing defined cultures under natural conditions.

A considerable number of bacteriologists have studied the method and Schultz and Gerhardt (1969) have published a very extensive review covering historical aspects, techniques, theoretical background, and possible uses of the dialysis culture. The use has, however, been very limited in bacteriology, and practically absent in algology up to three years ago, when we in our ignorance rediscovered the method.

In their theoretical treatment of the dialysis culture Schultz and Gerhardt enumerated four systems: (1) batch reservoir–batch fermentor, (2) continuous reservoir–batch fermentor, (3) batch reservoir–continuous fermentor, and (4) continuous reservoir–continuous fermentor.

The batch reservoir–batch fermentor system is simply a normal batch culture in which the microorganisms have been restricted to a small part of the nutrient medium, which they, however, have access to.

We have during most of our studies worked with the continuous reservoir–batch fermentor type, and I shall discuss some of the characteristic properties of this system.

Schematically the microorganisms in the fermentor have more or less limited access to a constant level of nutrients in a reservoir with continuous replenishment. The culture will, after having passed the lag phase, go into logarithmic development, and the growth of the organisms is not limited by nutrient concentration. Compared with a conventional batch culture, the logarithmic phase of the dialysis culture is extended, since diffusion also brings in nutrients during this phase in the latter system. The growth curve of the logarithmic phase is expected to follow

$$\frac{dN}{dt} = k \cdot N \tag{1}$$

where N is the number of organisms, t is time, and k is a constant.

At some critical population density, however, growth becomes nutrient limited, and the dialysis culture enters a unique phase of linear increase in cell number (or biomass) with time. Since the nutrients have to diffuse through the diaphragma and this process is a linear function of the concentration potential over the membrane, growth will also become more or less linear dependent upon concentration differences between the outer medium and the culture liquid. In this situation the growth curve of the culture may be described by

$$\frac{dN}{dt} = Y_N \cdot P_m \cdot A_m \cdot \frac{S_r - S_f}{V_f} - Y_E \cdot Y_N \cdot N \tag{2}$$

which is linear when $(S_r - S_f)$ is constant and $Y_E \ll Y_N$, where N is the number of cells/ml, t is the time, Y_N is the yield coefficient for conversion of substrate to cells, P_m is the overall membrane permeability coefficient for substrate, A_m is the geometric area of membrane, S_r is the substrate concentration in reservoir, S_f is the substrate concentration in fermentor, V_f is the fermentor volume, Y_E is the specific maintenance rate.

Sooner or later the cell number in the fermentor should reach a level at which the nutrients that diffuse through the diaphragma are consumed in the basal metabolism of the cells. There is nothing left for growth and division. In this phase, the stationary one, $dN/dt = 0$ and $Y_E \cdot N$, becomes highly significant. This situation is described by

$$Y_E = \frac{P_m A_m \cdot (S_r - S_f)}{V_f \cdot N} \tag{3}$$

For heterotrophic microorganisms and for animals it is easily understood that there must exist a maintenance requirement for energy (food). This concept may, however, seem less obvious when applied to photosynthetic organisms which obtain the energy from light.

All the above-mentioned growth phases were detected in our dialysis cultures of planktonic algae. Good correlations were obtained between the concentration of nitrate (the limiting nutrient) in seawater and the transition point between logarithmic and linear growth for several algae. The growth rate in the linear phase was also well correlated with the nitrate content of the outer medium, and finally, the population density in the stationary phase was determined by the concentration of the same nutrient in the surrounding seawater. The data obtained allowed estimations of Y_N and Y_E to be carried out.

Our conclusion was that dialysis cultures were well suited for estimating the capacity of seawater for supporting growth of algal populations.

USES IN INTEGRATED AQUACULTURE

Monitoring Growth Conditions and Capacity

In any integrated aquaculture system there must be a primary producer upon which the whole system rests, and a more or less continuous monitoring of the growth conditions of the primary producer is highly desirable. Dialysis cultures seem to be especially suitable for such controlling activity. The assay can be carried out directly in the medium to be used, under conditions nearly identical to those felt by the primary producer in situ, and one may of course use the primary producer itself as test organism. Its growth and development can be followed for 8-12 days in the case of marine phytoplankton, which means that one receives continuous information on the growth condition over the whole of this period by means of one and the same assay. Furthermore, this method should allow normal operation of the concentration mechanisms of the alga or test organism. Conventional bioassays are usually carried out in quite small volumes of medium, and accumulation of trace compounds is limited by the small total amount present in the culture vessel. In nature, turbulence and general mixing expose the microorganisms to new bodies of water and accumulation may be and is frequently brought much further than it can develop in the batch bioassay. Similar mechanisms will operate in large-scale recycling systems. The dialysis technique allows renewal of most bioregulators and trace compounds which are taken up by the test organism, and any accumulation process should express itself more or less freely.

We have for a couple of years followed the capacity of seawater for supporting growth of phytoplankton using a simple device constructed in our workshop (Jensen et al., 1972). This gave the acceptable correlations cited above between algal development and the limiting nutrient ($-NO_3$). Responses to variations in the nitrate content were also satisfactory, and accumulation by the test algae of heavy metals through the dialysis membrane was clearly demonstrated in experiments on zinc and copper tolerance of planktonic algae (Jensen et al., 1974, 1975). The copper content of cells of the diatom Phaeodactylum tricornutum in dialysis culture corresponded to a quantitative uptake of the copper in a volume two to four times that of the dialysis bag. In the case of zinc uptake, the algae must have had access to a volume of water at least 20 times the fermentor volume.

It seems that dialysis-type bioassays form the most suitable and relevant technique for the monitoring of growth conditions and uptake phenomena of the primary producer in integrated aquaculture.

Production of Inocula of Primary Producers

The dialysis technique will make possible growth of starter populations of the primary producer under exactly the same conditions which prevail in the large-scale growth units. This should eliminate shock reactions which frequently cause extended lag periods when a separately cultivated inoculum is introduced into the large-scale medium.

In principle, one may also grow the primary producer in the tank where the first order consumer is, providing grazing is controlled by a dialysis system. This should allow influx of nutrients, partly excreted by the grazing organism, and should keep out the grazer so that feeding may be controlled externally. No application of this type of dialysis culture has taken place, but studies of the impact of grazing upon phytoplankton population using dialysis cultures have been started at our institute. By placing dialysis bags in natural phytoplankton populations in the sea and comparing growth in the bags and in the blooms, some idea of the influence and dimension of grazing may be obtained.

Large-Scale Dialysis Cultures

Large-scale dialysis cultures of algae have not been carried out yet. The most difficult problems to be solved are those connected with the membrane. To secure satisfactory speed of nutrient influx, we have hitherto based our studies on a very high ratio of membrane area to culture volume. This forced us to use transparent membranes in order to bring enough light to the algae. The choice of membranes was thereby seriously limited, and we have had to resort to regenerated cellulose. This material has several disadvantages. It is weak; it is attacked by a number of bacteria; it has relatively low permeability for most nutrients; and it is relatively expensive.

The first improvement seems to be a separation of the photosynthetic from the dialysis step in the system. This will open up the way for more materials to be used in the dialysis membrane.

There are at least two reasons why we want to explore large-scale dialysis cultures. First, this system makes it possible to grow organisms in media with very high nutrient concentrations—concentrations which would kill the population or completely inhibit its growth had the organisms been brought in direct contact with the nutrients. In other words, dilution of a concentrated medium is not necessary, and this has further consequences. It gives much smaller culture volumes and denser populations compared with conventional cultivation in dilute media, and the cost of harvesting is reduced significantly.

Second, manipulating the microorganisms becomes easier in dialysis cultures, in comparison with either batch or chemostate systems. Due to the membranes, washout does not occur in dialysis cultures. The organisms have to remain in the reactor irrespective of the dilution rate and the composition of the medium. Cooperative microorganisms may be put in the desired growth stage—exponential, retarding, or stationary—and kept there for considerable time by proper regulation of the composition of the medium and suitable adjustment of the population density. This will mean good control of the composition of the microorganism or of the type of product it exudes into the medium. Since most organisms give the desired product only during a short period of its life cycle, and frequently in the stationary growth phase, means of forcing the organism into this situation will be very valuable.

REFERENCES

Jensen, A., B. Rystad, and L. Skoglund. 1972. The use of dialysis culture in phytoplankton studies. *J. Exp. Mar. Biol. Ecol. 8*:241-248.

Jensen, A., B. Rystad, and S. Melsom, 1974. Heavy metal tolerance of marine phytoplankton. I. The tolerance of three algal species to zinc in coastal sea water. *J. Exp. Mar. Biol. Ecol. 15:* 145-157.

Jensen, A., B. Rystad, and S. Melsom, 1975. Heavy metal tolerance of marine phytoplankton. II. The tolerance of three algal species to copper in coastal sea water. In preparation.

Metchnikoff, E., E. Roux, and T. A. Salimbeni. 1896. Toxine et antitoxine cholérique. *Ann.
Inst. Pasteur*, T. Pp. 257–282.

Schultz, J. S., and P. Gerhardt. 1969. Dialysis culture of microorganisms: Design, theory,
and results. *Bacteriol. Rev. 33*:1–47.

DISCUSSION

PERSOONE: Are the dialysis membranes commercially available, and what are they made
of, and secondly is there no danger of bacterial fouling or fouling by other organisms to
interfere with the exchange?

JENSEN: These dialysis bags are available commercially. They have been used for many
years for ordinary dialysis purposes and they are made of regenerated cellulose in various
sizes. They are definitely attacked by bacteria. In our area there is an abundance of
cellulose discharged to the sea from the paper mills, which we believe has increased the
bacterial flora that will digest cellulose. Usually the bags will develop holes in the course
of 10 to 14 days.

PERSOONE: Do you have an idea when the first holes appear?

JENSEN: You can, actually, if you inspect the growth curve. Deviation from normal
growth curve will often indicate hole formation. Very quickly, also, the holes will manifest themselves if you squeeze the bag—you will see streams of water coming out from the
holes. Bags without holes are able to withstand a moderate pressure from the hand.

PERSOONE: Are there bags available with different porosities?

JENSEN: There are a number of different qualities available, and also a lot of membranes
with different porosity or permeability. The majority of these are rather expensive, and
we have confined ourselves to the sausage tubing, which is easy to operate and cheap.
Membranes in sheet form would require a different system from ours to be handled
conveniently.

CHRISTENSEN: One advantage of the dialysis culture was said to be that one need not
deal with very dilute solutions. This is to me not completely convincing, because as far as
I can see not only should the solution and the volume of the solution be suitable for the
organisms, but in addition you must also have a concentrated solution for replenishment.
I would also like to get an idea of the economics involved in such a system when it is
enlarged.

JENSEN: Our experience shows that with the organisms we have used, we are able to get
very fast growth even under conditions where there is no nitrate detectable by chemical
means in the outer volume. With the systems exposed to the water from the fjord, we
always had very fast growth, and regularly there was a gradient of nitrate from the outside
to the bag, which would secure sufficiently rapid transport of nitrate for the algae to give
rapid growth. One reason for this is that the water on the outside is very rapidly exchanged, so there is no net depletion of nitrate on the outside. On the other hand the
dialysis system is self-regulating, for if the concentration of nutrient on the outside be-

comes too high, the permeability of the bag will effectively diminish the transport of nutrient from the outside to the inside to a rate which is compatible with the growth of the organisms. In industrial practice this aspect has been put to use to control the dosage of one or more of the components in chemical reactions.

As far as the commercial utilization of this process is concerned, I think the potential will be in the production of relatively expensive chemicals, e.g., antibiotics. In this case the microorganism can be put in the desired stage of its life cycle in the dialysis system, where it can be exposed to chosen conditions which encourage antibiotic production. For such applications I think it is still premature to speak about costs.

CONOVER: You have said that you are beginning a program where you are comparing the growth in dialysis bags with the population growth outside in a natural environment. Have you gained enough experience to say whether the population composition is affected by the conditions inside the bag? Is there some density-dependent factor involved?

JENSEN: We have no extensive data so far. Our intention is to assess the total capacity of the water for algal growth. In the actual case one would have to use a number of test organisms. We propose to do such experiments in connection with anticipated blooms, and correlate the growth rates of, say, *Skeletonema* and other selected algae inside and outside the bag with respect to growth rates and density. In the bag you will determine the potential growth capacity for the particular alga, and on the outside you will see that the growth is in the natural conditions, where you have competition from all the organisms and also grazing.

We have not got so far yet, because of technical problems. It is rather difficult to run these drums in the free water masses. We have tried a system that just consisted of a buoy floating on the surface, with the bags suspended underneath in a drumlike construction. With the wave movements on the surface the drum would rotate and thus agitate the dialysis bags. In the first trial run the drums moved too rapidly, but this is now corrected.

In the later trials the system has operated satisfactorily and we have obtained good growth curves, also in the open fjord. However, at the time of the trials we had no bloom to compare with; this is something that will have to be done later.

CONOVER: Why do you need to agitate the bags; Isn't there enough water movement past the bag when it is just suspended in the sea?

JENSEN: We really do not need much movement, but it has to be sufficient so that the culture does not settle at the bottom of the bag. We have found the best way to do this is to place the bag horizontally with a small bubble in it, so that the bubble moves gently from one end of the bag to the other.

PERSOONE: Are these membranes really tough, and is there no risk of mechanical disruption?

JENSEN: For the first week they are quite tough, after that they weaken. You can even hang them in the wake of a boat if you want to.

PERSOONE: Do metabolites or high molecular weight compounds pass through the membranes?

JENSEN: With ordinary regenerated cellulose it is said to have a cut at the ten thousand molecular weight or so, but this can, of course, be regulated by choosing the right kind of membrane.

Food Chains and Their Use: Mussels, Mollusks, and Bivalves

Integrated Systems of Mollusk Culture

John H. Ryther and Kenneth R. Tenore

Woods Hole Oceanographic Institution
Woods Hole, Massachusetts

Mollusks have two environmental requirements for their growth and commercial culture that may often be mutually exclusive. First, the water temperature must be in the proper range for the animals to pump and filter water; second, the water must contain enough microscopic food organisms of the proper size and composition to provide food for the shellfish. These two prerequisites are often not present simultaneously in the same environment. Tropical and semitropical waters are naturally poor in nutrients and normally lack the level of primary productivity (i.e., phytoplankton growth) for substantial mollusk growth. The more eutrophic temperate and boreal waters have temperatures too low for feeding and growth of bivalves for at least part and often as much as half the year.

Generally speaking, the best environment for molluscan growth is found in temperate areas in summer, and this indeed is when and where most commercial shellfish culture takes place. Some familiar examples are culture of the Pacific oyster (*Crassostrea gigas*) in Japan, the American oyster (*C. virginica*) in the United States, and the flat oyster (*Ostrea edulis*) and the blue mussel (*Mytilus edulis*) in Europe. Precise information concerning the temperature at which these species do or do not pump water, filter food microorganisms from it, assimilate their food, and grow is usually lacking, but it is probably safe to say that most of them, for one reason or another, do not grow significantly at temperatures that are much below 10°C and that their temperature optima are in the 15° to 20°C range. Thus, in the southern part of their respective ranges (e.g., *C. gigas* in the Inland Sea of Japan and *M. edulis* in the Galician Bays of Spain), where winter temperatures rarely fall below 10°C, year-around growth is possible and a marketable crop may be produced in just one year. In northern Europe and the United States, however, the temperatures of coastal waters and estuaries approach and sometimes exceed freezing levels in winter and fall below 10°C for as much as half of the year (Bumpus, 1957). Under these conditions, a commercial crop of shellfish may take 3-5 years for its production (Bardach *et al.*, 1972). Clearly, the use of heat under these circumstances would enhance the growth of the mollusks during the unproductive winter months and improve the prospects for their artificial cultivation. It is also obvious to anyone who cares to look into the economics of the problem that artificially heating water in the volumes that would be needed for commercial mollusk production would be prohibitively expensive.

However, waste water from industrial cooling systems, particularly steam-generating electric plants where once-through cooling water systems are commonly used, is an obvious source of free heated seawater. Many power-generating stations are located in temperate estuaries and coastal marine waters just because of the availability of an abundant, if not unlimited, supply of cool ocean water. Frequently, such installations are located in natural shellfish-producing areas, where other conditions are favorable for the growth of the mollusks.

The new, relatively large, steam-generating power plants, whether nuclear or fossil-fueled, if they utilize a once-through cooling water system, produce a flow of about 1000 ft^3/sec (648 million U.S. gallons or about 2.6 million m^3/day). Disregarding, for the moment, the possible beneficial effects of the warm water, how much of a molluscan culture industry will that quantity of pumped water support?

Let us assume that the amount of particulate organic matter of suitable size and composition for molluscan food that will be entrained in the power plant cooling water averages about 1.0 mg wet weight per liter (Ryther, 1969). Then the power plant discussed above would provide, not only water of elevated temperature, but also a flux of 2.6 metric tons of food per day or about 1000 tons of food per year. If the mollusks can filter, consume, and assimilate that food with a conversion efficiency of 20% (Tenore *et al.*, 1973), an annual production of 200 metric tons, fresh meat weight, of bivalve mollusks could be supported by the cooling water pumped from one modern power plant. To put this in other terms, 200 tons meat weight is equivalent to about 1500 tons of whole oysters or about 800 tons of whole mussels, shells included. It is also equivalent to 10-20 million individual, medium-sized shellfish, assuming that the average meat weight is 10-20 grams per animal. In other words—depending, of course, upon the species and its intrinsic value at the time and place in question—a single power plant could conceivably produce the proper combination of warm water and food for a small but commercially viable shellfish industry.

Let us now consider the temperature of industrial cooling water with respect to its suitability for and potential usefulness in molluscan culture. Depending upon the design of the cooling water system (as may be required for environmental protection, for related legal reasons, or for economic or engineering reasons relating to the availability of water), the condensers of a once-through cooling water system will increase the temperature of the intake water anywhere from about 5° to 25°C. Very commonly, the cooling water Δt is about 15°C.

Obviously, the usefulness of power plant cooling water for molluscan culture, as for other forms of aquaculture, depends upon the Δt of the particular power plant in question, the annual temperature cycle of the ambient or intake water, and the thermal optima and tolerances of the species of animal to be cultivated. To some extent, the choice of species provides some flexibility. For example, American oyster can not only tolerate but continue to grow at temperatures close to 30°C, while the thermal maximum for *Mytilus edulis* is about 20°C. Generally speaking, however, it is unlikely that optimal growth conditions for any species of mollusk can be provided throughout the year by using undiluted treated effluent. Cooling water that is 15°-20°C in winter (i.e., when ambient water temperatures are 0°-5°C) will generally be of the order of 30°-35°C in summer, temperatures too high for most species of mollusks that would find the ambient temperature at that time of year to be near optimal. Thus, it is probable that the use of heated cooling water for mollusk culture will, if practical, prove to be a seasonal practice. Alternatively, provision could be made in a mollusk culture facility for diluting the cooling water with ambient seawater following its discharge or, by providing a second source of pumped, ambient seawater to the culture facility, providing any desired mixture of cooling and ambient seawater and, as a result, achieve near optimal temperature for the organisms throughout the year. While the latter could represent an ideal solution from a technical point of view, the economics of such a practice could be prohibitive.

Assuming, however, that the commercial culture of mollusks is potentially viable on both a technological and an economic basis, there are many risks and constraints to such an enterprise. First, utilities are often prone to discharge into their cooling water a variety of other wastes associated with the operation and maintenance of the plant. The cooling water system is readily available and the volumes and flows of water involved are deceptively large with respect to the dilution and removal of toxic or nuisance-type plant wastes. These may include fly ash (in fossil-fuel plants), nuclear wastes (in nuclear plants), chemical wastes from regenerating ion-exchange columns, sanitary wastes, and washdown water including oil, grease, and other objectionable and toxic substances. Many power plants use chlorine or other biocides to prevent fouling of the condenser tubes or other parts of the cooling water system, and the effluent may contain residual quantities of these substances. While some of these may be discharged only occasionally, they may be highly toxic and need only be present once to destroy the cultivated organisms or render them unsuitable as human food. The management, engineers, and operators of any power plant that enters a joint venture with an aquaculture enterprise must recognize these dangers and accept the fact that the culture operation will impose some conditions and constraints upon the power plant and its normal operation.

With respect to the release of toxic substances in the cooling water, legal constraints may prevent the marketing of the product whether or not dangerous levels of such toxicants are actually picked up and concentrated within the mollusks. For example, in the United States, the so-called Delaney amendment to the U.S. Food, Drug, and Cosmetic Act of 1938 prohibits the sale of any food product containing an "additive" that is a known carcinogenic agent. Since the latter includes any radioactive substance, the amendment has been interpreted to include any food organisms cultivated in the effluent from a nuclear power plant.

Another problem and risk associated with the utilization of industrial cooling water is that the plant may shut down suddenly, removing the source of heated water. This may occur regularly, for routine maintenance, which may be acceptable, particularly if it can be scheduled in the summer when warm water is not needed and when, in fact, there may be a wish to avoid it. But often shutdowns are sudden and unpredictable, due to accident or malfunction, giving the culturist little opportunity to plan or prepare for the loss of the warm-water resource.

Aquaculturists may, to a large extent, avoid the problem or reduce the risks of shutdowns by judicious choice of the power plant. Many utilities have expanded their capacity by adding new independent units, which are separate operating entities as far as generating power is concerned, but which usually share many common services and facilities. Among the latter is the cooling water system, which for a multiunit utility normally consists of a single intake and discharge system, with common pumps, pipes, canals, etc., in between which the water is dispersed to the separate units. It is seldom, except in the most drastic emergencies, that all sections of a multiunit power plant would be shut down at one time, and hence very rare that no heated effluent would be available.

One further consideration for the culturist to take into account in planning an aquaculture system utilizing the warm water effluent from a power plant is the present and projected use of the utility. Is it a *base load* plant, which is intended to be operated continuously, or is it a *peak load* facility, used only at times when there are maximal power requirements? Is it an old plant, use of which will gradually taper off to peak or emergency use, or a new plant scheduled to operate at or near full capacity for many years?

Finally, in this connection, it should be pointed out that mollusks in general, and some species such as oysters in particular, are among the organisms most capable of withstanding heat and cold shock. If the temperature of their ambient water is suddenly reduced by 15°C (i.e., from 20° to 5°C), there may be some mortality, but in many cases the bivalves will simply close up and withstand the change without damage. Further, they can remain closed and in an inactive condition for days, weeks, or even months, as they normally do in winter in north temperate climates, or they may open up, increase their metabolism, and recommence pumping, feeding, and growing as soon as the heated water again becomes available. It is doubtful that any species of fin fish, crustacean, or other food organism is as hardy and adaptable in this respect as are many of the mollusks.

In addition to the problem of radioactivity, discussed above, other legal constraints may affect the possibility of aquaculture in thermal effluents. The United States Environmental Protection Agency has established guidelines for the individual states (which ultimately regulate environmental protection), recommending that cooling water discharges not be permitted to increase the temperature of estuaries or coastal marine waters by more than $1.5°F$ ($0.8°C$) in summer or $4°F$ ($2.2°C$) in winter outside of some mixing zone to be defined by the states. Some of the latter (e.g., New York) have defined the mixing zone as a circular area 100 meters in diameter or less at the water surface. In practice, meeting this regulation requires that the utility in question discharge its heated effluent through a multiport submerged diffuser whereby the cooling water is diluted with ambient seawater immediately upon its discharge through the process of jet entrainment mixing. Where such practices are required and implemented, the cooling water is diluted and cooled so rapidly and so thoroughly that it could not conceivably be used for culture. While the preceding example represents an extreme case, the general tendency, at least in the United States, is to reduce if not eliminate (i.e., through closed-cycle cooling towers or other devices) the presence and hence the environmental impact of heated cooling water—a philosophy that is unfortunately incompatible with that of the beneficial uses of cooling water, including aquaculture.

Calculations were made earlier concerning the amount of molluscan food that may be expected to occur in the cooling water of a modern power plant and that could serve as nourishment for a molluscan culture industry utilizing this cooling water. The food concentration referred to (1.0 mg/liter wet weight) is an average, geographically as well as seasonally, and can be expected to vary considerably with respect to both time and space. In particular, the amount of phytoplankton in coastal and estuarine waters may be considerably reduced in winter, particularly at high latitudes where low levels of solar energy as well as low temperature and other unfavorable environmental factors may significantly reduce photosynthesis and hence the primary production of organic matter.

If cultivated mollusks are subjected to optimal temperatures and their metabolism, water pumping, and feeding activity are functioning at high levels, but the water does not contain adequate food for their growth or even for their metabolic requirements, the animals may fare worse than if they were allowed to *hibernate* at a low level of metabolism in the unheated water. This was demonstrated by W. Foster, Maine Department of Sea and Shore Fisheries, who cultured American oysters (*Crassostrea virginica*), European oysters (*Ostrea edulis*), hard clams (*Mercenaria mercenaria*), and mussels (*Mytilus edulis*) in the warm water effluent from the Central Maine Power Co., Mystic Station, Wiscasset Harbor. Temperatures in the area range from about 0°C in winter to 15°C in midsummer. Because the cooling water is rapidly diluted upon its discharge, water of only about 10°C above ambient could be used in the experiments. In late spring, summer, and early fall, Foster observed marked enhancement of growth of all species. However, in late fall, winter, and early spring the mollusks clearly were actively pumping and filtering, but due to the paucity of food in the water, they not only did not grow but actually lost weight and their condition deteriorated. In a subsequent winter, Foster reared marine phytoplankton artificially and fed this food to the mollusks, which then responded by growth throughout the winter season (Maine Dept. Sea and Shore Fisheries, 1972).

Food organisms, of the proper kind and in adequate concentrations, are probably seldom present in nature at optimal levels for the best possible growth of mollusks. In the preceding discussion, a concentration of 1.0 mg/liter (wet weight) of food microorganisms was assumed as a mean level that might be expected in estuaries and coastal waters. However, concentrations as high as 100 mg/liter have been achieved in outdoor cultures where nutrients are not limited (Oswald, 1970). It is well recognized that environments enriched with organic or inorganic wastes (from domestic, food processing, agricultural, or similar sources of pollution) are, despite their other objectionable characteristics, notoriously beneficial for mollusk production. Only when such environments become so eutrophic that they are unstable and become occasionally or permanently anoxic does the productivity of shellfish and other marine organisms decline or cease.

In addition to the risk of their becoming anoxic and abiotic, however, such highly enriched situations tend to be highly inefficient with respect to the utilization of food by mollusks. There is an optimal concentration of unicellular algae that is most efficiently filtered and assimilated by these animals (Tenore and Dunstan, 1973) and above which an increasing proportion of the food is rejected as pseudofeces and/or ejected as unassimilated fecal material. At still higher food concentrations, the mollusks actually reduce their pumping and filtering rates (Loosanoff and Engle, 1947). In view of this dilemma, the authors have conceived of the concept of *controlled eutrophication*, in which maximum potential production of both algae and mollusks is obtained without the risks and instability of highly eutrophic natural systems by compartmentalizing the systems for growth of the mollusks and their food separately (Ryther *et al.*, 1972).

As indicated above, maximum production of phytoplankton, by as much as two orders of magnitude above average production in nature, requires enrichment of the water with essential nutrients, particularly nitrogen and phosphorus. The amounts of these substances needed are substantial, for example, 70 tons/hectare/year of fertilizer containing 10% nitrogen to maintain algal productivity of 100 grams (wet weight) of organic matter/cm^2/day (i.e., 10 g carbon/m^2/day). Economically, this could be prohibitive, particularly in open situations where complete control and recovery of the added nutrients would be difficult.

Other alternatives to enrichment with commerical fertilizers have been considered. Roels and Gerard (1970) are conducting experiments in the artificial rearing of marine phytoplankton for mollusk food in which the source of nutrients for algal production is nutrient-rich, deep (i.e., 700 m) ocean water. However, the cost of pumping sufficient water (e.g., 20 million m^3 or over 5 billion gallons for the above-mentioned 7 tons of nitrogen) would also be substantial. Furthermore, the concentration of essential nutrients in deep ocean water, though considerably higher than surface water, is not great, but is in fact only about one-tenth that needed to sustain the maximum potential production of algae. This means that 10 times the area is needed to produce the same crop of phytoplankton that could be grown by artificially fertilizing seawater. Finally, easy accessibility of deep ocean water close to shore and to a potential mollusk culture industry is relatively rare and restricted to a few specific locations on earth.

A final alternative source of nutrients is that mentioned above, which is responsible for the uncontrolled and usually considered undesirable eutrophication of natural waters—the wastes of our society. Treated domestic wastes contain some 20 g/m^3 of nitrogen as ammonia or nitrate, an order of magnitude higher than the concentration of nitrogen in deep ocean water and sufficient to maintain algal production at its maximum potential. Initial experiments have demonstrated sustained growth of marine diatoms in a culture medium consisting of 50% secondary treated sewage effluent and 50% seawater with 75% of the 4000-liter continuous-flow culture turned over each day. Maximum algal yields from this system have reached 5 g carbon/m^2/day (ca 50 g wet weight of organic matter/m^2/day) (Oswald, 1970). The yield from these continuous cultures was fed into 2000-liter mollusk tanks, each of which were stocked with 2000 juvenile oysters. The algal culture was diluted approximately 10:1 with filtered seawater as it was fed into the oyster tanks, thereby reducing its concentration to the optimal concentration for filtration and assimilation by the mollusks. The diluting seawater also provided the flow of water and supply of oxygen needed by the animals. Under these conditions, in experiments carried out during the summers of 1972 and 1973, the oysters were able to remove an average of about 85% of the algae from the food suspensions and to assimilate this food with a conversion efficiency of 10–20%.

One of the possible constraints to a mollusk culture system using human wastes as a basic source of nutrients is the risk of the shellfish concentrating such trace contaminants as heavy metals, organic compounds, and human pathogens from the sewage effluent. To date, we have had no evidence of unacceptable levels of heavy metals or organic trace contaminants in the cultured mollusks.

Chlorinated sewage effluent may be used in the system, so that enteric bacteria are also not a problem. However, human viruses are known to survive chlorination and therefore represent a potential problem. At present we are initiating studies on both the pretreatment of the effluent with high-energy electron bombardment to kill or deactivate viruses before their introduction to the system, and the monitoring of mollusks which have been allowed to purify themselves in clean seawater for various periods of time following their growth on sewage-cultivated algae.

Beginning in the fall of 1973, the above project was moved to a new facility consisting of six 16 X 16 X 1 meters (130,000-liter or 35,000-U.S. gallon capacity) algal ponds and eight 13-m-long cement-channel raceways. This facility is capable of utilizing almost 500 cubic meters/day of treated sewage effluent and producing over a metric ton per year of oysters or other bivalve mollusks. A limited capacity to heat the algal ponds and the seawater flowing through the mollusk cultures will permit use of the facility throughout the year and simulate a culture system using heated industrial cooling water. With this pilot scale of operation, it is expected that both the technology and the economics of a mollusk culture system using both waste heat and nutrients may be evaluated.

ACKNOWLEDGMENT

The authors wish to acknowledge the support from the following research grants: NSF RANN Grant GI-32140, NOAA Sea Grant 04-4-158-5, and NSF GA-39159.

REFERENCES

Bardach, J. E., J. H. Ryther, and W. O. McLarney. 1972. *Aquaculture*. Wiley-Interscience, New York. 868 pp.

Bumpus, D. F. 1957. Surface water temperatures along Atlantic and Gulf Coasts of the United States. Special Sci. Rep.–Fisheries No. 214, Fish Wildl. Serv., U.S. Dept. Interior. 153 pp.

Loosanoff, V. L., and J. B. Engle. 1947. Effect of different concentrations of microorganisms on the feeding of oysters (*O. virginica*). *U.S. Bur. Fish., Fish. Bull. 42* (51):31–57.

Maine Dept. Sea and Shore Fisheries. 1972. Mason Station aquaculture, 1972. Fourth Annual Report Environmental Studies, Maine Yankee Atomic Power Co., Augusta, Maine.

Oswald, W. J. 1970. Growth characteristics of microalgae cultured in domestic sewage: Environmental effects on production. Prediction and measurement of photsynthetic productivity. Proc. IBP/PP Tech. Mfg. Trebon. Sept. 14–21, 1969. Wageningen Center for Agric. Publ. and Document. N.V. Noord-Nederlandse Drykkers, Merdel, Wageningen.

Roels, O. A., and R. D. Gerard. 1970. Artificial upwelling. *Mar. Technol. Soc. Proc. Conf.* Food-Drugs from the Sea, 102–122.

Ryther, J. H. 1969. The potential of the estuary for shellfish production. *Proc. Nat. Shellfish. Assoc. 59*:18–22.

Ryther, J. H., W. M. Dunstan, K. R. Tenore, and J. E. Huguenin. 1972. Controlled eutrophication—Increasing food production from the sea by recycling human wastes. *BioScience 22*:144–152.

Tenore, K. R., and W. M. Dunstan. 1973. Comparison of feeding and biodeposition of three bivalves at different food levels. *Mar. Biol. 21*:190–195.

Tenore, K. R., J. C. Goldman, and J. P. Clarner. 1973. The food chain dynamics of the oyster, clam and mussel in an acquaculture food chain. *J. Exp. Mar. Biol. Ecol. 12*:157–165.

DISCUSSION

PERSOONE: How fast will the detrital layer of the raceways build up? Will a six months' continuous operation lead to excessive accumulation? Do you have figures on the energy left in the detritus and how much is utilized by the worms.

TENORE: There are two different systems based on *Nereis* and *Capitella*, respectively. Lab experiments with C-14 tracer techniques have suggested that *Nereis* cannot assimilate detritus per se, but that the activity of microfauna is necessary.

When nematodes were added to the system, there was a significant incorporation of the labeled detritus, presumably through the food chain. We are now working on a separate project on detrital utilization.

The *Capitella* have a tremendous turnover capacity of the sediments. From the experiments with tracer techniques it seems that the *Capitella* are unable to assimilate fresh detritus, it is probably utilizing the microfauna associated with the detritus. As you age the detritus, you observe increased incorporation of the detritus.

As to the accumulation of bottom deposits during the summer, conditions on the bottom of the tank become quite eutrophied and to avoid anoxic conditions we had to, at times, put air stones near the bottom. The worms didn't die because the *Capitella* is a stress organism anyway. They just migrated to the surface on the sides until better conditions. When the water becomes better oxygenated they go back into the sediment.

KORRINGA: In the raft culture of oysters in Japan similar conditions are encountered. All the feces and pseudofeces of oysters and epibionts and ascidians will settle on the bottom underneath the raft in such large amounts that the organic material cannot readily be assimilated by the normal bottom fauna. At times this might lead to anoxic conditions so that the undermost oysters on the rafts may suffer from these adverse conditions. So it is a difficult problem to secure good conditions for the shellfish, at the same time as obtaining a good utilization of the organic debris. It might of course be said that the formation of large amounts of biomass will be a benefit, because it gives a better nutrient supply for the fishes, but it is doubtful whether fishes will eat the *Capitella*. This is particularly so because the *Capitella* live in the bottom and are difficult to get at.

TENORE: This is not always so. *Capitella* in high densities will leave their burrows and inhabit the sediment surface. Moreover, Rowe and Haedrick, working on the Mystic River—a downtown polluted harbor river in Boston—have shown winter flounder congregating through the year, and gut analysis showed that they were feeding on *Cap-*

itella. Flatfish are well adapted for feeding on benthic polychaetes, even burrowing forms.

KORRINGA: Do they really crawl out from the sediment?

TENORE: Yes, when the density is high.

CARSTENS: Another alternative for scavenger animals on the bottom might be lobsters. Would they suffer from a lack of oxygen at times?

TENORE: In the culture of lobsters or crabs or any of the other what I would call the pugnacious species, the major problem is to keep them separate from one another. We have worked with small lobsters, which will feed on the detrital food source. What you eventually end up with is one large lobster who has preyed on his fellows when they were molting. There is a need to design habitats where, like in mazes, there is a minimum probability of two lobsters meeting one another. The use of such systems, however, will tend to increase the labor costs unduly, a major drawback in our part of the world.

PILLAY: It was most interesting to hear about the progress of the investigations in Woods Hole, one of the centers where some scientific investigations have been carried out on the use of sewage for aquaculture. Although sewage is used on a very large scale for aquaculture in many countries, scientific investigations have not been as frequent as one would like.

One reason why the use of sewage for aquaculture is not as extensive as it might be may be twofold: One is the aesthetic objection that consumers have to anything grown in sewage; the second is the public health aspect. Although it seldom has been proved, there exists a suspicion that the use of sewage involves danger to public health. Would you care to comment on these two aspects, on the basis of your experience?

TENORE: In our work we are exclusively using the *effluent* of secondary treated sewage, i.e., the organic matter has been mineralized. In a summary I think that most of the metals and organic pollutants have been absorbed and chelated out in the sludge.

The problem of bacterial and viral contamination of shellfish from the sewage effluent is a possibility, and we are presently studying this problem.

KORRINGA: There is a great difference between the use of fishes like carp, which is boiled before eating, and shellfish, which are eaten raw. And, especially in the case of shellfish, great care should be taken not to contaminate your shellfish with water which may contain viruses and pathogenic microorganisms from discharged sewage.

TENORE: The other problem of social acceptability of sewage-grown food is a bigger problem. The oysters grown in the experimental system have shown no accumulation of metals or PCB above levels acceptable for the Commonwealth of Massachusetts, but we fear that prejudice might limit the use of such food.

KORRINGA: In part such prejudices are without basis, as long as we find it completely normal that industry should use cow dung and pig dung to manure the fields for raising vegetables.

TENORE: Another type of prejudice is customary eating habits. For instance, mussels are not eaten in any significant amount in the United States.

KORRINGA: It seems that one of the reasons why the public in the United States does not like mussels is that they are often full of small pearls. It is also striking that neither in Japan will people eat mussels, and it takes a long time to change the food prejudices of the people.

CONOVER: Have you considered using the mussels as supplementary food for a possible flounder population?

TENORE: We have thought about it, but the shorter you keep the food chain, the more production you will have in such a system. Another point against the growing of mussels is that there are no commercial stocks of seed available in the United States.

CONOVER: Have any estimates been made of production costs in such an integrated system of aquaculture?

TENORE: The Woods Hole project is presently at a size scale where we cannot automate and in this way assess the costs of large-scale production. The economists have figured out that in the present system, and given the present price of shellfish on the market, the system could break even. The possibility of making a profit will, of course, depend upon how efficient and optimized the polyculture system is operated.

Another point is that in the United States cities and towns are eventually going to be forced to include a tertiary treatment of the sewage because the systems of bays on the east coast with relatively low turnovers cannot stand the constant influx of the large amounts of secondary treated sewage effluent without suffering adverse effects. With the necessity of a tertiary treatment enforced upon the towns, the possible revenue from running an aquaculture system might be attractive.

The use of a polyculture system is attractive in terms of economics. If the price of one species should fall because of overproduction, the system might be restructured toward another trophic level. As one instance, we can refer to the oyster spat prices in the United States, which have fallen drastically because the hatcheries have done too good a job on production.

OPPENHEIMER: Referring to the overall scheme, are you at the point where you can put carbon values on the flow efficiencies?

TENORE: We can give values up to the detrital production. We are now looking at this problem, and we have some data for the *Capitella*. One of the real problems with *Capitella* is that the population cycle fluctuates heavily in biomass, which means that we need correspondingly long terms of observation to obtain dependable figures.

We are also doing experiments with flounders as predators of the *Capitella*, and when we get these data I think we will be in a better position to give a complete picture.

The weak point is the data on production of seaweed. Culture techniques have been difficult. *Ulva* is the great problem, because of the high production of gametophytes during higher temperatures. Last summer, every six or seven days the culture just shred with gametophyte production. We are therefore looking to *Rhodomania* as a possible alternative. We hope to accumulate good growth data for different seaweeds to assess their potential economic role in the polyculture system.

OPPENHEIMER: What is the trophic level efficiency for your oysters?

TENORE: This depends on food type and concentration, etcetera. We have found a value of 20% to be a good rule of thumb. Mussels showed the traditional 10%.

NESTAAS: What would be the efficiency of the system with regard to phosporus and nitrogen removal from the sewage?

TENORE: The algal culture in the present system removed 98% of the ammonia, while residual phosphorus goes through the system. The invertebrates regenerate roughly 20% of this nutrient.

On a per-gram basis, the seaweeds are more effective in removing the nutrients than the phytoplankton population. So you could have just a seaweed/abalone culture.

KORRINGA: They do a better job than the diatoms.

SOEDER: But isn't this at a lower concentration?

TENORE: It is at the same concentration as phytoplankton, 20 parts of sewage to 80 parts of seawater.

CHRISTENSEN: In the lecture it was mentioned that Spain had an annual production of 160,000 metric tons of mussels, which would correspond to about 4 kilograms per year capita including children and babies. Is this really so?

KORRINGA: This figure is according to official statistics, but only 20% to 25% of the tonnage is edible meat, and at least 50% of this quantity is exported. What is most impressive about this figure is that 25 years ago there existed no mussel production at all in this area. It is doubtful, however, whether this growth can continue, because it is now noted that the mussels take a longer time to reach a marketable size as compared to some years ago when the number of parks was smaller.

TENORE: It is apparent that the latest expansion of the mussel culture in Galicia has shown that the productivity of the rafts differs very much according to where they are placed in the bay. It has also been shown that the nutrients which these mussels utilize to a large extent derive from Atlantic water that has been pushed in due to displacement of Mediterranean water coming up the coast.

KORRINGA: Yes, and there seem to be definite signs that the productivity of the rafts is declining, possibly because the crop of food is fully utilized locally.

TENORE: We have had the opportunity to follow the growth of some rafts in the bay of Galicia. Some of these rafts are situated in relatively shallow water behind a large island, and the productive rafts are situated so that they will feed on the nutrient-rich water coming from the outside.

KORRINGA: We have exactly the same situation in our country with oyster farms in the Oosterschelde area, which is continuous with the North Sea. The oyster beds are situated between the sea and the site of discharge of raw sewage from the city of Bergen op Zoom. It was found that the best oysters are obtained in an area a good distance away from the sewage outlet, where the water rich in phytoplankton comes directly from the North Sea. Practically no positive effect can be traced to nutrients directly derived from the sewage outlet of Bergen op Zoom. On the other hand, in areas still further west the bottom of mixing sand is unsuitable for oyster farming, whereas the currents are too fast there for the oysters to feed and grow well.

TENORE: In the rafts in Spain we have been looking at the tremendous production of detritus from the mussel and the associated epifauna, and we actually have found a community of fish like plaice, turbot, and sole associated with the areas underneath the rafts. We have also observed a change in feeding behavior; for instance, the cod will feed on the small crabs growing on the ropes of the mussels. This is a good example of a polyculture system, where secondary fauna will develop as a consequence of the production of mussels.

PILLAY: In Woods Hole there is a fair supply of untreated sewage or semitreated sewage that can be used for aquaculture as described in your presentation. Is there a possibility of expanding this type of culture activities to other parts of the United States?

TENORE: There seems to be a number of cities in the States where such a system might be applicable. We have had many inquiries from communities, but we would like to have a few years' experience with the present setup before we expand the activities.

KORRINGA: In a number of cases the sewage system is designed to take care of both the domestic sewage and the runoff from the streets from rain, and so on. As a consequence of heavy rain, the sewage cleaning system can cease to function because of the overloading of the system, resulting in virtually untreated sewage entering the culture beds. Another drawback is that in resort towns like Woods Hole the population is possibly four to six times as high in summer as in winter, and it will then be very difficult to keep a purification plant on a biological basis running satisfactorily.

TENORE: It is obvious that our system is not an alternative to secondary treatment; it should be added to the existing secondary treatment system. I believe that the difficulties you refer to can be surmounted. And a great advantage with this kind of a system is that it does not require costly land areas that otherwise would be utilized for the construction of buildings, etcetera. The real limitation might appear when a number of small townships having installed such plants, to the effect that the market will be saturated by the output of such plants.

CONOVER: Has ozone been tried for treatment to eliminate viruses and pathogens and is it effective?

WALNE: Ozone is used in shellfish purification plants, particularly in France but in some cases also in Britain. This is a quite satisfactory method of sterilizing seawater, but apparently a little more expensive than using either ultraviolet light or chlorination.

CHRISTENSEN: A common practice is to let contaminated oysters rinse themselves in unpolluted chlorinated water, and this seems to be the most efficient method to remove viruses and pathogens.

KORRINGA: We have ample experience with such purification. There is a choice of either keeping the shellfish in quarantine in clean water for a sufficiently long time—a matter of weeks—to make sure that the procedure is sufficient. The other alternative is to operate purification plants where the water is sterilized by chlorine, ultraviolet light, or ozone. As an index for cleaning, the thermotolerant coli might be used. For viruses the procedures are not yet so well developed. The practical experience is limited. Particularly, hepatitis seems to be dangerous.

SOEDER: The eating of raw shellfish seems in any case to be a rather unsafe thing to do. Especially with the large number of people visiting sea resorts in the summer time, the risk of spreading the hepatitis virus will be considerable. In the case of viruses there is a definite need for further experience with systems of removal of such agents.

TENORE: One alternative might be to submit the sewage to primary and secondary treatment; another alternative might be electric bombardment like the system being tried out at the M.I.T. on a pilot scale.

PILLEY: It is necessary that the public feel assured of the safety of the product. It is therefore essential to develop suitable techniques for providing a safe product.

DEVIK: The latitude of Woods Hole is comparable to the Mediterranean. Further north one major problem will be the phytoplankton growth in the wintertime, when there might be practically no growth. How can we manage the cycles so as to get the most out of the summer season, and what can be done during the winter to utilize the nutrients of the sewage?

TENORE: This aspect of the problem has not been so important in the Woods Hole experiment because we have a spring bloom in Woods Hole at the end of January, which means that there is ample phytoplankton productivity at least 10 months of the year.

KORRINGA: But there will be the added difficulty that at the time of the spring bloom the temperature in the water is too low for the mussels or shellfish to make efficient use of this bloom. In other words, there is a great risk of the spring bloom going to waste. How is that avoided in your system?

TENORE: In the terms of the Woods Hole system, the temperature is sufficiently high throughout the year because of the use of waste heat.

RAYMONT: Are there problems in keeping the diatom culture growing? You mentioned in passing that by adding silica you were getting diatoms throughout, though of course your species would vary. However, if you get a big dinoflagellate burst, you may have the wrong dinoflagellates and this could really cause problems.

TENORE: There are several factors besides just the silica addition, such as the intense harvest rate, 50% to 75% of your culture, together with the proper concentration of the sewage. With too high a sewage addition dinoflagellate might come in, but then the combination of intense harvest and proper sewage level will keep the faster growing diatoms in dominance, particularly with silica enrichment.

KORRINGA: Yes, we have the impression that a relative lack of soluble silica may promote a dinoflagellate bloom which is unwanted.

TENORE: One of the problems again is that we might have diatoms of the wrong kind.

OPPENHEIMER: The answer to the question that Dr. Devik posed may be the answer to the fundamental question: Which way should mariculture go? The traditional way seems to be to go from the sewage to a bacterial cycle, to the regeneration of tissue from nitrogen and phosphorus. One might alternatively take the straight sewage and pass it through a trophic level which might be reconstituted by chemical technology. In this way you might get around the problem of needing sunlight. This might then mean to take the raw sewage and feed it, for example, to *Tilapia,* or possibly to some other kind of fish, and then, by chemical technology, reconstitute the fish for further use.

KORRINGA: This corresponds to actual practice in India, where the fish will eat from the sewage all kinds of partly decomposed particles, thus avoiding the long way around mineralization and phytoplankton.

PILLAY: There are also experiments in using pelletized sewage sludge as a fish feed.

TENORE: It may be that the answer is that we have to develop more than one type of system, each suited to the particular area.

CONOVER: Would a system like the one tried out in Woods Hole be considered a threat to the normal livelihood of the oyster growers, clam fishermen, and others? I might give the example of the raising of Irish moss, *Chondrus,* which at present provides at least a partial income for about 4000 fishermen in the Canadian maritimes. Indications are that, by efficient management, the corresponding amount of *Chondrus* might be grown in a limited area, and development of such a method might throw 4000 fishermen out of work. Have these problems been evaluated, and have you had unfavorable reactions in your dealings with these other people?

TENORE: So far the relationships have been very good.

KORRINGA: Personally, I would consider that whenever it is economical, we should replace laborious and complicated methods with efficient ones. We should always aim at the creation of economically sound units that do not require considerable subsidies to be able to operate.

CONOVER: In a sense I would agree, but we have to take into account the socioeconomic adjustment that takes place when you bring about such a necessary transformation.

TENORE: If such developments were imminent, then one might find the shellfish grower less adaptable. We consider, however, that the trophic levels other than the shellfish will not constitute mass culture. The main outlet will be in the form of shellfish.

JENSEN: What will be the total amount of sewage going into aquaculture in the future? With such a large proportion of sewage in aquaculture, we might envisage a surplus production of seafood, which might be unacceptable to man because he wants other things to eat besides seafood. Possibly we should aim, not at 100%, but at a much lower proportion of the sewage going back into aquaculture.

TENORE: I think it is a good idea that we pretend to think that we can use all the sewage in aquaculture, but we might not have only one system. There might be a range of systems suited to the various countries, and in this sense the output might not be complete in the form of fish or fish products.

BRATTEGARD: In Woods Hole you have developed a system suitable for the recycling of the sewage from smaller townships. There are a number of such towns, from Texas up to the north, that might be interested. Would you introduce the same producers, herbivores, and so on in all the different sites, or would you develop a particular community for each particular site?

TENORE: Given the differences in ecology between the Gulf of Mexico and the Gulf of Maine, different species would probably be employed. However, the trophic positions and basic model of the food chain is considered to be fairly universal. In some cases you might end up with animals that at present are not utilized, and we have to get the food technologists to work on these problems to get them into useful products. There is, for example, the case of a clam found in the estuaries of the southeast, *Rangia,* that is soaked in *Mercenaria* juice to give it some flavor and cover up the muddy taste.

KORRINGA: In Holland there is the example of *Cardium,* which is now collected with suction dredges in large amounts. That is not eaten in our country, but after shucking it is frozen and shipped to Spain, where it is canned and to a large extent reexported. Some of these might even get back to Holland.

Factors Affecting the Relation Between Feeding and Growth in Bivalves

P. R. Walne

Fisheries Experiment Station
Conwy, Gwynedd, U.K.

Bivalves have a number of natural advantages over other animal species when the problems of utilizing eutrophicated waters are explored. Basically herbivores, they have, on the one hand, a feeding mechanism which is specifically adapted to dealing with small particles (1-10 μm) and, on the other, they are themselves large enough to act as food items for man. A number of species, particularly oysters, are highly prized for food in many parts of the world and this has resulted in a substantial body of knowledge about large-scale culture systems. These comprise a whole spectrum from the highly controlled hatchery rearing of the larvae, through well-developed methods of catching natural spatfalls, to mechanized methods of cultivation on the seabed, raft cultivation, harvesting, purification, and processing.

The effect of food supply on growth can be considered in two stages: the quantity of food consumed and the quality of the food. In this chapter the effect of a number of factors which influence the quantity of food consumed are examined first and then the quality of food and its conversion into meat are considered.

QUANTITY OF FOOD CONSUMED

Bivalves feed more or less continuously by filtering particles from the water which is pumped through the gills by ciliary activity. This leads to two terms:

pumping rate, which is the volume of water passing through the gills in unit time; and filtering rate, which is the volume filtered in unit time. If filtration rate is 100% efficient, then the two terms are equal. The rate of water filtration by bivalves has attracted considerable experiment attention. Factors which affect the rate of filtration will, in turn, affect food consumption and growth, but few studies have directly linked filtration and growth. At present the data can be used as a guide. The pumping rate can be modified by changes in ciliary activity and by alterations in the width of the opening from the outside into the mantle cavity. The filtration rate can be modified by changes in the *mesh size* of the gill. Since estimations of the percentage of oxygen removed during the passage of water are usually in the 3–10% range (Ghiretti, 1966), most of the pumping activity is usually thought to be primarily for feeding.

On a unit weight basis small animals filter at a greater rate than large animals. Table I (from Walne, 1972) shows the specific filtration rate in ml/g dry weight/ min for animals of three dry meat weights at 20–22°C. The animals were kept in individual boxes and each received a flow of 300 ml/min/animal.

A further instance of the relative inefficiency of larger animals is given by Vahl (1973), who suggests that in *Cardium edule* larger animals consume more oxygen for each liter of water pumped than small animals.

The rate of filtration is adjusted according to the abundance of particles in the water. At low concentrations filtration rate is increased, and at high concentrations it is reduced and surplus food is rejected as pseudofeces (Winter, 1970; Walne, 1972). The result of this adjustment is to allow a constant level of ingestion over a fairly wide environmental range. Presumably, as the particles become too sparse then pumping is reduced, since the energy expended would not compensate for the food obtained, but this point remains to be elucidated.

The production of pseudofeces occurs when trapped particles are not carried to the mouth but are rejected through the inhalent aperture. In the natural environment it is usually a response to water containing large amounts of inorganic matter, and it is possible that some sorting occurs so that a relatively higher proportion of the organic fraction is ingested. Several authors have described this in laboratory studies using pure cultures of algae. For example, in *Modiolus* it occurs at between 40 and 60×10^6 cells/liter of *Chlamydomonas* (Winter, 1970), while in *Crassostrea virginica* (Loosanoff and Engle, 1947) it occurs at $1–2 \times 10^6$ cells per milliliter of *Chlorella*.

Table I. Specific Filtration Rates

Species	Dry meat weight		
	0.5 g	1.0 g	2.0 g
Ostrea edulis	250.6	146.6	85.7
Crassostrea gigas	287.4	173.2	104.4
Mytilus edulis	98.4	63.5	41.0

The effects of various levels of food supply on growth have been demonstrated in experiments with recently metamorphosed juveniles (Walne, 1970a). In these trials frosted glass plates to which *Ostrea edulis* were attached were suspended in tanks of filtered water enriched with various levels of unicellular algae. Examples shown in Figure 1 illustrate how too little food and too much food can both lead to a reduction in growth.

Food particles have to be large enough to be trapped by the filtration mechanism. Haven and Morales-Alamo (1970) found that in *Crassostrea virginica* particles 1-3 μm in diameter were only removed one-third as efficiently as larger particles. Vahl (1972a, 1972b, 1973) reported that *Mytilus* efficiently removed 2-3 μm particles, *Cardium* was rather less efficient, and *Chlamys* full efficiency was not reached until a diameter of 6 μm was reached.

There is evidence that a current of water promotes filtration activity (Walne, 1972) and in a growth experiment in which small *Ostrea* and *Crassostrea* (0.12-0.13 g dry organic matter) were kept individually in boxes, growth was significantly greater at flows of 183 ml/min compared with 69-74 ml/min. Some similar experiments also suggested that filtration was reduced when the water had already passed over feeding animals.

Elevated temperatures enhance the rate of feeding but, as will be shown later, this may have other effects which may not be so beneficial.

QUALITY OF FOOD

The great majority of experiments on the value of different diets have been concerned with unicellular algae, since some of these have been shown to support good growth in the laboratory. In the sea bivalves eat a wide assemblage of microorganisms as well as dead organic matter or detritus. The difficulties involved in defining detritus and preparing reproducible samples have made it impracticable to study the importance of this fraction in bivalve diet.

There are considerable differences in the food value of different species of algae to bivalves. Most of the work which has been reported deals with the larval stage, but the data on juveniles suggests that the two stages have similar requirements. From the data provided by a series of trials with *Ostrea edulis* (Walne, 1970a) it was possible to arrange the species tested in order of the food value (Table II). Since each of these trials included several concentrations of the diet, it is probable that the maximum growth rate was obtained with each species. In these experiments growth was assessed by the increase in shell length over about 21 days.

Further studies have shown that different diets can produce animals of different composition. For example, in a series of eight 21-day experiments with the quahog, *Mercenaria mercenaria*, carried out in the period June–September,

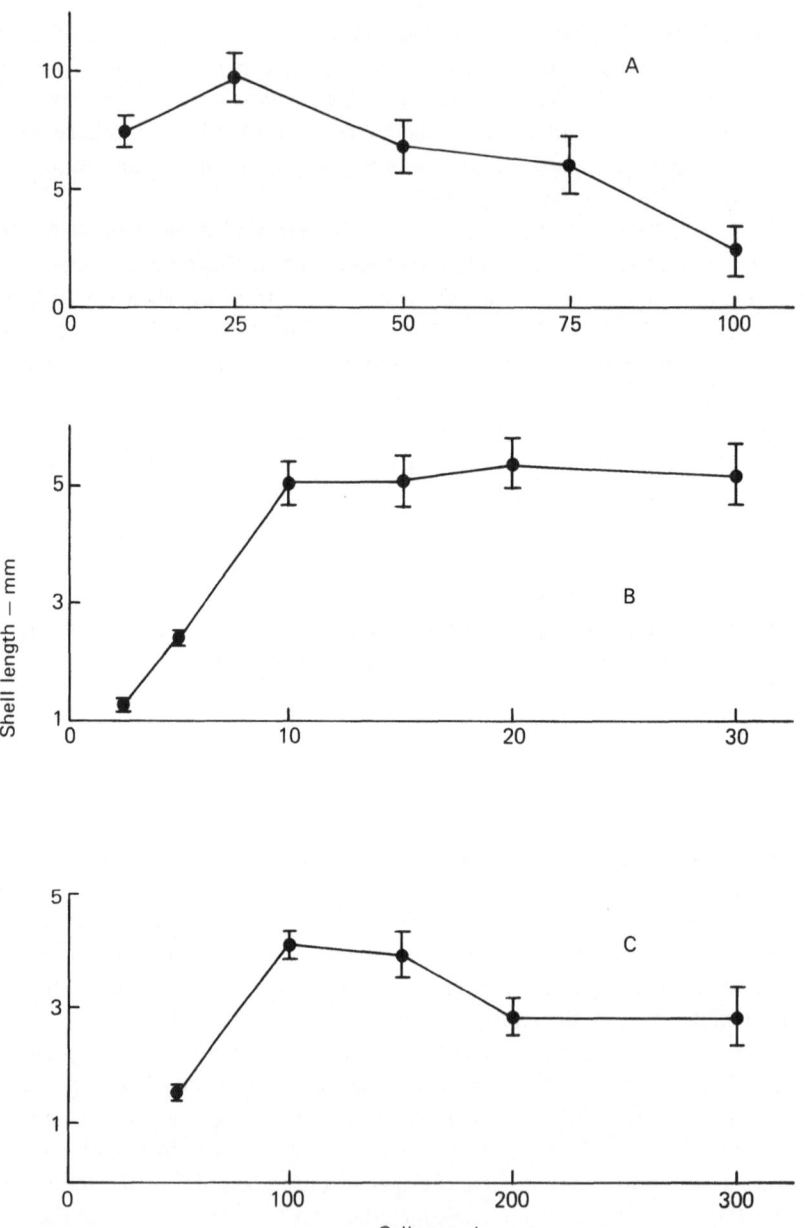

Fig. 1. The mean size reached by juveniles of *Ostrea edulis* when grown from metamorphosis in different concentrations of algae. The 95% confidence limits of the mean are shown. (A) *Isochrysis galbana* for 49 days; (B) *Tetraselmis suecica* for 33 days; (C) *Dicrateria inornata* for 26 days.

Table II. The Growth of *Ostrea edulis* Spat When Fed on Various Foods Compared with the Growth of Controls on *Isochrysis*[a]

Species	Index of food value	
	Individual experiments	Average
Monochrysis lutherii	1.70, 1.03	1.36
Chaetoceros calcitrans	1.28	1.28
Tetraselmis sueica	1.42, 1.06, 1.12	1.20
Skeletonema costatum	0.93, 1.09	1.01
Isochrysis galbana	–	1.00
Dicrateria inornata	0.85, 1.04	0.94
Cryptomonas sp.	0.54, 0.74	0.64
Cricosphaera carterae	0.41, 0.83, 0.61	0.62
Chlorella stigmatophora	0.65, 0.56	0.60
Phaeodactylum tricornutum	0.62, 0.43, 0.73	0.59
Olisthodiscus sp.	0.47, 0.71	0.56
Nannochloris atomus	0.60, 0.60, 0.41	0.54
Chlorella autotrophica	0.37, 0.66	0.52
Pavlova gyrans	0.69, 0.32	0.50
Micromonas pusilla	0.50, 0.40, 0.42	0.44
Dunaliella euchlora	0.40	0.40
Dunaliella tertiolecta	0.36, 0.42	0.39
Chlamydomonas coccoides	0.26, 0.33	0.30

[a]The index has been calculated by dividing, for each experimental series, the largest mean size in the series at the measurement made on the nearest to 21 days from the commencement of the test, by the mean size of the control on that day.

the growth of meat and shell was determined in six feeding regimes. The results (Table III) show how, in the absence of additional food, some growth took place in natural sea water, but the proportion of dry meat expressed as a condition index dry meat/dry shell declined. In the presence of extra food the growth rate was considerably enhanced, and the condition index also tended to improve.

Table III. The Mean Shell Weight of *Mercenaria* Grown for 21 Days in Various Feeding Regimes[a]

Additional food	Natural sea water		Filtered sea water	
	Shell (mg)	% meat	Shell (mg)	% meat
Initial	36.8 ± 2.4	4.7 ± 0.7		
None	53.4 ± 10.7	3.5 ± 0.2	41.9 ± 6.8	3.2 ± 0.2
20 cells/µl *Tetraselmis*	92.1 ± 32.3	5.8 ± 0.3	81.6 ± 28.1	4.8 ± 0.3
100 cells/µl *Isochrysis*	97.1 ± 36.2	5.2 ± 0.5	94.6 ± 28.6	5.2 ± 0.4

[a]The dry meat weight is expressed as a percentage of the shell weight. The 95% confidence limits are shown.

In experiments with the butter clam, *Saxidomus giganteus* (Walne, 1973), the organic content was analyzed in more detail. The nitrogen content increased at about the same rate as the live weight on the three diets used—*Tetraselmis chui, Isochrysis galbana*, and *Phaeodactylum tricornutum*—and remained at about 6-7 μg/mg live weight. The condition index increased in all the experimental conditions, with a mixture of *I. galbana* and *T. chui* yielding a significantly higher index than either species on its own. The relative gain in carbohydrate (glucose) was substantial on all diets but with a diet of *Tetraselmis* the glucose/nitrogen ratio was about 1.8 to 2.0, while on *Isochrysis* it was about 0.9 at the higher cell concentrations where the best growth occurred.

Some recent studies at Conwy have examined the value of diets consisting of a mixture of algal species. With oyster larvae, growth was only slightly better when the diet was a mixture of good food species, but a larger proportion survived metamorphosis and the spat produced were more vigorous compared with those fed on a single species (Helm, 1969). In similar trials with juvenile *Ostrea edulis* the growth obtained could be related either to the total volume of the food in the water or to one of the components of the diet. No poor food—for example, *Chlamydomonas*—was improved by the addition of small amounts of a good food—for example, *Isochrysis*. There have been no reports of the effect on growth of mixing useful species of algae with inorganic particulate material.

CONVERSION EFFICIENCY

There are few reports of estimates of the conversion efficiency in situations where the animals grow rapidly. Walne and Spencer (1974) estimated the conversion efficiency of *Tetraselmis suecica* by *Ostrea edulis* kept in culture trays. They obtained 1-3 mg increase in live weight of juveniles of 3-15 mg live weight for every 10^6 cells fed in a 21-day period and calculated that between 10 and 20% of the food offered was converted into dry meat. As expected, the less efficient conversion was obtained with abundant food when each spat was fed about 10^7 cells over the 21-day experimental period. The higher efficiencies were obtained when the spat received about 20% of this amount.

While more rapid growth was obtained at 24°C, the condition index was about 4.0 compared with 5.5 at 14°C. The conversion efficiency in terms of increment of organic matter per 10^6 cells fed was similar at 14, 19, and 24°C.

In these experiments the daily ration was 4 to 40% of the dry organic matter of the juveniles per day. Helm *et al*. (1973) examined the effect of adding *Tetraselmis* at about 6% per day of the dry organic weight to seawater flowing over breeding stock oysters. This ration was below the maintenance level, since the stock lost condition, but it had a beneficial effect as the larvae produced had greater food reserves than those from stock which had not received a supplement of cultured algae.

SUMMARY

It is clear that a number of conditions have to be fulfilled before high growth rates of bivalves are obtained and before the stock is in good condition. Laboratory trials suggest that there are wide differences in the food value of different species of algae and that the use of mixtures of species does not lead to an improvement in the value of poor species. Simple enrichment of seawater may not lead to a bloom of good food species. However, the species used in laboratory tests are those which are available in culture and may not be representative of those occurring in the natural environment.

The use of dead foods—and no good diet has yet been defined—leads to difficult presentation problems, since the small particles required are readily attacked by bacteria which foul the water.

Given an adequate diet, it is necessary to present it to the animal in a manner leading to an adequate level of ingestion. Too high a level leads to waste through the production of pseudofeces, low conversion efficiency, and poor growth. A suitable water current is required to stimulate feeding and carry away feces. The accumulation of the latter is a major problem in areas of intensive bivalve culture where water currents are weak.

Increasing the temperature will hasten growth but it may divert energy from somatic growth to gonad growth. It also appears to hasten shell growth more rapidly than meat growth, since spat grown at high temperatures have less meat in proportion. This may explain the observation that oysters at low latitudes tend to be in poorer condition than those from high latitudes (Walne, 1970b).

REFERENCES

Ghiretti, F. 1966. Respiration. Pages 175-208 in: K. M. Wilbur and C. M. Yonge, eds. *Physiology of Mollusca*, Vol. 2. Academic Press, New York.

Haven, D. S., and R. Morales-Alamo. 1970. Filtration of particles from suspension by the American oyster, *Crassostrea virginica*. *Biol. Bull. Woods Hole, Mass. 139:* 248-264.

Helm, M. M. 1969. The effect of diet on the culture of the larvae of the European flat oyster, *Ostrea edulis* L. ICES CM 1969 E:7, 7p (mimeo).

Helm, M. M., D. L. Holland, and R. R. Stephenson. 1973. The effect of supplementary algal feeding of a hatchery breeding stock of *Ostrea edulis* L. on larval vigour. *J. Mar. Biol. Assoc. U.K. 53:* 673-684.

Loosanoff, V. L., and J. B. Engle. 1947. Effect of different concentrations of microorganisms on the feeding of oysters. *Fish. Bull., U.S.*, no. 42, 51: 31-57.

Vahl, O. 1972a. Efficiency of particle retention in *Mytilus edulis* L. *Ophelia 10:* 17-25.

Vahl, O. 1972*b*. Particle retention and relation between water transport and oxygen uptake in *Chlamys opercularis* (L.) (Bivalvia). *Ophelia 10:* 67-74.

Vahl, O. 1973. Porosity of the gill, oxygen consumption and pumping rate in *Cardium edule* (L.) (Bivalvia). *Ophelia 10:* 109-118.

Walne, P. R. 1970*a*. Studies on the food value of nineteen genera of algae to juvenile bivalves of the genera *Ostrea, Crassostrea, Mercenaria* and *Mytilus. Fishery Invest. Lond.,* Ser. 2 *26*(5): 62 pp.

Walne, P. R. 1970*b*. The seasonal variation of meat and glycogen content of seven populations of oysters, *Ostrea edulis* L., and a review of the literature. *Fishery Invest., Lond.,* Ser. 2 *26*(3): 35 pp.

Walne, P. R. 1972. The influence of current speed, body size and water temperature on the filtration rate of five species of bivalves. *J. Mar. Biol. Assoc. U.K. 52:* 345-374.

Walne, P. R. 1973. Growth rates and nitrogen and carbohydrate contents of juvenile clams, *Saxidomus giganteus,* fed three species of algae. *J. Fish. Res. Board Can. 30*(1): 825-1,830.

Walne, P. R., and B. E. Spencer. 1974. Experiments on the growth and food conversion efficiency of the spat of *Ostrea edulis* L. in a recirculation system. *J. Cons. Perm. Int. Explor. Mer. 35*(3): 303-318.

Winter, J. 1970. Filter feeding and food utilisation in *Arctica islandica* L. and *Modiolus modiolus* L. at different food concentrations. Pages 196-206 *in:* J. H. Steele, ed. *Marine food chains.* Oliver and Boyd, publishers.

DISCUSSION

PERSOONE: Have you tried *Cyclotella nana*, or *Thalassiosira pseudonana* (which seem to be quite hard to grow in our countries as compared to the United States)? And did you try *Dunaliella viridis*?

WALNE: We have tried *Cyclotella* quite a lot, particularly as a food organism for *Crassostrea gigas* larvae in the early stages. We found that the production of an algal culture is not quite as good as *Chaetoceros*, so we normally use *Chaetoceros* in preference to *Cyclotella*. Which was the other one?

PERSOONE: *Dunaliella viridis*, which is much better apparently for culturing other species, such as, for example, crustacea.

WALNE: Undoubtedly crustacea are different. The *Chlamydomonas* and *Dunaliella* that are in this table are a particularly poor food. We have used it quite a lot for *Artemia*, and it is the food now being used on the Isle of Man for *Artemia* and for rotifers. Good food for one species of bivalves seems to be good for the others, but does not apply going from bivalves to crustaceans.

PERSOONE: Do you think that the difference in nutritional values of the species of the table which you have given here is due to morphological differences or to the chemical composition of the species, for example, the absence of a micronutrient or a vitamin?

WALNE: There is probably not one answer, but morphological differences are quite important. In some circumstances morphological factors are decisive, particularly related to the cell wall construction and the ability of the animal to break down the cell wall in the quite rapid transit time that bivalves seem to have.

The gross biochemical composition does not seem to be decisive. A number of people have looked at amino acid composition of the algae. The concentrations are not very different from algae with a good food value as compared to those with a bad one, but the whole thing is very complex because we do not know if the animal can digest amino acids more readily from some species of algae. There is very little information on the digestibility of proteins related to bivalves. Neither have our experiments with micronutrients given indications as to their importance.

PERSOONE: Do you think that the metabolities which might be present in high density cultures could interfere with the growth of the larvae, for example, ammonia at an alkaline pH?

WALNE: Yes, but I don't know where you would get ammonia from in an algal culture.

PERSOONE: Well, if the algae are past the log phase there might be ammonia present and besides, the excretion of your bivalve larvae could also give some ammonia in high density cultures. Recent work indicates that ammonia can have very bad effects, particularly on sensitive organisms like oyster larvae.

WALNE: Yes, water quality might be a factor. We have done a number of experiments at Conwy in the last three or four years, and our technique at present is to rear larvae on either filtered seawater or on artificial seawater as a control. We find that in the winter months the growth we obtain over four days is better in natural water than it is in our artificial water. When the spring outburst is present, the conditions are reversed. It is accentuated in the last *Phaeocystis* outbreak, usually at the end of May or June. Toward the end of July, the natural water comes back and is better than the artificial water. So far we have not been able to establish any chemical component that will correlate with this phenomenon.

PERSOONE: I am not only thinking of the natural waters, but rather of metabolites and the quality of the water of the algal cultures which we use as a food. Did you ever try to separate the algae from their medium before feeding, and would there be a different result on the food efficiency?

WALNE: We have not done such experiments. Normally we take great care to grow our algae so that they are growing pretty sharply—in particular, we do not use declining cultures and senescent cultures.

KORRINGA: Were you concerned with the effect of zinc at some time?

WALNE: We have a lot of zinc in the water sometimes, but we seem to be able to continue to grow larvae in such water, even if the zinc level measured by chemical methods suggests that we should not be able to.

KORRINGA: But what is the explanation that the water in which *Phaeocystis* is growing is less suitable than your artificial water? Could that be the effect of some metabolites?

WALNE: It seems so. There are indications that declining *Phaeocystis* cultures do produce organics in water. There is some work in the Antarctic by Lewin, where they demonstrated that guts of king penguins were completely sterilized by the antibiotic activity of *Phaeocystis* cultures. In the waters of the Northern Hemisphere, with *Phaeocystis* present, the waters contain a very reduced diversity of bacterial flora. But the question we cannot answer is: Has something been added to the water or has something been removed?

SOEDER: The table of your prepared paper indicates a utilization of less than 100%. Does this mean that the animals are excreting more of the food in the feces or is the food not taken up at all?

WALNE: The index of that table is obtained by comparing the size of the oyster at the end of the three-week experiment with the size of equivalent oysters in the control. There is no reference to the amount of food ingested, or to the filtering efficiency.

SOEDER: Would you think that the same amount of food in terms of grams per gram by weight of the animals has been taken up?

WALNE: The experiments were designed so that the amount of food available to the animals was equal.

SOEDER: But not the food taken up?

WALNE: I have some evidence that the bivalves did not filter so actively on some of the unfavorable foods, and I think this could be part of the explanation.

SOEDER: This table is quite remarkable, especially if we consider that, for instance, for *Daphnia Chlamydomonas* is an excellent food. It may be a suggestion that it has been found that acid polysaccharides from algae are able to inhibit some proteinases in the gut of monogastric vertebrates, so it might seem that the cell content cannot be attacked by the enzymes.

WALNE: But that is not the whole story. It might be an explanation for *Chlorella*, but, for example, *Dunaliella* does not have an obvious cell wall. *Pavlova gyrans*, for example, is a very soft sort of flagellate which you would expect would be digestible.

SOEDER: Well, in the case of the green algae, I can say for certain that there are differences in digestibility. Kraüt, Jekat, and Pobst (1966) have compared raw and homogenated green algae, and also algae processed in other ways. In all cases the digestibility of intake was very poor. Correspondingly in bivalves, we can see green algae coming out intact. We cultivated silver carp with green algae of the *Chlorella* type. They eat *Chlorella* like mad, but as long as the organisms are raw they are unable to digest them. But when the cells are broken or denatured, they are completely digestible. In such cases we found no difference in digestibility for all algae, or no conspicuous difference as large as indicated in your table. This would point to the importance of the breakdown of cell walls.

PRITCHARD: Is it possible to investigate the chemistry of the carbohydrate, lipid, and protein structures of these foods and then begin to directly formulate and produce cultures to specifications? Secondly, do we have any indication at all as to what energetic systems operate in these bivalves, and what the critical energy ingredient is? Is it a short chain fatty acid as in a ruminant, is it glucose, or is energy being obtained from a protein source. What are the critical pathways?

WALNE: The reason why these things have not been done is the practical difficulty in working with bivalves. The objectives of these experiments was to find good algal food for juvenile bivalves; that was the end of the road as far as these experiments were concerned.

But there are a number of experimental difficulties in working with bivalves, and in particular the problem of ensuring that what is given to the bivalves are really algae and not partially decomposed algae contaminated with bacteria. This might necessitate working in a bacteria-free environment in the case of bivalves. In trying to use definite formulations this problem will be greater still because of secondary growth on these nutrients. For these purposes encapsulations have been tried out. I agree with your question that this is an extremely interesting and important area to try and solve some of the basic nutrition in bivalves.

PRITCHARD: Apart from the need for a purified diet approach, there seems to be a lack of carbohydrate, protein, and lipid analyses on the cultures that you are growing. It should be possible to take these cultures and do an analysis of them at the same time as doing an analysis of the nutrients added.

WALNE: I have explored two lines. The amino acid spectrum has already been mentioned. The other thing we have done is to measure carbohydrate and protein content of *Isochrysis* cultures over various periods of time, and feeding these to larvae and to spat. In our apparently standard system we got quite big differences over a time period in the carbohydrate and protein content on a per-cell basis of *Isochrysis*, but there was no correlation between that and the growth of larvae. The growth was good whether there was a high protein per cell or a low one.

One student in our lab is at present looking at the standardization of the major biochemical constituents of our major foods by culture techniques, in the hope that she will be able to standardize these algae and obtain some of those data you are asking about. But what I really think requires tackling is the use of diets of a known composition.

PRITCHARD: Is there any reason that one cannot do studies on metabolic pathways using isotopes?

WALNE: No, not really. We have used phosphorus as an indicator. We are thinking of using labeled glucose and studying the incorporation of the label to determine the essential amino acids. But there is still a big problem of the digestibility of the proteins. There seems to be no way around the problem except feeding the protein to the animal and seeing how the animal gets on.

KORRINGA: Some forty years ago a biologist called Gilbert Ramson was confident that he had demonstrated that the oysters could feed on dissolved organic matter. What is your view on that?

WALNE: It has been demonstrated that oyster tissues could take up dissolved organic matter and dissolved glucose from water, but again to interpret the experiment one needs to work in a bacteria-free environment.

RAYMONT: In analyzing algae for carbohydrates there have been difficulties in distinguishing between the wall carbohydrates and the cell carbohydrates. There is also the possibility of considerable differences depending on culture conditions, not only nutrient concentration but also light regimes. All these factors can affect the carbohydrate composition of the algae. The same probably holds for lipid materials. This technique of growing algae under controlled conditions is presumably then rather difficult, since algae appear to be particularly labile.

MATTHEWS: Dr. Walne mentioned the importance of temperature in changing the quality of food. In your case shell length has been used as the index for measuring the nutritional value of the food. Is this a reliable estimate? Could the algae at the top of the list be good for producing shell, but not necessarily good for producing quality oysters?

WALNE: Shell length has been used as an estimate because it is the easiest thing to do. Supplementary information of dry meat weight and dry shell weight indicates definitely that the ones at the top of the list are bigger whatever component you take, compared with the ones at the bottom of the list. But in terms of lesser differences, there might be some interchange of position.

In some recent experiments I have investigated the proximate composition of butter clam spat fed on two different diets, *Tetraselmis* and *Isochrysis*. *Tetraselmis* was not a very good food, neither in shell length nor in overall weight increase, and it led to a considerable increase of carbohydrate per animal. With *Isochrysis* there was much less accumulation of carbohydrates in relation to protein. Part of your comment, then, might well be correct in so far as the animals at the top of the list or in any part of the list may have had rather different proximate composition in terms of protein and carbohydrate ratio, although they were not very different regarding the total weight or overall length.

BØHLE: Related to Dr. Walne's paper we had some preliminary results of the behavior of mussels in heated seawater and the nutrient level necessary for growth. In these experiments small juvenile mussels, 13 mm across, were put in different boxes and under a continuous flow of seawater with a continuous addition of algae. The experiments were performed at 8° for five weeks, and then repeated at 12°, 16°, 20°, and 24°, respectively, with different concentrations of algae. The experimental setup was as indicated on the first slide. The next slide shows that the weight increases with the algal concentration up to a certain level when the temperature is relatively low, with an optimum temperature at around $16^\circ C$ at the highest algal concentration. The next slide shows the same with the length as indicated. These experiments were done with *Isochrysis galbana* supplied by Dr. Walne. These results will serve to underline the importance of adding sufficient food to heated seawater, as the first prerequisite for efficient aquaculture.

WALNE: May I make a point in relation to the food value of algae? We know from experimental basis that there is a very considerable difference in the food value of algae. This has the most important application in the aquaculture sphere in the sense that the production of algae by whatever methods we use is by no means the end of the road. It will be necessary to have a system which is not only efficient in terms of carbon fixation or uptake of nitrogen, it also has to be precisely defined in terms of species composition. Reading about and observing various experiments of this sort in different parts of the world, it seems rather clear to me that the algal population that are built up in these very semicontrolled environments are very different in different places, and they are very different from time to time in the same place. We have an example of the Long Island bivalve hatcheries which were originally developed on the principle of passing through a continuous centrifuge and allowing it to bloom for 24 hours; this water was then used for the larval culture.

An important point is that it is not just a matter of producing a large amount of algal material per se, but the quality of the algae has to be controlled in itself in some way. I have discussed this at various similar meetings, and there does not seem to be a great deal of suggestion as to how you can influence culture conditions in large volumes of water so as to ensure that you are producing just the right diatom or just the right flagellate. As one example, Dr. Tenore just mentioned that at Woods Hole last year they did not have good results in terms of growth because the algal culture with which he was provided was predominantly *Phaeodactylum*.

TENORE: In the switch from semibatch to continuous culture you will inevitably obtain a predominance in *Phaeodactylum*, and the efficiency of energy flow for the year before was 21% for oysters, but last year was 8%—threefold decrease—so that is very true.

CONOVER: Do you consider that the *Nitzschia* or the *Phaeodactylum* might be a symptom of eutrophication in the culture system?

TENORE: Dr. Dunstan has published several papers on the phytoplankton component of the system. *Phaeodactylum* being a fast grower, eventually outcompetes other species. In old cultures there is probably a eutrophied system functioning. Dr. Dunstan discussed the idea of using what he called pioneer communities of large mass algal cultures. In this system there would be a series of ponds where one pond is used for one week or so and then switched off, if it is not exhausted, and then going on to another one. In this fashion the bivalves are always fed cultures dominated by more favorable algal species.

PERSOONE: That is the system that we have been working with.

WALNE: In the domination of *Phaeodactylum* in a mixed culture of *Phaeodactylum* and *Isochrysis*, it might not only be the question of the growth rate of *Phaeodactylum*, but also of a possible antibiotic effect on the *Isochrysis*.

TENORE: Whatever the reason is, it takes over in the end. I think it would be hard to envisage mass culture even at a thousand-liter level or at our system of two thousand liters where you can keep specific monocultures going in a marine environment.

SOEDER: There are, of course, differences which make your culture turn to specific species, as for example *Phaeodactylum*. Once you know the reason for that you can avoid it.

WALNE: I would like to make reference to some experiments with which Dr. Raymont was in particular associated in the south of England, where he was growing *Phaeodactylum* in tanks. These experiment tanks become badly overgrown with a species which we call *Monas*, and they say it ate up very nearly all the *Phaeodactylum*.

RAYMONT: May I elaborate a little? We had a system of static tanks where we added CO_2 to the air supply used for mixing, but which also stabilized the pH. We found that once we had gone past the log phase with *Phaeodactylum*, the *Monas* suddenly bloomed, even though it was so sparse as to be almost uncountable initially, and the *Phaeodactylum* count fell rapidly. In a few laboratory experiments we could literally see *Phaeodactylum* being eaten, but the rate of consumption did not seem to be sufficient to explain the rapid decrease. In other experiments we filtered off the *Monas*, and upon addition of this filtered water to the *Phaeodactylum* culture, the *Phaeodactylum* declined as before. The culture soon settled on the bottom, so there seemed to be an extracellular effect of the *Monas*, quite apart from actual consumption.

TENORE: In our cultures with the *Phaeodactylum* we observed essentially the same phenomenon.

CONOVER: Is there any real reason why we cannot go from a pure test-tube culture to a four hundred-liter tank? I really think we are ultimately going to have to work with single cultures, at least of cells. I think that we are going about as far as we can by just pouring sewage into batches of natural seawater. Where the ecosystem is altered, it is going to become unbalanced one way or another ultimately. It seems to me that the type of problem we have been discussing here is inevitable if we continue to take this approach. It is probably the cheap approach, but in the long run I'm not certain that it is the reliable approach.

TENORE: I think Dr. Soeder's remark was important in the sense that there must be some limiting factor which you can exploit in order to keep a mixed culture. I think we have had success in the 500-liter cultures growing *Cyclotella nana* for periods of time, but it is prohibitive in means of cost to try to autoclave 200 liters of water or Millipore filter the water.

I think that in terms of a larger system, if you can have cultures with change of dominance of species composition as long as you are within a field of acceptable species, I think this is an alternative way of thinking about it.

WALNE: Once you start having more than one species, in my experience, you wind up with *Phaeodactylum* or *Chlorella* or the nettles of the field, so to speak. Our commercial hatcheries are operating up to batches of about 300 liters with single species cultures using quite simple equipment, but they estimate that the cost is between 3 pence and 5 pence per liter for a fairly thick culture. This is not an economic food for a growing bivalve. It is all right for the juvenile stages. The other thing is that we want a real understanding of the causes of succession in the sea.

CONOVER: When we are putting a mixed population of diatoms into multiculture with bivalves, we are encouraging disproportionate growth of some species, because presumably the bivalves are not equally able to sample all of the forms of organic matter that are there. It seems then that we have to break the system into components. We have to do it all the way down and end up with monocultures.

PERSOONE: I really wonder if one starts a quite large culture with green algae and one adds a relatively small volume of a natural plankton community, whether one of the natural phytoplanktons will take over. I am not sure that the algae present in the normal seawater will take over.

CONOVER: I doubt it.

PERSOONE: So this could be a way of approaching it.

TENORE: In the experiment two years ago we were adding a natural stock each time, and we did see a change in dominance from *Skeletonema* to *Chaetoceros*.

PERSOONE: It depends what you are starting with. I think if you can start either with a green algae, such as a good growing *Dunaliella* or *Tetraselmis* or something like that, the normal phytoplanktons present in the natural seawater will not take over.

WALNE: I have done this with *Chlorella* in a 10,000-liter tank. I've had a rich culture of *Chlorella* standing outside quite unprotected and we have drawn it off and made up with ordinary seawater. You can do that for a long period with species like *Chlorella*, but who wants *Chlorella* anyway? So far I have not found a good oyster food yet which is resistant to these other species.

CONOVER: One of the problems one finds in culturing copepods is that the concentrations of food which one must offer to these organisms are always very much greater than those which are commonly found in the sea. I therefore believe that what we use is not the really ideal food for the copepods in question, and I suspect that for bivalves, too, there are probably some more suitable cells than we are using now. I don't think we are really working on improving the strains to any extent. There does not seem to be such work going on.

OPPENHEIMER: Has anyone used one of the green photosynthetic bacteria, for instance, the rhodospirillum group? They grow vigorously, one can genetically control them, and they require sunlight for their energy. Their size is about $1 \times 7\text{-}8$ μm, long enough to get hung up in your feeding web. They are also relatively easy to grow in pure culture, to separate out, and to grow in vast batch cultures like penicillin, even to grow in 50,000-liter containers of pure culture if you want.

There are several species of interest, such as *Rhodospirillum rubrum*, which utilizes sulfur, but nonsulphurous species can also be used. They might lend themselves to some research because one can really produce them in very dense cultures.

WALNE: How do the proteins of the bacteria compare to algal and plant foods in general?

OPPENHEIMER: It depends on the species. One can usually regulate the proteins in most bacteria. By changing the medium condition one can, for example, change the lipid content from 2% to 50% in some of the cells.

CONOVER: This is very interesting, although I don't believe anyone has ever had much success feeding bacteria to bivalves.

OPPENHEIMER: That is because biologists don't attend to the problem.

CONOVER: As a sort of general food chain ecologist, I should like to believe that there are bacteria which will grow sufficiently rapidly and efficiently to provide food for higher planktonic levels. However, we cannot really demonstrate any utilization of bacteria by most of the common species so far studied.

The Introduction of New Species in Habitats of Heated Effluents

J. E. G. Raymont

Department of Oceanography
University of Southampton
Southampton, England

INTRODUCTION

There are few genuine examples of the conscious introduction of marine animals
to new areas. Immigrant species have usually been accidentally introduced; ships
arriving from foreign ports are regarded as the most likely source of immigrants
either as ballast or on hulls. The history is usually unknown and the spread of
the immigrant species is largely conjectural, one of the major difficulties, of
course, being that even bottom-living forms frequently have planktonic larval
stages and can spread relatively rapidly with tides and currents.

There are still fewer examples of immigrant species related especially to
habitats of heated effluents, though the biology of such species, especially growth
and reproduction, may be of particular interest. Generally, investigations of this
type tend to be related to measures for the control or eradication of unwanted
or harmful marine animals. Species of shipworm for example, have long been
cited as harmful immigrants, some of them essentially warm water forms which
have benefited from the artificially warmed conditions of certain docks. Sudden
and very damaging outbreaks of such shipworms are known, and though it is
often difficult to correlate these outbreaks with environmental factors, the damaging attack by *Teredo navalis* in Swansea Dock during the recent war, as well as
the occurrence of the ship worm *Lyrodus* at Shoreham Harbour, may almost
certainly be associated with unusual artificially warm conditions in those areas.

Harmful marine species include the imported pests, particularly associated with the oyster industry, such as the American drill (*Urosalpinx cinerea*), and the competitor species, *Crepidula*. But neither of these introductions can be associated with warmed water. Moreover, the precise effects of immigrants on native species are difficult to quantify, especially where nonbiological parameters have also changed.

Examples of the introduction of marine species that may prove beneficial in the sense of providing the basis for commercial exploitation in the future are very limited. In discussing marine immigrants I propose to limit myself to species introduced into Great Britain, and essentially to consider the effects of heated effluents arising from electricity power stations. This limitation stems largely from the fact that the electricity industry has tended to cooperate in this type of marine research, and some fairly long-term records are available.

I shall refer to accidental unwanted or harmful introductions at some length, since such colonizers might greatly affect the success or failure of inoculations or implantations of marine key species desired for commercial exploitation. New "weed" species can flourish and then might rapidly modify a marine environment.

I do not propose to discuss any fish farming experiments in warm effluents, since most of these involve local fish species in Britain. Neither shall I refer to the experiments with foreign species of prawns, since warm water experiments have, as far as I know, been confined to tank cultures.

OBSERVATIONS

One of the best authenticated accounts of immigrant species related to warm waters comes from studies of Queen's Dock, Swansea, due primarily to Naylor (1965*a*, *b*). The dock received warm water used for cooling purposes from an electricity power station and has only a limited exchange to the outside sea. Temperature records indicate temperatures of some 14° to 26°C, depending on season, inside the dock; this is approximately some 10°C in excess of the outside area.

Naylor points to some peculiar features of the fauna of the dock. Throughout 1956-1958 he records a number of immigrants, several of which appear to be warm water species.

A species of Kamptozoan, *Loxosomella kefersteinii*, new to Britain, settled abundantly inside the dock. *L. kefersteinii* is a warm water species—a native of the Red Sea, the Adriatic, and the eastern Mediterranean—and probably came in with a Mediterranean tanker. Settlement in Swansea docks commences in July and reaches a peak by September, the animals reproducing by budding. The animals reach maximum size from buds in only two to three weeks, so a very rapid colonization of the area can occur (cf. Ryland and Austin, 1960).

Two species of *Bugula*, also settling organisms, occur abundantly in the docks. *B. neritina* is essentially a warm water species, widely distributed in the warmer waters of the world. It is found apparently only in the south and south-west of Britain, in docks and harbors, especially where artificial warming occurs. In Queen's Dock it is abundant, settling from May to October, with a peak about July or August. A related species, *B. stolonifera*, also occurs in Queen's Dock and has a similar breeding time. It is essentially a Mediterranean species (Ryland, 1960).

Two species of worms exemplify some of the problems in trying to identify genuine immigrant species benefiting from warmed water. Both occur in Queen's Dock. There is some doubt about whether the first *Hydroides incrustans* should be a separate species. In any event it is a warm water form which was first observed in another warm water area (Shoreham Harbour canal). The other species, *Mercierella enigmatica*, can occur in considerable quantity but only sporadically. Tebble (1956) refers to its presence in Weymouth Harbour, where it can suddenly appear in very large quantities. Markowski (1962) also refers to the species as occurring sporadically but in great numbers in Cavendish Dock, an artificially warmed dock in the north of England which receives power station effluent. To some extent its appearance appears to be partly dependent on lowered salinity, but the species seems to flourish with raised temperatures.

A very much clearer example of a warm water immigrant is the barnacle *Balanus amphitrite* var. *denticulata*—the only barnacle found in Queen's Dock near the outfall of the power station over the four years, 1956 to 1960. Naylor believes that its occurrence is associated with the raised temperature. This is supported by the work of Crisp and Molesworth (1951) who found this species, a recognized subtropical/Mediterranean form in Shoreham Harbour canal—an area which receives effluent warm water from a power station. The barnacle could occur in large numbers—500 to 1000 per square meter. Outside the warmed area of Shoreham Harbour canal *B. amphitrite* occasionally made temporary settlements in the estuary and even as far as three miles distant, but these colonies were never permanent. In summer the temperature of Shoreham Harbour canal frequently was of the order of 25°C. Crisp and Molesworth believe that successful breeding of the barnacle occurs every year, followed by settlement from about May to August sufficient to maintain the species. Patel and Crisp (1960) found that *B. amphitrite* commences breeding at temperatures of 16-18° and it breeds readily at temperatures exceeding 32°C. It is not found naturally in Europe on open coasts north of the Charente area of the Bay of Biscay. Temporary records are at Plymouth near a power station outfall, and a dock at Llanelly, South Wales, which also received a warm water discharge. In Queen's Dock it appears to be an obvious warm water immigrant, breeding in April–May and again during autumn, with water temperature remaining sufficiently high. In any transplantation of key species to areas of warmed water the dangers of introducing a barnacle such as this are only too obvious.

Two immigrant species of crab are also listed by Naylor as occurring in Swansea Dock. The first is *Brachynotus sexdentatus;* this is a Mediterranean species, not found elsewhere in Europe but quite common in Queen's Dock. Several species of crabs have been suggested as being carried in the ballast of ships, and establishing themselves as immigrants in European countries (e.g., *Rhithropanopeus*). But *Brachynotus* appears to be very much restricted to artificially warmed areas and breeds only in the summer. Another foreign species found in Swansea Dock, *Neopanope texana sayi*, occurs normally on the eastern coasts of the United States from Florida to the Gulf of St. Lawrence. Although it probably benefits from the warmer conditions of Swansea, its widespread distribution in North America suggests that it could colonize British waters more widely.

Many of the immigrant warm water species in Queen's Dock, and indeed *Balanus amphitrite*, appear to have a fairly long breeding period, mainly over the summer months. This is a fairly common pattern with warm water forms. However, from early 1961, since the time that the temperature inside Queen's Dock has fallen somewhat, some faunal changes have occurred. For instance, a native species of barnacle (*B. crenatus*) previously absent from the dock, has appeared; this species is presumably now competing with the warm water immigrant.

An especially interesting species found in Swansea has been identified as the wood-borer, *Limnoria tripunctata*. The history of identifying species of *Limnoria* in Britain is complex.

In 1957 Eltringham commenced a study of *Limnoria* in our own locality—Southampton Water. This was partly related to the construction of a new power station along Southampton Water at Marchwood which would contribute considerable quantities of warm effluent seawater. Until 1953 only a single species of *Limnoria* (*L. lignorum*) was known in British waters. Eltringham and Hockley (1958) discovered that three species existed in the area; indeed, all three could occur in the same piece of wood piling. Jones (1963) carried out a geographical survey of *Limnoria* species throughout a large area of Great Britain. *L. lignorum* occurs all around British coasts, though it is rare in southwest England. It is a typical boreal species in both Europe and North America. It breeds in Southampton Water in early spring (possibly even from November) to as late as May–June. The second species (*L. quadripunctata*) is essentially a warm temperate species. Menzies (1959) regards it as confined to areas where the water temperature averages between 11.4 and 16.2°C for at least five successive months of the year. It is fairly widely distributed around the coasts of Britain, though it appears to avoid the colder coasts. In Southampton Water, in the warmer areas at Marchwood near the outlet of the power station, a large part of the population is *L. quadripunctata*. In Southampton it has a prolonged breeding period, when temperatures are rising and falling through about 10°C. The most interesting species however is *L. tripunctata;* this is a temperate/tropical species, probably near the northernmost limit of its range in Southampton. It is presumably an immigrant. Menzies suggests a minimum temperature for breeding of 14°C; this temperature must presumably be maintained for some period of time.

The species migrate as adults with their breeding. *L. lignorum* migrates first, roughly in March–April; *L. quadripunctata* migrates later, starting in April and proceeding during the summer until about August–September. *L. tripunctata* migrates still later beginning in May. At Marchwood, however, the migration of *L. quadripunctata* commences in March and lasts until October. Thus, it is two months longer and, in fact, does not entirely cease even during winter. This extension of the breeding migration time is presumably linked with the warm water around Marchwood.

While these investigations were in progress, Naylor reported that *L. tripunctata* was the only species of *Limnoria* found in Queen's Dock. It breeds there throughout the year, and would appear to be another example of an immigrant warm water species.

Investigations of *Limnoria* were also carried out at Poole, Dorset, where a power station outfall carries warm water into a comparatively restricted shallow harbor area. Near the power station outfall at Poole an actively breeding population of *L. tripunctata* was found in April but further away from the power station, outside the influence of the warm water, no breeding females were found at that time. The effect of warm water on breeding populations of this species thus seem clear.

The problem of defining an immigrant species and of relating its success to any influence of warm water can be applied to a study by Holmes (1968) on the ascidian *Styela clava.* This species was first reported from Plymouth in 1953 and was found in Southampton Water in 1957–58. It has since been found in such areas as Shoreham Harbour and Poole Harbour and more widely in Great Britain. It is a native of Japanese seas and possibly was transported on military craft from Korea.

In Southampton Water it can be very abundant. Holmes found up to 200 adults per square meter intertidally. Indeed, it is the only solitary ascidian in mid-tide and low-tide level. No study of the life history appears to have been made in Japan. The fact that it has colonized some areas of Great Britain affected by warmer waters might suggest that this is a species benefiting from effluent warm water. Holmes suggests that breeding is restricted to summer, approximately July to September, when water temperatures exceed 16°C. But it would be difficult to assess how far warm water is essential to this immigrant.

In Southampton Water investigations over many years have been concerned partly with the effects of cooling water from an electricity Power Station sited at Marchwood and commissioned in 1955. Effluent seawater is returned across the shore and reaches Southampton Water on the surface at a temperature of about 6° above ambient. From 1954 to 1959 a fairly intensive study was concentrated at an area near Marchwood, close to the effluent water, and at a station at Calshot (Figure 1) some seven miles further down Southampton Water.

A considerable "natural" temperature fluctuation occurs in Southampton Water; for example at Calshot, away from power station influence, a range of 4-20°C (in one year 23°C). At Marchwood a rather greater range, of the order of 0-22°C existed, but in the latter years of the investigation the warm effluent water further up the estuary at Marchwood tended to raise both summer and winter temperatures. In 1957 and 1959, the summer maximum has increased to 26-27°C. Although generally over the area there is no marked *surface* increase in temperature, since cooling water of full salinity is taken in at depth, the warm returned effluent may be of sufficient density to pass beneath the surface waters of lower salinity. A warmed pool of water originating from the outfall, therefore, may remain, at an intermediate depth, drifting to and fro with tidal flow. Such a body of water gives up its heat only slowly.

Raymont and Carrie (1964) conducted a zooplankton survey over the six years to try to assess the effect of the thermal changes, comparing Calshot as a control station with changes near the power station at Marchwood. Whatever the effects of thermal change, there appears to have been no reduction but, if anything, an increase in the zooplankton population in the area.

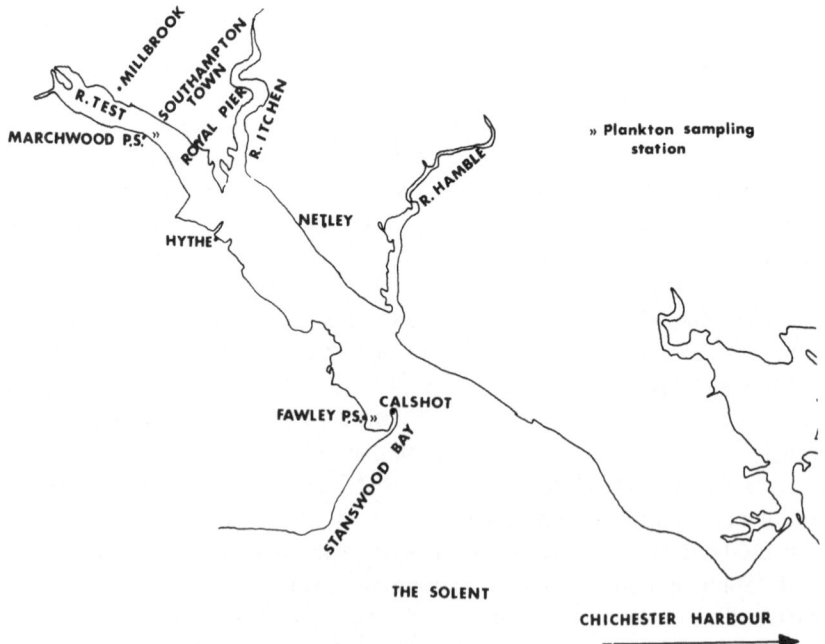

Fig.1. Map of Southampton Water showing plankton sampling stations.

Changes in the composition of the zooplankton, however, are much more important from the point of view of immigrant species. The most obvious was the great increase in the late spring, summer, and autumn populations of barnacles dominated by *Elminius*. *Elminius* is known as a species which was accidentally introduced to Great Britain, probably in Southampton Water, during the war years. Since that time it has rapidly spread around the British Isles, and though an Australasian species, it has successfully colonized wide areas, even including the colder east coast of Britain. Although it is a warm temperate species, its success cannot be attributed markedly to warm water influence. According to Crisp and Davies (1955) *Elminius* will breed in a natural environment at temperatures exceeding 6°C. In England it breeds throughout most of the months of the year. Its spread is largely accounted for by its ability to have a large number of broods per year and to utilize almost any type of phytoplankton food available, so that any marked positive effect of warmed effluent in Southampton Water is unlikely.

Another species in the plankton survey of Southampton Water, especially during the late summer of later years, was the small copepod *Acartia tonsa*. This was first recorded for Britain from Southampton in 1957 by Conover. It is an American species, widely distributed in coastal waters, but some work would suggest that it is a species favored by slightly higher temperatures. This, then, may be a factor in Southampton. Since our record in Southampton it has been spotted in one or two other areas. Davis (1967) comments that it occurs rarely in the plankton of the Blackwater estuary. This estuary receives effluent from the electricity power station of Bradwell. Perhaps the most interesting record, however, is that of Markowski (1962), from Cavendish Dock, which receives effluent from a power station. Up to 1955-56 other copepods were present in the plankton, but thereafter *Acartia tonsa* appeared in great numbers to the exclusion of other species. Markowski says that all stages of *Acartia* are found through the whole year; breeding continues year-round. He says that the populations are such that sometimes it appears virtually as a pure population.

So far nothing has been said of the introduction of foreign species of possible commercial importance. Portuguese and American oysters that have been repeatedly laid on British beds would undoubtedly benefit from warm conditions, since their breeding temperatures are not normally attained in British waters. Even our native oyster is near its breeding limit (normally about 16°) in Great Britain and so many are artificially laid that they might show a marked response to warm effluent. A striking example comes from Southampton Water, where a few years ago, another large power station (Fawley) was commissioned near Calshot. Some three years ago very young oysters were dredged from Stanswood Bay, around the effluent discharge. Two years ago commercial oystermen took large quantities. A survey revealed a remarkable new population around the outfall, estimated at some nine million oysters, all of approximately the same young age group. This represents a colonization of the area with the artificially warmed conditions.

The most remarkable accidental introduction of a new species to Southampton Water, which has undoubtedly responded to the locally warmer conditions, is that of the American hardshell clam, *Venus (Mercenaria) mercenaria*. We first recovered a single specimen in 1957 and Heppell (1961) records the species in the area from 1956 onward. The species has now so extensively colonized Southampton Water that in many areas it is the dominant bivalve and a commercial fishery is operating.

Venus mercenaria is a warm temperate species occurring on the eastern United States seaboard from Texas to Cape Cod. Though it can grow at lower temperatures, spawning, according to American authorities, hardly occurs below a temperature of 24°C. Loosanoff and his colleagues demonstrated that clams normally spawn in summer when a spawning stimulus (warming the seawater to about 30°) caused the clams to spawn rapidly. But clams can be conditioned to spawn during winter by first feeding with phytoplankton and gradually raising the temperature. Ansell (1967) has shown that this is also possible with clams from Southampton Water. Though the total egg numbers produced were rather smaller than in America, the behavior is reasonably similar. The number of eggs laid ranged from about 30×10^6 to as few as 0.5×10^6 eggs per female. However, in the United States the gonads begin to grow again during the autumn immediately after spawning; in Southampton no major growth of the gonad occurs after summer spawning until the following spring.

A further very marked difference of the Southampton clams, however, is that the critical temperature for spawning appears to be about 18°C—several degrees below the minimum suggested for American clams. Apparently already there is an acclimatized race. In Southampton some ten years ago, it appeared that only certain years were favorable for spawning; for example 1955 and 1959 were conspicuously "good" years and these happened to be very warm years. Apparently a warm summer coinciding with warm effluent water in Southampton Water contributed to a successful spawning. Later investigations, however, now suggest that probably some spawning occurs every year, though only certain years are really successful.

The occurrence of the clams in Southampton Water is still a matter of conjecture; the annual rings on the shells of the first specimens taken from 1957 onward show that these were relatively old clams which date from the 1930s. Even in 1962 a very large clam (10.8 cm) had at least 26 rings! It is remarkable that the area to the west of Southampton Water is marked by a complete dearth of records of *V. mercenaria*. Even in Southampton Water the species does not seem to occur naturally below Hythe. The warm effects of the effluent waters from the Southampton area are rapidly cut off to the west. On the east side of Southampton Water the clams extend well down, and outside the area to the east live specimens have been taken in small numbers in Stokes Bay, with an unconfirmed report even as far as Chichester Harbour. One or two introductions in the nineteenth century into estuaries in the north of Great Britain were completely unsuccessful. Although some artificial layings of American clams have been made in some French estuaries in the Charente and Brittany, it seems highly unlikely that viable larvae could have been carried across the English Channel. Most likely, the species has been accidentally introduced from the "kitchens" of liners from the United States docking in Southampton Water. With the warm water effluent from power stations successful spawnings and colonization of the area has now occurred.*

A remarkable feature of the Southampton Water population is the high density of clams. Carriker (1961) quotes three clams per m^2 as a reasonable commercial density. In the Marchwood area 100 clams/m^2 have been recorded. In the outfall of Marchwood, with the highest density of commercial clams, up to 160/m^2 occur. Young clams are, of course, far denser; one of the highest figures recorded was 4750/m^2 of a mean size of 10 mm.

V. mercenaria seems to be remarkably unselective with regard to its substratum. Around Marchwood the shore consists mainly of stone and shell gravel with mud; along the Netley shore the bottom is of soft mud; toward the end of Southampton Water the substratum is hard gravel and sand. This may be a major factor in successful colonization.

Difference in growth rate occur with clams even in Southampton Water. At Millbrook Point, for instance, the clams grow more slowly and reach a smaller size than near the outfall of the Marchwood Power Station, and on the whole the rates are below those in the United States. However, in the clams more directly exposed to the warm effluent waters of the power station, the growth rate approaches that in the United States. This effect has been tested experimentally by re-laying clams from Southampton Water in Poole Harbour, a relatively enclosed shallow area, near the warm water of the outfall of a power station. A number of other layings were made in parts of Poole Harbour well away from the warm water effluent. The results showed that animals started to grow earlier in the warm outfall water (approximately one month earlier) and they continued to grow longer during the autumn. The average growth increments were also greater. Ansell and his colleagues demonstrated that, whereas in the Marchwood outfall growth of clams proceeded for about 173 days of a year, at the Poole power station outfall growth continued for 204 days, as opposed to only 123 days in cooler water in the same area. The percentage length increment for a standard clam of 4 cm varied from 54% at Poole power station to as little as 14% in the cooler water.

Most of the growth in British waters occurs in the warmer months, especially

*Recent reports suggest that a few clams were laid in Southampton Water, circa 1926.

in June, July, and August, but exposure to power station effluents prolongs the season as well as increasing growth rate. American workers have suggested that growth rate is related to the abundance of food, particularly small planktonic diatoms. We were able to carry out a few laboratory experiments which suggested that addition of phytoplankton cultures to the water flowing past the clams increased growth but only when algae were in moderate concentration. During winter, feeding was also possible in laboratory experiments, providing the temperature was raised. There are a number of favorable arguments, therefore, for attempting to introduce this species to areas of warm effluents.

Ansell (1967) and his colleagues studied changes in the condition of *Venus mercenaria*, especially in the Marchwood power station outfall. The body weight of a clam rises from a winter minimum to a maximum about May–June; there is then a sharp fall which they interpret as due to spawning. The readiness to spawn (*spawning potentiality*), can be measured by testing clams with spawning stimulus (warming to 30°C) and noting the percentage of the animals which respond. Ansell found that spawning potentiality was at a peak in May–June. After the decline with spawning, in almost every year the condition rises again to another peak about August–September; spawning potentiality also rises, and a second spawning period apparently occurs.

Biochemical changes in clams from Marchwood also suggest that a first spawning occurs about May–June. Thus the water content of the clam, high during the winter and early spring, falls about May to reach a low value immediately before spawning occurs. After spawning the water content rises again. The nitrogen content as an indication of protein also rises to a peak during spring but falls on spawning. Carbohydrate increases from a low winter value to a maximum in the early summer. It may fall about the time of the first spawning but usually increases during the summer and is more affected by a general decline with the onset of autumn. The utilization of reserves during the winter period seems to be marked especially by a utilization of carbohydrate.

In the United States, in Long Island Sound, the clams appear to spawn normally in August. However, to the south in Carolina a spawning occurs in June with only a second minor peak in September–October. This would appear to be more similar to Marchwood.

The spawning in Southampton Water at Marchwood in May–June is believed, however, to be unsuccessful generally as regards settlement; probably the temperature outside the immediately warmed area is too low. In any event, outside the Marchwood area, in other parts of Southampton Water, the clams apparently do not spawn in May–June; instead they appear to spawn only in the summer (July–August).

Later work by research students at Southampton (Mitchell, Rodhouse, and Hibbert, unpublished) have included studies on the occurrence of clam larvae in Southampton Water. The larvae are limited to the summer; Mitchell suggests that a rising temperature of about 18°C is necessary to start spawning. In 1970 a minor peak of larvae in May–June, taken near the Royal Pier, probably represents larvae from the stock at Marchwood outfall, where the temperature had already passed 18°. A second peak probably represents later (July–August) spawning of other clam stock around the Royal Pier itself. In 1972 Rodhouse recorded densities of larvae in the same months, approximately comparable to the 1970 figures apart from the large peak. Two peaks of larvae occur; he believes that a high burst of phytoplankton may be responsible for depressing the amount of larvae in the intervening period. In any event, although probably larvae are spawned each year, only some years are followed by successful settlements. Apart from the very early years of 1955 and 1959, other successful years are 1964, 1968, 1970, and possibly 1971. Reasonably high temperature for settlement is essential. However, this is not the only factor, since 1973 was one of our very warmest summers, but it does not appear to have been a very successful year.

The colonization of this important commercial species, however, is surely largely dependent on the beneficial results of warm water effluents. The accidental colonization of a number of harmful or unwanted foreign warm water marine species in a few British ports, however, is sufficient warning to us of the problem of "weeds" in marine aquaculture.

REFERENCES

Ansell, A. D. 1967. Egg production of *Mercenaria mercenaria*. *Limnol. Oceanogr. 12:* 172–176.

Ansell, A. D., and K. F. Lander. 1967. Studies on the hard-shell clam, *Venus mercenaria*, in British Waters. III. Further observations on the seasonal biochemical cycle and on spawning. *J. Appl. Ecol. 4:* 425–435.

Ansell, A. D., K. F. Lander, J. Coughlan, and F. A. Loosmore. 1964. Studies on the hard-shell clam, *Venus mercenaria* in British Waters. I. Growth and reproduction in natural and experimental colonies. *J. Appl. Ecol. 1:* 63–82.

Ansell, A. D., F. A. Loosmore, and K. F. Lander. 1964. Studies on the hard-shell clam *Venus mercenaria*. II. Seasonal cycle on condition and biochemical composition. *J. Appl. Ecol. 1:* 83–95.

Carriker, M. R. 1961. Interrelation of functional morphology, behaviour, and autecology in early stages of the bivalve, *Mercenaria mercenaria*. *J. Elisha Mitchell Sci. Soc. 77:* 168–241.

Conover, R. J. 1957. Notes on the seasonal distribution of zooplankton in Southampton Water with special reference to the genera *Acartia*. *Ann. Mag. Nat. Hist.*, Ser. 12, *10:* 63–67.

Crisp, D. J., and P. A. Davies. 1955. Observations *in vivo* on the breeding of *Eliminius modestus* grown on glass slides. *J. Mar. Biol. Assoc. U.K. 34:* 357–380.

Crisp, D. J., and A. H. N. Molesworth. 1951. Habitat of *Balanus amphitrite* var. *denticulata* in Britain. *Nature, London 167:* 489–490.

Davis, D. S. 1956. The marine fauna of the Blackwater Estuary and adjacent waters, Essex. *The Essex Naturalist 32,* Part I: 1–61.

Eltringham, S. K., and A. R. Hockley. 1958. Coexistence of three species of the wood-boring *Limnoria* in Southampton Water. *Nature, London 181:* 1659–1660.

Heppell, D. 1961. The naturalization in Europe of the quahog, *Mercenaria mercenaria* (L.) *J. Conchol. 25:* 21–34.

Holmes, N. 1968. *Aspects of the biology of* Styela clava *Herdman*. Doctoral thesis, University of Southampton.

Jones, L. T. 1963. The geographical and vertical distribution of British *Limnoria* (Crustacea: Isopoda). *J. Mar. Biol. Assoc. U.K. 43:* 589–603.

Markowski, S. 1962. Faunistic and ecological investigations in Cavendish Dock, Barrow-in-Furness. *J. Anim. Ecol. 31:* 43–51.

Menzies, R. J. 1959. The identification and distribution of the species of *Limnoria*. Pages 10–33 *in:* D. L. Ray, ed. *Marine Boring and Fouling Organisms, Friday Harbour Symposia*. University of Washington Press, Seattle.

Mitchell, R. 1974. Studies on the population dynamics and some aspects of the biology of *Mercenaria mercenaria* (Linne). Doctoral thesis, University of Southampton.

Naylor, E. 1957. Introduction of a Grapsoid crab, *Brachynotus sexdentatus* (Risso), into British waters. *Nature, London 180:* 616–617.

Naylor, E. 1965a. Biological effects of a heated effluent in docks at Swansea, S. Wales. *Proc. Zool. Soc. London 144:* 253–268.

Naylor, E. 1965b. Effects of heated effluents upon marine and estuarine organisms. Pages 63–103 *in:* F. S. Russell, ed. *Advances in Marine Biology*, vol. 3. Academic Press, London and New York.

Naylor, E. 1960. A North American Xanthoid crab new to Britain. *Nature, London 187:* 256–257.

Patel, B., and D. J. Crisp. 1960. The influence of temperature on the breeding and the moulting activities of some warm-water species of operculate barnacles. *J. Mar. Biol. Assoc. U.K. 39:* 667–680.

Raymont, J. E. G., and B. G. A. Carrie. 1964. The production of zooplankton in Southampton Water. *Int. Rev. Ges. Hydrobiol. Hydrogr. 49:* 185–232.

Ryland, J. S. 1960. The British species of *Bugula* (Polyzoa). *Proc. Zool. Soc. London 134:* 65–105.

Ryland, J. S., and A. P. Austin. 1960. Three species of Kamptozoa new to Britain. *Proc. Zool. Soc. London 133:* 423–433.

Tebble, N. 1956. The control of *Mercierella enigmatica* (Polychaeta) in Radipole Lake, Weymouth, England. *Proc. Int. Congr. Zool. XIV (Copenhagen):* 444–446.

DISCUSSION

CHRISTENSEN: May I ask a few economic questions in regard to this clam? What roughly is the first hand price, and what are the harvesting costs?

RAYMONT: It is difficult to give the harvesting costs. The clams are harvested, put through cleaning tanks, and then mainly exported to France.

KORRINGA: This particular species sell at present in France for 5 francs each, I have seen.

PILLAY: Is there much difference in the condition of clams growing in these two different types of areas?

RAYMONT: The condition is, of course, best just before spawning time and the pattern is a bit different from the American pattern. In America the gonads appear to start developing immediately after spawning, so that the condition is fairly rapidly built up. In Southhampton Water the gonad doesn't regenerate quickly and it is only in the early spring the next year that the condition appears to begin to come back again. The commercial growers have to be aware of this in their harvesting.

PILLAY: Does the clam compete with any of the local species?

RAYMONT: It has probably pushed out the local *Cardium*. The number of this species seems to have declined but not much else has changed.

PILLAY: Have you any observations on the benthic algae, particularly microalgae, of this area?

RAYMONT: We have one study. We certainly have very extensive mud flats which are dominated by a few species of benthic diatoms. These are very rich, but I'm sorry I cannot quote you species. Presumably the clams make use of this to a very considerable extent.

WALNE: One of the interesting points about this area is that it is not a particularly good area for clams. We have planted them out in various parts of the country and there are quite a number of areas without heat where the clams will grow very much better than they will in the Southampton water when it is unheated. It definitely seems that the temperature has had a triggering action.

BRATTEGARD: Could you please specify more about the hazards of introducing such species, and also compare the situation in for example the Southampton Water estuary with a Norwegian fjord where the topographical barriers for spreading of faunal elements are more difficult to overcome.

RAYMONT: There are several problems that may arise. When you try specifically to introduce a species to farm it, there may be competition for suitable food. This is particularly so if some competing species are capable of very rapid reproduction and of reproducing for a considerable part of the year. Another hazard in accidentally introduced species is the actual coverage of surfaces. An example (not in Southampton) is the barnacle *Balanus amphitrite*. But perhaps the best example is the invading barnacle *Elminius modestus*, which came into Southampton Water during the last war. This barnacle may be characterized as a somewhat warm temperate species, but it can also benefit enormously by a wide variety of food. This barnacle has really swept across the Southampton area and in places literally covers exposed surfaces. The most striking instance was in a new dock region built as an unloading area for the new power station. Within a relatively short time the whole of that surface was covered by barnacles, virtually 100% *Elminius*.

BRATTEGARD: Perhaps I can tell a little about the conditions along the Norwegian coast. In many of the smaller fjords or fjords we call *polls*, we have distribution "pockets" of both northern and southern species. If you select a poll according to basin depth, to threshold depth, etc., then it is often possible to say whether you might find some southern species or not. An example is the "southern" sea anemone, *Anemonia sulcata*, which is present in some polls in the Bergen area. In polls with cold deep water the benthic communities are often similar to certain subarctic communities or similar to communities known from colder parts of the North Sea. Such polls might serve as local centers for the distribution of animals which can start sending out animals when temperature and salinity conditions change.

RAYMONT: I quite accept what you say, I think, particularly for deeper areas. The Southampton area is a shallow sort of semiestuary, although fairly saline with a salinity exceeding 30%oo. Generally, it is really a drowned valley; in such an area the possibility of colonization is more real. A very good example of colonization is the immigrant ascidian *Styela clava*, which does not seem to breed unless the temperature goes above 16°C. It has colonized the area since about 1960 and it is now one of the common ascidians. When we consider how efficient ascidians can be as feeders, such an organism could be a serious competitor when it is spreading.

PILLAY: There is certainly going to be a change in the fauna and the flora in an area which is exposed to a discharge of heated water. If there are no controls, would you be against the introduction of a species that is tolerant to the temperature conditions that will be built up there?

RAYMONT: Personally, I cannot see anything wrong. In the particular case of this clam it has turned out well, and given tangible economic benefits. We should require, however, that when such introductions are made intentionally, and not accidentally like the barnacle described, they should be preceded by sufficient investigations of the breeding, feeding, and other requirements, to have a picture of the biology. As an example of introductions with more adverse results, we can refer to some of the early introductions of oysters to various areas, and what they brought in the way of pests. Dr. Korringa can speak far better on these aspects.

KORRINGA: This happened in 1883 or '84 I believe, when people became enthusiastic about the possibility to introduce the American Atlantic oyster, *Crassostrea virginica*, in southern parts of England. The introduction was done rather crudely: The oysters were not carefully cleaned and one consignment of oysters brought along the predator *Urosalpinx*, the eastern drill, which killed enormous amounts of young oysters. It also brought along the slipper limpet, *Crepidula fornicata*, which was a competitor for space and food, and which, after crossing the North Sea, nearly put us out of business. In this same consignment of oysters other things came which were at that time unobserved, as they did not do economic damage. But one consequence was that the boring clam, *Pholas candida*, which was very numerous in the North Sea, was almost completely replaced by *Petricola pholadiformis*. This is one of the cases of a foreign species which behaves very vigorously on coming to a new environment where it can breed.

FENGER: In the case of the Kalundborg Fjord, which I referred to in my previous lecture, the species composition of the area was investigated before the power plant started. We described the fauna by dividing it into cold water species and warm water species. Cold water species were defined as those living north of the border formed by the English Channel. Looking at the diagrams on the blackboard, when we compare the temperature above the ambient for the various locations with the percentage of cold water species, we find quite a good correlation, and moreover we find that the distribution of the warm water species corresponds with the distribution of the cooling water after mixing in. This is then an example of how the former community changes because of discharged heated water.

NYMAN: I would like to put a few questions to Professor Raymont. An *Elminius* species was said to be damaging Southampton docks. Were they able to withstand a sudden drop in temperature? The second question mostly concerns fish. It has been recorded, for instance, that guppies have been established in a stream in England. Another example where an *exotic* species was established can be given from Sweden. Off a power plant in Lake Mälaren west of Stockholm a very large oligochaete of the species *Branchiura sowerbyi* was recently found. This species has a normal distribution in tropical Africa and southeast Asia; it has been found now and then in the British Isles under unheated conditions, but then dwarfed in size. In this Swedish locality it had attained a maximum size of around 7 cm, and of course nobody knew how it got there. Also in the Konin District in Poland pronounced secondary effects of introduction of fish into fresh water systems have been reported. In Konin three power stations use as cooling ponds five interconnected lakes, into which grass carp and other nonindigenous species were introduced. These carp have grazed down the littoral flora to the effect that certain species of fish like tench, bitterling, and stickleback have been virtually eliminated since their habitat was destroyed. Do you have any further examples or comments, Professor Raymont?

RAYMONT: On the question of lowering the temperature from a power plant, there are instances of decreased load on the power plant with a consequent sharp decline in temperature. It does not seem that the invading species have been eliminated, but as far as I can recollect, some species have ceased to breed, resulting in a sort of scattered remnant which is gradually going out. This was perhaps particularly shown in the slide for Swansea Docks describing the development of *Balanus amphitrite*. I think it is also true that the budding of *Loxosomella* shown on the first slide may have now ceased. It may be passing out quickly because the species is what I would call fairly "delicate."

Regarding fish, I can only cite the example of *Atherina boyeri*, which was found during the warming-up period at Swansea, and which is regarded as a tropical species. The commonest change is that mullet come in great shoals, but as to what you may call *foreign* fish, we have no records.

Of fresh water conditions I have little knowledge but I might mention the example from near the town of Peterborough in England, of what they call the "Cut," an area of water which receives most of the power station's warmed water. They have compared the growth of a number of what you might call coarse fishes which you get in the "Cut" and in the natural river which flows nearby. They have shown that there is a considerable population and that feeding continues for a longer part of the year in the "Cut," but that annual growth is not very different from the river.

KORRINGA: Perhaps I can add that it is not always so that species which come from warmer countries will perish at low temperatures in midwinter. Take, for instance, the case of *Mytilicola* which came from the Mediterranean, and can stand the severest winters in Holland, conditions which are never encountered in their original habitat.

NYMAN: Also on the west coast of Sweden mullet have been attracted by hot-water effluents.

KORRINGA: But these fish are sensitive to low temperatures in winter, and if they cannot get away in winter they perish.

Safeguards in the Exploitation of Domestic Effluents for Aquaculture

P. Korringa

Netherlands Institute for Fishery Investigations
IJmuiden, The Netherlands

Just as in agriculture and animal husbandry, one should consider that the primary production is always the basis of aquaculture, and that sunlight, nutrient salts, carbon dioxide, and water in liquid form are the well-known prerequisites for primary production. If one farms aquatic organisms of the first link of the food chain, the principles are virtually the same as in agriculture and horticulture. The yield per hectare will depend, in the first place, on the availability of sunlight and of nutrients. Nutrients can be administered in some form or other to raise the production.

If one farms aquatic organisms of the second link of the food chain, one can harvest a modest yield per hectare only if the organisms under cultivation feed exclusively on the vegetation growing naturally on the site of farming. There are two ways to obtain a much better yield per hectare: (1) When the organisms under cultivation are filter feeders using plankton and detritus, an intensive flushing will lead this type of food to the farmed organisms from a much larger acreage than that on which they actually grow. This explains the high figures for mussel production with rafts moored in bays flushed on the tide. (2) When one grows herbivores feeding on particulate food in stagnant water, one can directly administer suitable particulate food, grown elsewhere, or fertilize the water with inorganic fertilizer or with some organic fertilizer, domestic sewage included.

If one farms aquatic organisms of a carnivorous nature, belonging to higher links in the food chain, one can never rely on the local primary production, since at the very most 1% of the primary product can ever be found back in the bodies of organisms of the third link in the food chain. It is a sheer necessity to administer extra food in some way or other to reach a remunerative level of farming. Pellet feed prepared on a scientific basis is ideal, but costly. Minced scrap fish can be used, provided it is either fresh or deep frozen. Slaughterhouse offal can be used in certain types of farming, but enrichment with fertilizer, dung, or domestic effluents requires considerable care. When the results of such enrichment are to be obtained via the primary production, it can, as such, hardly lead to really high yields when one operates in stagnant water.

Domestic effluents are in certain cases efficient in enriching the aquatic farm areas, because it will only partly require the long way via complete mineralization to step up the productivity. The domestic effluent contains particulate matter, partly in the process of decomposition by microorganisms, which can directly be utilized as food by some fishes. They can convert decomposing proteins and carbohydrates into their own tissues, thus making a shortcut. One should therefore not speak of biodegradation but rather of bio(de)gradation, indicating that part of the material can be upgraded directly. This is the great advantage of using domestic effluents in certain types of aquaculture.

In recent years experiments have been made with another shortcut to avoid the long and energetically costly way of complete biodegradation of organic waste. It was proved possible to grow yeasts on certain components of mineral oil and on fecal material. If one can control such a yeast production efficiently, one can in its turn use the yeast, rich in proteins, as food in some farming operations, aquaculture included. However, this does not mean an entirely new source of food that has nothing to do with primary production, for somewhere at its base one will find the old photosynthetic processes. When mineral oil is used for growing yeasts, concern has been expressed about the possible incorporation of carcinogenic constituents.

Economically the most important case of farming the first link of the food chain in the aquatic environment is the *nori* culture in Japan. The redweed *Porphyra* is farmed on such a large scale in the coastal waters of Japan that its proceeds are eight times those of oyster farming in Japan: the annual production amounts to 6 billion sheets of the processed product, worth 12 yen each. Attempts have been made to increase the nori production by spreading fertilizer, since nutrients containing nitrogen and phosphorus may be the limiting factor in primary production in coastal waters. Experiments have demonstrated that one can thus obtain a higher production but that most of the nutrients administered will be washed away by the tide, not to return. Other than in terrestrial monocultures, one cannot expect that a high percentage of the fertilizer administered will find its way to the cultivated crop. "Weeds" of various description, including the microscopic plants of the plankton, will take advantage of the administered fertilizer in an area much larger than where the crop of nori is actually grown. For nori farming, it is more economical to utilize a coastal area slightly polluted with domestic effluents. The rapid biodegradation in shallow coastal water increases its nutrient content and thus promotes the growth of *Porphyra*. Since the farmers need not pay for the domestic effluents, this system is much cheaper than spreading inorganic fertilizer, and the fact that here, too, part goes to weeds—all the plants other than *Porphyra*—does not matter economically.

Domestic effluents would entail the risk of contaminating the edible product with pathogenic bacteria and viruses. *Porphyra*, not being a filter-feeding organism, does not accumulate bacteria or other particles discharged in the domestic effluents. Neither in Japan nor in England has one recorded spreading of contagious diseases via consumption of redweed. The way it is processed and cooked apparenlty forms a reasonable safeguard against man's contamination with bacteria and viruses. *Porphyra* does, however, accumulate certain elements, among which particular attention has been paid to some radioactive isotopes contained in industrial effluents. In fact, discharge of radioactive waste in coastal waters of Great Britain reaches man's food via the same redweed *Porphyra*, locally called *laver*. Industrial effluents therefore constitute a greater risk in aquaculture dealing with the first link of the food chain than do domestic effluents, and safeguards should focus on these risks.

Among the organisms of the second link of the food chain there exist especially filter feeders of the realm of the molluscan shellfish which are farmed on a large scale. Oysters, mussels, and clams filter the water in search of food particles, and in doing so may ingest and accumulate bacteria and viruses discharged with domestic effluents. Outbreaks of typhoid, cholera, and hepatitis can sometimes be traced to consumption of raw shellfish grown or stored in contaminated water. It has been known since the turn of the century that farming and storage of molluscan shellfish entail health hazards for the consumer when the water is polluted with domestic effluents. Therefore, many countries have set up a system of bacteriological control of both water and shellfish, and certificates stating that the consignments concerned are free of potentially dangerous bacteria and poisonous chemical constituents should accompany the shellfish on its way to the consumer. This clash between shellfish farming and sewage discharge has led to abandoning many formerly productive waters, and to "red line" districts in which shellfish may be farmed and fished, but from which it should not be marketed. Such contaminated shellfish has to be relaid on clean beds for a sufficiently long time (expressed in weeks) to produce a bacteriologically impeccable product.

Slightly contaminated shellfish can pass through a special purification plant where they sojourn long enough in sea water sterilized with ozone, chlorine, or ultraviolet light to become decontaminated. All this entails extra work and extra costs for the shellfish farmer, which costs are not refunded by the pollutors, i.e., the cities discharging domestic waste on shellfish grounds.

Farming molluscan shellfish is possible in slightly polluted water where proper quarantine measures and purification plants are used under strict control, but a heavier influx of domestic sewage invariably leads to other deleterious effects. It is clear that where lack of oxygen and presence of hydrogen sulfide is found during part of the tidal cycle, oysters, mussels, and clams will not be able to survive. But beyond that area the water may be apparently clean and rich in oxygen, and display a green hue to be ascribed to a dense stock of phyto-plankton. From a biological point of view such water is still polluted, for it contains usually, or at least periodically, a flora of dinoflagellates which may be harmful for the shellfish and for its consumer. Eutrophication, leading to too rich a plankton of an aberrant composition, is dangerous for oyster larvae, which may starve amidst such overly rich plankton and be hindered by toxic metab-olites. In serious cases of eutrophication the health of the consumer of shell-fish may be endangered even when the latter is bacteriologically safe to eat. Such may be caused by accumulation of poisonous metabolites of certain dinoflagellates such as *Gonyaulax* and *Dinophysis*. Cases of paralytic poisoning by mussels collected in a polluted canal have been reported from Belgium, and gastrointestinal complaints have often been heard after consumption of mussels collected in the Oslofjord in a period of bloom of dinoflagellates, to be traced to eutrophication caused by domestic pollution. To safeguard the consumer of shellfish, one should therefore oragnize a good system of bacteriological control and in addition make regular observations on the type of phytoplankton prevailing in the area where the shellfish is farmed and stored. In case of doubt, assistance of white mice and white rats should be invoked to determine whether the shellfish contains dangerous constituents.

It will be clear from the foregoing that microbiological purification of domestic sewage followed by discharge of a clear effluent does not guarantee that shellfish can safely be farmed and stored in the water receiving such effluents. The clear fluid may seem harmless but can contain so much nitrogen and phosphorus compounds that a bloom of a dangerous type of phytoplankton may ensue. Third-degree purification or discharge by a pipeline to areas far away from shellfish farming are the only appropriate measures to safeguard the consumer of shellfish. The toxins produced by dinoflagellates are not destroyed by boiling, whereas the effects of bacteriological contamination are not observed when the product is properly cooked.

Industrial pollution of shellfish beds and storage basins may lead to other effects: tainting (making the product unsaleable) or accumulation of poisonous compounds such as heavy metals, radioactive isotopes, and chlorinated hydrocarbons.

In some cases one farms fish species belonging to the second link of the food chain. As such should be mentioned the milkfish (*Chanos chanos*) farmed on a large scale in brackish water in tropical countries in Asia, and various species of mullets (*Mugil* sp.). The farmers enrich the water of the ponds to increase the amount of food available for the fishes. In the case of milkfish, one may apply cow dung or urea prior to stocking the ponds, or vegetable matter to decay after stocking, to promote the growth of the blue-green algae milkfish feed on. Domestic sewage is not used. Mullet farmers in Israel administer a given dose of chicken dung and superphosphate in the ponds and offer pellet feed in addition. Domestic waste is not used.

In farming fish of higher links of the food chain, feed is an item which weighs heavily on the budget. Therefore, one has to study, not only which type of food will be best (e.g., fresh fish or formula feed), but also which is economically sound to administer. Since olden days fish farmers in Asia have found that it is good to enrich their ponds with organic fertilizer. To that end they build latrines over their carp ponds or lead domestic effluents from sewers in controlled quantities into the larger carp ponds for extensive farming. One should administer these domestic effluents carefully, for microbial degradation requires a considerable amount of oxygen and even a fish like the carp cannot survive where oxygen becomes depleted and the poisonous hydrogen sulfide appears. But since the partly decomposed organic materials contained in sewage effluents are either eaten directly by the carp to be digested and converted into their tissues or follow the long way via mineralization, photosynthesis, and herbivores such as *Chironomid* larvae, much of it finally leads to better growth and fattening of the carp. Since carp are duly cooked in the countries under consideration, there is no health hazard in enriching the pond with domestic effluents.

An extreme case of farming fish in domestic sewage is the so-called caramba farming in Indonesia. One places bamboo cages with young carp in a stream receiving a fair dose of domestic sewage and it is amazing to observe how fast these carp grow, grabbing all the particles that flow through their cages, and feeding in addition on *Tubifex* and *Chironomid* larvae that they find on the bottom of their cages. The water must flow rather rapidly, for in stagnant water such a high percentage of domestic sewage would rapidly lead to oxygen depletion. There is no health hazard ensuing from consumption of fish farmed that way, for one cooks the carp thoroughly so that any germs of contagious diseases are adequately destroyed. One should, however, not sell the carp immediately after taking them from the cages in the rivulets, for their flavor may turn out to be rather disappointing. General practice is to keep them a few weeks before marketing in a cage placed in unpolluted water and to administer some high quality feed to improve their organoleptic qualities. The caramba system may seem a somewhat unpalatable way of farming at first sight, but one should consider that European cauliflower, when boiled, often reveals markedly whether it has been manured with fish or with muck, and that there is no way to improve its flavor by placing it in a clean soil prior to marketing.

DISCUSSION

OPPENHEIMER: What kind of productivity did they get for the carp, how fast did they grow, and to what size?

KORRINGA: It is a question of two months from the introduction of the fish to the time it reaches marketable size. Small carp are produced in the typical carp ponds to be transferred to the sahwas (rice fields). There are special producers growing small carp for sale and the usual practice is to buy such carp to place in the cages. There are a number of such seedling centers in Indonesia which breed their particular type and which people buy for the cages.

CONOVER: How many fish are there per cage?

KORRINGA: In one cage there may be 50 or 60 at one time. I do not have exact figures on growth and conditions such as oxygen levels. For such data I would refer to a paper on caramba farming produced by Dr. Vaas, now the director of the Delta Institute in Yerseke, The Netherlands.

CONOVER: I find the problem of the dinoflagellate toxicants particularly intriguing. You probably know of the chronic shellfish toxicity problem in the Bay of Fundy. This is an area which would be considered relatively unaffected by human activity, but peculiar hydrographic and biological conditions seem to result in a situation where the shellfishing at certain times of the year has to be prohibited because of occurrence of toxic shellfish.

KORRINGA: Especially in warm weather and when the water is rich in nutrients this will occur. For intensive shellfish farming there is a need for water rich in food, but as soon as you get really rich water by upwelling or sewage or other reason, under certain weather conditions you may see that suddenly things run out of hand. As one example I might cite the conditions of the coastal water from the Belgian coast up to the German coast. The coastal stretch of water contains three or four times as much phosphates and nitrates as further offshore, but the plankton content in the coastal water is not three or four times as rich as further into the North Sea. This is explained by the preponderance of silt in the coastal water, which reduces the light intensity sufficiently to lower the plankton population. Once in a while conditions will be suitable for a large bloom, which in practice means that we have to keep the waters under constant supervision to know when such things will occur.

CONOVER: Last year was unusual insofar as the toxicity spread much further south than had been anticipated.

KORRINGA: The causes and the circumstances of the paralytic poisoning from *Gonyaulax* are quite well known and there is a lot of literature. The gastro-intestinal effect of dinoflagellates is for the medical profession something new. You get quite different symptoms from the same group, albeit from different species.

PILLAY: There are, of course, many cases of dinoflagellate blooms in areas which receive large quantities of wastes, but are there reports of the occurrence of such blooms in areas where there is controlled discharge of wastes?

KORRINGA: I only know the dinoflagellate blooms from the sea water. I do not know if any occur in really fresh water and domestic sewage is usually discharged in fresh water.

PILLAY: They do occur in brackish water and coastal bays like the inland sea in Japan.

KORRINGA: In these areas you sometimes have dinoflagellate blooms.

PILLAY: Yes, but these are areas where there is control of all waste discharge.

KORRINGA: Well, in the oyster area there is control. The oysters and the water are sampled for control. For example, in the Hiroshima area there is control, but there occurs at times a red tide. In such cases the oyster farmers are advised to move their rafts to safer water devoid of dinoflagellate blooms.

PILLAY: The oysters are controlled, yes, but not the discharge of waste into the inland sea.

CONOVER: It is more or less true of the eastern Canadian waters in the Fundy area that there is a minor amount of domestic sewage entering the system in an uncontrolled fashion. But I am quite certain the shellfish toxicity antedates most of the human proliferation in the area. From past records, we know it occurred 50 or 75 years ago. It seems fairly certain that these conditions are not related to domestic pollution, but we have not resolved what the particular conditions are contributing to these occurrences of toxicity.

KORRINGA: I think we need more research on the biology of dinoflagellates, and especially to study the conditions of outbreak of such blooms, not only from the point of view of nutrients which play a part in it, but also including weather conditions. There was, for example, a case in England about two years ago in the month of May where bad weather on the east coast of England brought a bloom of *Gonyaulax* such as we have never seen in the North Sea. The first indication of the bloom came from bird watchers who observed vomiting birds; only later were cases of human poisoning reported.

PILLAY: We also know of such cases happening in other situations, for example in the lagoon of Tunis, which receives considerable quantities of raw sewage. It seems then to be a question whether we can really control the discharge effectively.

KORRINGA: Where you have extensive farms of shellfish designed for the market, I think the officials should have good control over the products to safeguard the health of the people. The control should not only be for the shellfish itself, but also for the development of toxic flagellates.

SOEDER: As a comment to Dr. Pillay's question, I would refer to the studies of H. H. Seliger and co-workers. They show that the hydrographic conditions can concentrate dinoflagellates enormously, especially if a layers of brackish water swims on top of the normal seawater, in which case the dinoflagellates will swim upward and then be trapped in that layer. Then the wind can transport enormous masses of these dinoflagellates to the shore, so it is not always the case with local bloom, but it might be a bloom due to a concentration caused by active migration.

KORRINGA: A bloom described in 1961 showed these features. The bloom was first traced near Ostend in Belgium where the color of the water was unusual. Next, bloom emerged in the Westerschelde, where people picked mussels from the breakwaters; next it came round the Isle of Walcheren, into the Oosterschelde, then it gradually went inside, and two weeks later the bloom was observed in the Waddenzee, and the bloom followed exactly the pattern of the tide transport. The cause of the bloom was not found, however.

CARSTENS: It seems, then, we have to accept the toxic dinoflagellates as an act of life regardless of whether the water is polluted or not, so it is important to develop some kind of predictive tool to estimate the hazard of such occurrences. My question is toward the other end of this problem: Are there any prospects of developing an indicator for poisonous shellfish that is better than the present biological test?

KORRINGA: At present the best way for prevention is a frequent inspection of the plankton in places where you grow your shellfish. That gives you the first indication, and if dinoflagellates really are proliferating, we employ the biological test. It would, of course, be better to have a chemical test, and there is a group in The Netherlands working on a colorimetric method to determine this kind of toxin. It seems to be a matter of trying the method out during the next bloom to see whether it is practicable.

RAYMONT: Do we know whether these dinoflagellate toxins vary in their toxicity? They are supposed to vary in culture; is it the same in nature as well?

KORRINGA: We suppose this is the case because we thought first of *Provocentrum micans*, which was the most numerous species. Of course, this does not constitute a proof; it could be that there are some 10% of other species which are far more dangerous, *Dinophysis*, for instance. We also know that pure cultures of *Provocentrum* from Heligoland are not poisonous at all, but there may be wide fluctuations for different populations. Unfortunately, we know too little about this.

RAYMONT: There is the case of the *Gymnodinium culture* in Plymouth. The species was cultured for years, but apparently it can lose its toxicity almost totally.

KORRINGA: This might be because it no longer had any need for the toxin when it was cultivated. I feel the toxicity is not something abnormal, but something the organisms make in order to win against other organisms in the race for nutrients or growth. We know that many plants produce products which make it impossible for the seeds of other plants to germinate. The succession one observes in nature in the plankton is partly due to certain populations blooming up and suppressing others, which is effective as long as there are sufficient nutrients. It is only when there are not enough nutrients that the species in question will die off and lessen its hold.

MATTHEWS: In connection with Dr. Soeder's remarks, mention was made of Wyatt's work postulating a model for red tide. He pointed out that by the time the red tide occurs it is often too late to look for the cause, and he produced a model which substantiates very much what Dr. Soeder mentioned in the sense that it is a hydrographic model. It seems that there is a certain stage in the development of the bloom which seems to be toxic, it is not by any means restricted to these notorious dinoflagellates. In most cases it appears to be a dinoflagellate and that the red tide is associated with the chemical warfare which comes into play with the question of succession. One simply cannot look at the plankton and detect or predict a red tide from the presence of *Gonyaulax*.

SOEDER: I should like to comment on Dr. Korringa's paper with regard to the contamination by heavy metals. We have analyzed some of the Japanese seaweeds with respect to cadmium, for which the maximum admissible daily intake is about 0.1 milligram per head and day. Seaweeds which have apparently been exposed to some kind of sewage were found to contain up to 9 ppm cadmium, and if you are eating some of these delicacies which might be thought harmless, you can easily consume more than your maximum admissible daily intake.

I would like to suggest that in the framework of the nations assembled here, we should propose something like a central analytical system for aquacultured products. The reason for this is that a trace component like benzpyrene, for example, is very difficult to determine properly—which is also the case with mercury, cadmium, DDT, and other chlorinated compounds, to mention a few. There is a need for laboratories that can undertake such analyses for the various countries, and do it in a reliable fashion.

KORRINGA: In The Netherlands we have a system of monitoring the conditions in chosen locations. The stations are chosen in areas where products for export are fished or produced, as, for example, sole, shrimps, oysters, mussels, and eels. Specimens of these and other animals are analyzed four times a year for heavy metals, chlorinated hydrocarbons and radioactivity. The main objective is to have data at our disposal at times when some adverse publicity concerning the safety of the export products occurs. It is important to have such data to protect our exportation, and of course such analyses are done in central laboratories with adequate resources.

Food Chains and Their Use:
Fish and Fish Fry

Heated Effluent for the Rearing of Fry—for Farming and for Release

Richard L. Saunders*

*Aquaculture Group, Department of the Environment
Fisheries and Marine Service, Biological Station
St. Andrews, New Brunswick, Canada*

The objective in aquaculture, as in traditional animal husbandry, is to produce a high-quality product in the shortest possible time and at minimum expense. One of the main constraints in achieving rapid growth of poikilotherms is maintenance of suitable temperature for feeding and growth. In Fry's (1947) classification of the environment, temperature is a controlling factor; it governs the metabolic rate, thus affecting feeding appetite, digestion, and growth rate.

In this presentation I shall illustrate how thermal effluent, and therefore thermal control, might be used to hasten the growth rates of some potentially valuable aquaculture candidates. These are the salmonids in general and the Atlantic salmon in particular. I shall not list the criteria (Gaucher, 1970) on which the choice of salmonids is made. Suffice it to say, the Atlantic salmon, like the lobster, is a high-priced commodity and in limited market supply. But Atlantic salmon (Figure 1), like lobsters (Figure 2), are slow-growing animals, at least under natural conditions in the Northwest Atlantic area. Under seasonally favorable thermal conditions in summer and autumn, we cultured Atlantic salmon and rainbow trout to the so-called pan size (225–340 g) in eastern Canada within a single season of postsmolt growth. But it would be preferable to raise Atlantic salmon to 2–5 kg and to command premium prices. Although seawater temperatures in most of eastern Canada are too low in winter to permit year-round operations, large Atlantic salmon are cultured in Norway (Saunders, 1973;

*Present address: North American Salmon Research Center, St. Andrews, New Brunswick.

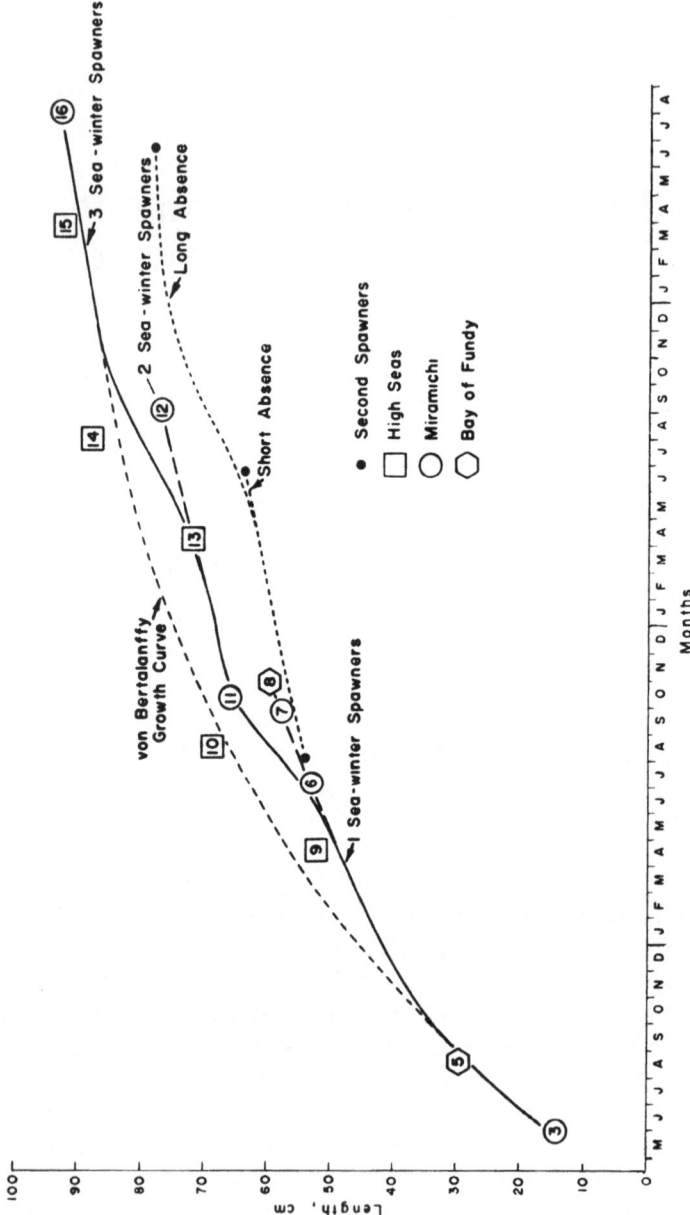

Fig. 1. Growth rates of Atlantic salmon in the Northwest Atlantic. From Allen *et al.*, 1972.

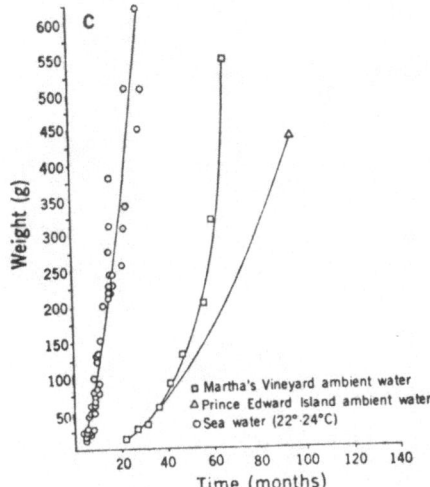

Fig. 2. Growth rates of the lobster (*Homarus americanus*) in relation to temperature. From Hughes *et al.*, 1972.

Fig. 3. Net enclosure for Atlantic salmon culture in Norway. From Saunders, 1973.

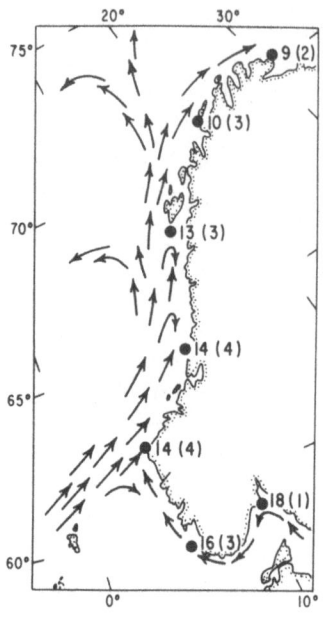

Fig. 4. Sea surface temperatures (°C) along the coast of Norway during summer and winter. Numbers in brackets are winter temperatures. From Møller, 1973.

Møller, 1973) and Scotland. Floating net cages (Figure 3) and enclosed coves are used year-round to rear salmon from the smolt stage to market size. This is possible along the outer coast of Norway because the Norwegian current, an offshoot of the Gulf Stream, provides reasonably warm water even during winter (Figure 4). Under good management, marketable sized (3-5 kg) salmon are produced in 14-16 months of postsmolt growth. In other words, stocks of salmon which normally produce 1.5-2.5 kg grilse in the wild state have the potential to reach twice this size during the same period in captivity where suitable temperatures and assured food supplies are provided.

Another example of the growth potential of Atlantic salmon is found in the British Isles where grilse (i.e., salmon which mature and return to spawning rivers after only one winter at sea) reach 4-6 kg. Went (1971) mentions exceptional individuals of 6.8 kg taken in Irish rivers. Similarly, large grilse are also taken in Scotland (Figure 5). A look at the annual temperature regime in the area west of the British Isles helps to explain this remarkable growth rate (Figure 6). An interesting question is: If some salmon avail themselves of this favorable temperature and grow rapidly, why do not all salmon in the Northeast Atlantic show such rapid growth? Many stocks of salmon in the Northeast Atlantic part of their range show growth rates no greater than their Northwest Atlantic counterparts. The answer to this question is likely to be found in genetic differences among stocks—both in respect to growth and developmental rates and migration patterns. For our mariculture developments, we should choose from among the many stocks of Atlantic salmon those that seem to have the genetic capability to grow fast as well as other heritable traits that would make them amenable to environmental manipulation.

Given suitable environmental conditions, Atlantic salmon can be made to grow much faster than they do in nature. The first serious attempt to provide a controlled environment for commercial production of salmonids in Canada was Sea Pool Fisheries in Nova Scotia. This bold endeavor failed owing to a number of problems involving management difficulties, inadequate fresh and salt water supply (with respect to both quality and quantity), poor nutrition, various diseases, and salinity stress. At Sea Pool there was provision for thermal control providing suitable temperatures year-round. Live steam was released in the water to maintain favorable temperatures in both the fish ponds and the water reconditioning units. The need to conserve heat required a low volume of cold makeup water and a reconditioning system to remove metabolic waste products. In the final analysis, the plant never achieved the planned level of production. Although no figures are available, the cost of production must have been very high owing largely to the high capital and operating costs of the water heating system.

Recently, we made a proposal to develop an aquaculture facility in conjunction with an oil-fired electrical generating station being built by New Brunswick Electric Power Commission near Saint John. Phase I is scheduled for completion in 1976 with a power output of 945 MW and *ca.* 1,000,000 liters/min of thermal effluent 16°C above ambient. Minimum winter temperature

Fig. 5. Large grilse taken in North Esk River, Scotland. Courtesy Dr. P. F. Elson.

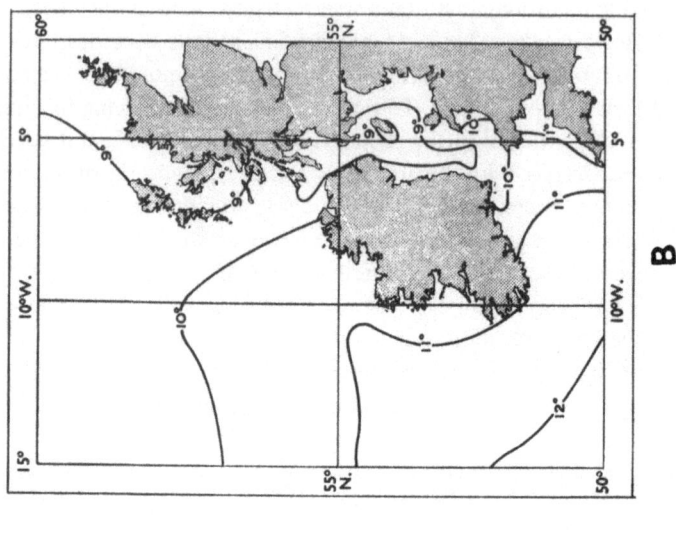

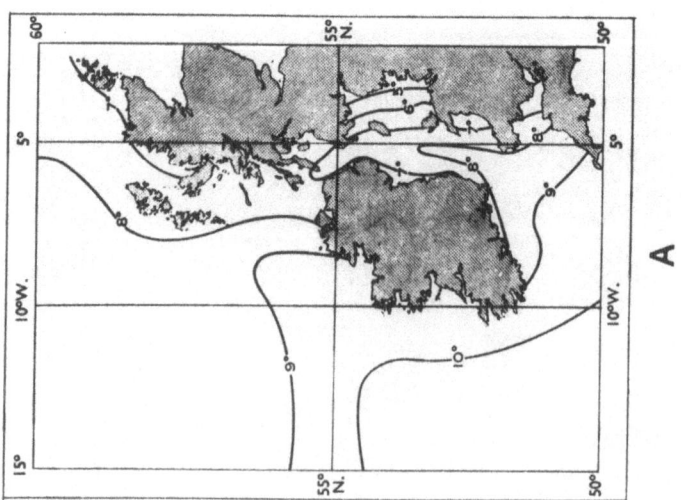

Fig. 6. Sea surface temperatures to the west of the United Kingdom. (A) winter; (B) summer. From *The Irish Coastal Pilot*, Anon., 1968.

at the cooling water intake will be about 2°C with a possible rise in the condenser to 17.6°C. It is unlikely this thermal effluent can be used directly because of toxic materials it might contain. Allowing for moderate efficiency of a heat exchanger and reasonable loss during transport, we might expect a minimum winter temperature of 12-13°C in the mariculture plant. At other times of year unheated sea water could be mixed with heated water to give desired temperatures higher than this. Ambient sea temperature at Lorneville is sufficiently low in summer that too-high temperatures would not be a problem at the mariculture plant. During much of the winter, the plant would depend almost entirely on heated effluent; at other seasons dependence on effluent would be reduced but probably never eliminated. This prospect of thermal control, together with an ample supply of good quality fresh water, offers attractive possibilities for culture of salmonids which grow well in the 10-16°C range. How could we use these thermal assets to optimize the growth of Atlantic salmon or other attractive aquaculture candidates?

HATCHING AND EARLY DEVELOPMENT

Canadian Atlantic salmon eggs are deposited in the gravel in October-November as the river water is cooling. By December their temperature is only slightly above freezing. Development is slow and hatching takes place in early spring. The alevins remain in the gravel and develop further while absorbing the yolk sac. Emergence and active feeding are delayed until the water warms sufficiently in April or May. In common hatchery practice, natural temperatures are approximated and development and hatching proceed only slightly faster than in nature (Belding et al., 1932). Recently, salmonid culturists have used relatively warm ground water or artificially warmed water to accelerate hatching and development to the active feeding stage. Marcus (1962) reports success with hatching Atlantic salmon at temperatures near 10°C and rearing at 16-18°C with growth to the smolt stage within one year. Gray (1929) incubated brown trout (*Salmo trutta*) eggs at temperatures from 2.8 to 13°C without high mortality. Above 15°C mortality was high. Gray found that trout hatched slowly at low temperatures are larger than those hatched more rapidly at high temperatures. McCawley and Trimborn (1968) reported good hatching and survival rates of rainbow trout at 11-14°C in recirculated water. Hayes et al. (1953) incubated Atlantic salmon eggs at temperatures between 1.3 and 14.5°C. At the higher temperature hatching and subsequent development were abnormal and the alevins died. Eggs reared at 11.5°C developed normally and without a high mortality rate at hatching. R. H. Peterson (personal communication) has recently incubated and hatched Atlantic salmon at 12°C at the Biological Station at St. Andrews.

Incubation and hatching took one month. In another month at that temperature, yolk sacs were absorbed and active feeding had begun. With judicious use of warm water it should be practicable to rear salmon to the active feeding stage in 2-3 months or by mid-January. Mathiasson (1970) describes how geothermal water is used in Iceland to hatch and rear Atlantic salmon. With this geothermal water, Gudjonsson (1973) reared Atlantic salmon to the smolt size in a single season.

PRESMOLT GROWTH

There is general agreement that growth of salmon fry and parr proceeds well in the 12-18°C range. Siginevich (1967) found in a Baltic stock of salmon that growth under natural conditions in the Daugava River was favorable in the range 16.6-18.1°C and the best growth was at 16.6°C. With seasonally changing temperatures between 8 and 14°C, Gudjonsson produced yearling smolts in Iceland. Mathiasson (1970) gives 12-14°C as being optimal for growth of smolts in an Icelandic hatchery. Peterson et al. (1972) generalize that water temperature should never exceed 18°C even during the hottest part of the summer and should remain above 10°C during the longest possible time to obtain optimum growth.

Presmolt growth can be further accelerated by rearing in brackish water. Canagaratnam (1959) demonstrated that coho salmon (*Oncorhynchus kisutch*) fry grew faster in brackish than in fresh water. Growth rates were directly related to salinity up to 18‰, the highest level he used. LaRoche (1950) found that growth of Atlantic salmon parr and smolts was faster in 26‰ salinity than in fresh water. Saunders and Henderson (1969a) reared Atlantic salmon to the smolt stage in water of various salinities up to 12‰ (Figure 7). Growth was better in 6 and 12‰ than in fresh water. Although the ambient salinity of 12‰ is close to the isosmotic level (10‰) of Atlantic salmon (Byrne et al., 1972), growth was better at 6 than 12‰. In some Norwegian aquaculture plants it is common practice to rear Atlantic salmon to the smolt stage beginning with fry in gradually increasing salinities (Saunders, 1973). Saunders and Henderson (1969a) found that salinities between 2 and 12‰ are effective in reducing mortality from fungus infections, particularly at the fry stage.

SMOLTING CONTROL BY PHOTOPERIOD AND TEMPERATURE

So far, we have seen that hatching, early development, and growth to the smolt stage can be hastened by using suitable temperatures. The production of yearling smolts is a considerable improvement over the minimum of two years

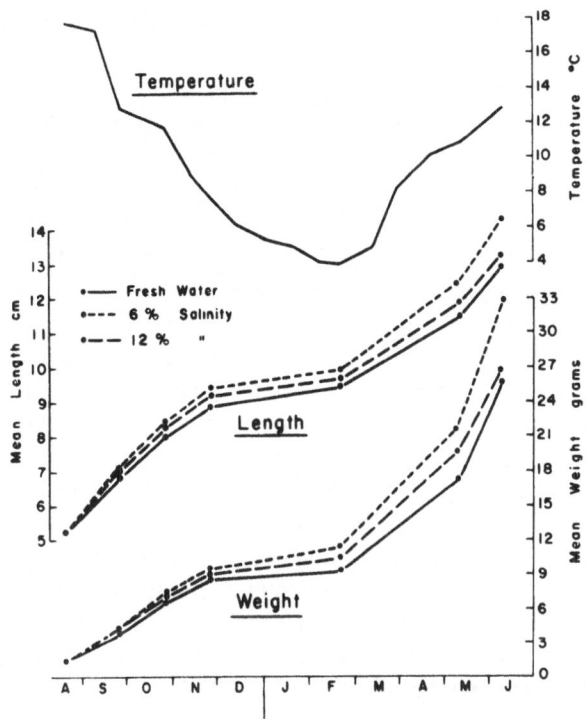

Fig. 7. The effect of salinity on the pre-smolt growth of Atlantic salmon.
From Saunders and Henderson, 1969*a*.

required to produce a smolt in nature. What are the possibilities of producing a smolt in less than a year?

Salmon undergo a number of physiological and behavioral changes during the parr/smolt transformation. The outward signs of smolting are a change from a generally brown to a silvery color. The fin tips become black. Smolts are schooling fish, whereas parr are solitary and territorial. In nature, smolts emigrate from rivers to the sea; in hatcheries they show restlessness and often change orientation, swimming with the current rather than against it. Physiologically, smolts have greater salinity tolerance than parr. Hoar (1965) suggested that the smolting process in anadromous fishes, and a number of physiological and behavioral changes that the process comprises, are probably triggered by changing photoperiod. Saunders and Henderson (1970) showed that photoperiod influences smolt development as indicated by growth rate, feeding appetite, and food conversion efficiency. Withey and Saunders (1973) showed that photoperiod affects the standard rate of oxygen consumption of smolts. When transferred to sea water in late summer, salmon which had been subjected to decreasing daylength in the spring had reduced standard rates of oxygen consumption and low growth rates and conversion efficiencies in comparison with control fish which had been subjected to increasing daylength in the spring. Wagner (1970) showed that changing photoperiod is the main environmental factor influencing the onset of smolting in steelhead trout and that by advancing or retarding the photoperiod regime it is possible to advance or delay smolting as judged by migratory behavior. In our laboratory we have recently caused presmolt Atlantic salmon to gain salinity tolerance during autumn or winter by subjecting them to long daylength at a time when similar sized fish under a short daylength regime had very low tolerance. Salmon subjected to long daylength in winter became silvery in February in contrast to controls, which did not silver till April. Growth rates of the long daylength fish were higher than those of controls. In agreement with these findings, Brown (1946) and Eisler (1957) reported that increased daylength enhanced growth rates of brown trout and chinook salmon. However, Johnston and Eales (1968) concluded that temperature had a greater effect on silvering of Atlantic salmon than did photoperiod.

Among the physiological changes taking place during smoltification is an increase in Na^+, K^+-activated ATPase activity in the gills. The enzyme ATPase is involved in the pumping mechanism whereby monovalent ions are excreted from the body via the gills. An increase in ion pump activity and in the concentration of ATPase has been reported for a number of euryhaline fishes during adaption to sea water (Epstein et al., 1967; Kamiya and Utida, 1969; Zaugg and McLain, 1970, 1971). Zaugg and McLain (1970, 1972) have shown that an increase in gill ATPase is associated with smoltification of rainbow trout and Pacific salmon and precedes their seaward migration. Recently, Adams et al. (1973) showed that the increase in ATPase concentration and consequent salinity tolerance of steelhead trout are temperature dependent. At temperatures of 6.5 and 10°C there was a twofold increase in Na^+, K^+-ATPase concentration which was associated with salinity tolerance. At 15°C or higher there was no increase in the enzyme concentration and little salinity tolerance. Assuming that this temperature effect on smolting is a generality for salmonids, it will be necessary to regulate both temperature and photoperiod very carefully in our efforts to produce smolts at any time of year. In our preliminary attempts to produce increased salinity tolerance and smolting by long daylength treatment during fall and winter, results were favorable at 10°C, the only temperature we used. It is reasonable to believe we can not only grow salmon to smolt size in less than a year but that we can also cause them to develop physiologically into smolts well before the natural May–June smolting season.

POSTSMOLT GROWTH

What environmental conditions are necessary for rapid growth of postsmolts during the period corresponding to sea life in nature? Saunders and Henderson (1969b) compared postsmolt growth of Atlantic salmon at constant temperatures from 10 to 18°C. Whereas 18°C is well within the diurnal range of temperature fluctuations in streams where parr grow well and within the optimum range for fry growth (Siginevich, 1967), it is above the optimum for postsmolt growth. Comparison of growth rates of postsmolts at 15 and 18°C during the summer showed best growth at 15°C. Comparisons in the winter between 10 and 14°C showed best growth at 14°C. These observations over a limited temperature range suggest best postsmolt growth at 14-15°C, somewhat lower than for presmolts.

Leggett and Power (1969) gave evidence that landlocked salmon in Newfoundland avoid temperatures over 14°C, moving into deeper water in July and in one case effectively ending the growing season. Hurley and Woodall (1968) found that pink salmon in vertical temperature/salinity gradients select lower temperatures and higher salinities with increase in fish size. Elson (personal communication) made some observations which point to different thermal optima between pre- and postsmolt Atlantic salmon. Salmon feed and grow in Greenland waters at 2-4°C, temperatures at which parr take little or no food (Allen, 1941; Saunders and Henderson, 1969c). Elson observed that, whereas parr in a New Brunswick river had stopped growing, as shown by their scales, when temperature fell below 4°C, salmon taken in Greenland waters at 2-4°C had full stomachs and had not yet formed winter bands on their scales. These observations indicate decreasing thermal optima during salmon development. Fine tuning of a controlled environment for aquaculture should take into account these as well as temporal or seasonally changing optima.

Saunders and Henderson (1969b) also compared postsmolt growth of salmon at various salinities from 0 to 30%₀. The highest salinity favored growth in early summer; at other times growth was faster in intermediate and low salinities. Hoar (1965) presents an attractive hypothesis to account for a changing physiology of salmonids whereby the fish are adapted to life in fresh water during autumn and winter and to life in the sea during spring and summer. Strong support for this hypothesis is found in studies with salmonids of temporal changes in various physiological–behavioral parameters: resistance to sea water (Conte and Wagner, 1965; Conte et al., 1966); changes in salinity preference (Baggerman, 1960a; McInerney, 1964); and variations in thyroid activity (Baggerman, 1960b; Eales, 1963, 1965). Hoar adds that cyclic changes in physiology are under hormonal control of pituitary products and that photoperiod and temperature are the environmental cues which maintain the cycle. Swift and Pickford (1965) showed striking changes in the hormone potency of pituitary glands from perch (Perca fluviatilis). Using hypophysectomized Fundulus for a bioassay, they found that the pituitary material from perch collected in the spring and early summer had more growth-promoting effect than that collected at other seasons. Gross et al. (1965) showed in a laboratory experiment that green sunfish (Lepomis cyanellus) grew best under a gradually increasing daylength, whereas decreasing daylength depressed growth. Their data also show a pronounced seasonal effect which potentiates the effect of photoperiod. In accord with Swift and Pickford's results, the maximum growth activity in the sunfish occurred in the spring, the period of increasing daylength. In view of the foregoing and our demonstrations of the effect of photoperiod on growth and metabolism of Atlantic salmon (Saunders and Henderson, 1970; Withey and Saunders, 1973), I suggest that we incorporate photoperiod in the environmental control scheme to optimize post-smolt growth to market size. It would be a simple matter to extend effective daylength by artificial lighting, but it might be impracticable to reduce daylength in a large outdoor culture facility.

ENVIRONMENTAL MANIPULATION FOR BEST GROWTH EFFICIENCY

Although feeding, nutrition, and food conversion efficiency are not, strictly speaking, part of the physical environment to be manipulated, their control in relation to temperature is an essential part of any scheme to speed salmonids to market size.

Brett and his co-workers (Brett, 1970a, 1971a,b; Brett et al., 1969), working with Pacific salmon, have done a lot to elucidate the principles underlying the interactions between temperature and feeding and growth rates. A good example is in the relations among gross and net conversion efficiency, temperature, and specific growth rate of sockeye salmon (Figures 8 and 9). Although the highest specific growth rate occurs with excess rations at *ca*. 15°C, the highest gross conversion efficiency occurs with reduced rations of 4% dry body weight/day at 11.5°C. If the maintenance ration is known—that amount of food required to maintain an animal without gain or loss in weight—it is possible to calculate the net efficiency of food conversion and to derive the efficiency of utilization of the food available for growth. In the case of sockeye salmon, the maximum net efficiency was at 9–10°C with a ration only slightly higher than the maintenance level. From their analyses of published results on feeding and growth of various fishes, Paloheimo and Dickie (1966) concluded that gross growth efficiency decreases with increase in rations.

There are many published data showing the importance of dietary considerations in relation to rearing temperature. I shall mention a few which show how we might match quality and quantity of food to particular temperatures. Atherton and Aitken (1970) working with rainbow trout found that 12°C is a compromise between a high absorption efficiency in the gut and high energy demands at 16°C, as opposed to inefficient absorption efficiency and low energy demands at 8°C. High-fat diets gave better growth at all temperatures tested with maximum values from 16 to 20°C. Maximum growth with low-fat diets was at

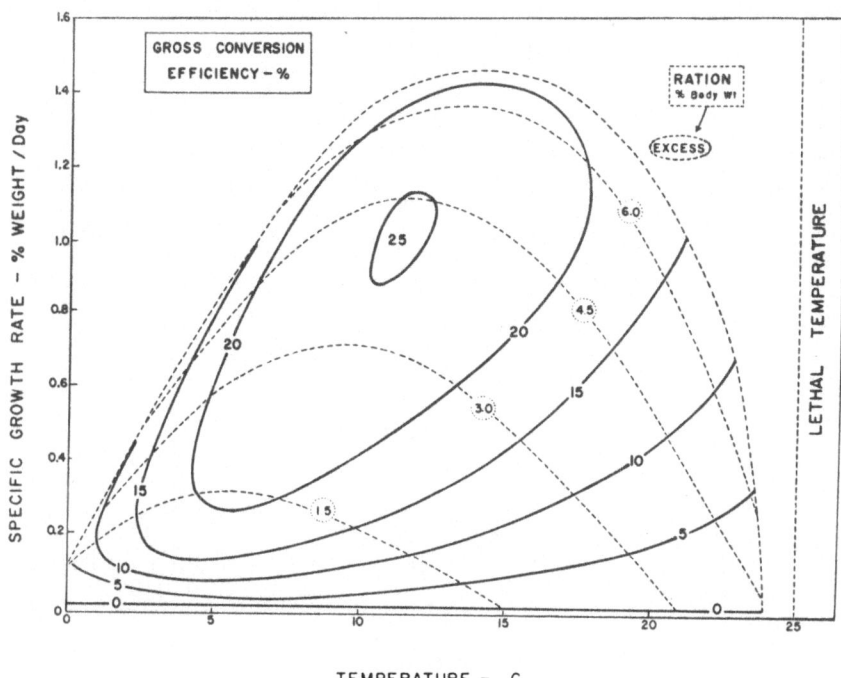

TEMPERATURE – C

Fig. 8. Gross food conversion efficiency in relation to temperature and ration level. Gross conversion efficiencies shown as isopleths overlying growth curves for different ration levels in relation to temperature. From Brett *et al.*, 1969.

12–16°C. At 16°C there was a greater percentage of fat absorbed, more ingested nitrogen was retained for growth, and less ammonia was excreted. Atherton and Aitken suggest that the yield of trout could be increased by using thermal effluent from power stations to maintain temperature around 16°C and by using a high-fat diet to meet metabolic demands. Bergström (1973) found that diets with as much as 16% fat content, composed of marine oils and lecithin, gave good growth and survival of Atlantic salmon underyearlings. She suggests that with such a diet growth to the smolt stage should be possible even without heated water. Siginevich (1967) points out that as long as the fat content in the viscera of Atlantic salmon fry is greater than that in the body and the overall fat content exceeds 13% of dry body weight, the food base is meeting individual requirements of energy necessary for growth. Phillips *et al.* (1966, 1967) and Poston *et al.* (1969) stress the advantages of using fat as a calorie source to spare protein in trout metabolism. A high-fat content meets the energy demands, particularly at high temperatures, and leaves a surplus to be stored. The protein component of the diet is then available for growth.

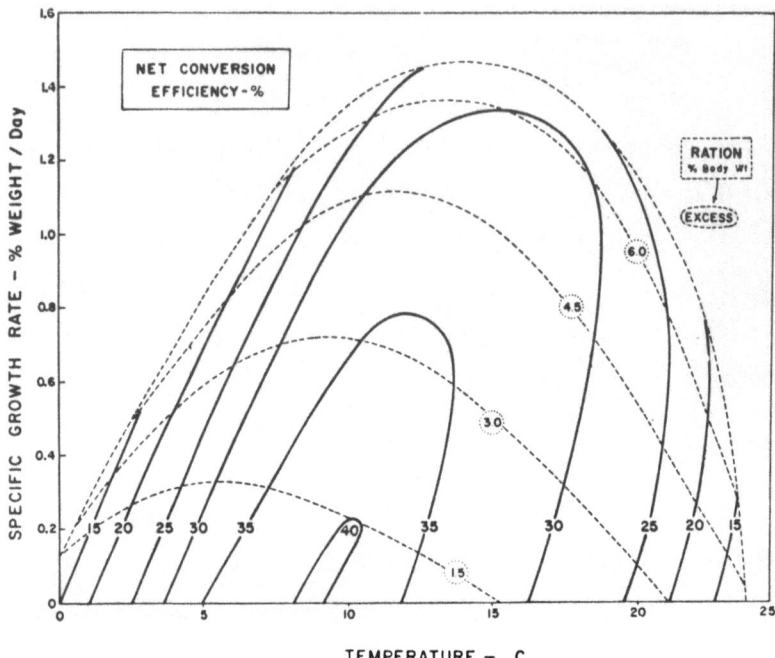

TEMPERATURE – C

Fig. 9. Net food conversion efficiency in relation to temperature and ration level. This relationship derived by subtracting maintenance ration from total ration. From Brett *et al.*, 1969.

Brett (1970*b*) makes the distinction between energy expended in total metabolism and that deposited in growth at different developmental stages (Figure 10). Nutritionally, there is great variation in the value of fats from different sources (Lee and Sinnhuber, 1972) and there is a need to choose the proper ones. Stickney (1972) found that channel catfish (*Ictalurus punctatus*) reared at temperatures from 18 to 34°C developed lipid contents increasing linearly from 24 to 44% of live weight. Owing to excessive fat content, catfish reared at optimum temperatures for rapid growth had lower dressed weights than those reared at less then optimum temperatures.

Peters and Boyd (1972) found that food intake by the hogchoker (*Trinectes maculatus*) increased between 15 and 25°C. However, maximum food conversion efficiency was at 15°C with *ad lib.* feeding but growth was slow at that temperature owing to low food intake. The best compromise in the hogchoker would seem to be at 25°C, where feeding and growth rates were maximum and conversion efficiency intermediate.

Doudoroff (1969) explains the need to match food supply to particular temperatures. He was referring to dietary needs in response to thermal pollution but this argument applies equally well to thermal manipulation for aquaculture. "The higher the temperature ... the higher is the metabolic rate of the fish, and therefore, the greater is the amount of food required for maintenance of its body weight, with no growth. The efficiency of utilization of food for growth, or its conversion to body tissue, may be reduced at high temperatures for this reason and others. Given a certain restricted daily ration, a fish living at a low temperature may be able not only to satisfy the relatively small maintenance-food requirement, but also to grow to a moderate rate. The gross food-conversion efficiency—that is, the ratio of the weight-gain or its caloric equivalent to the weight or caloric equivalent of food consumed—may be quite high. But [with] a higher temperature that is optimal for food intake ... and for growth when rations are unrestricted, the same restricted daily ration may be barely sufficient or insufficient for maintenance of body weight. The gross food-conversion efficiency thus may fall to zero at this 'optimal' temperature."

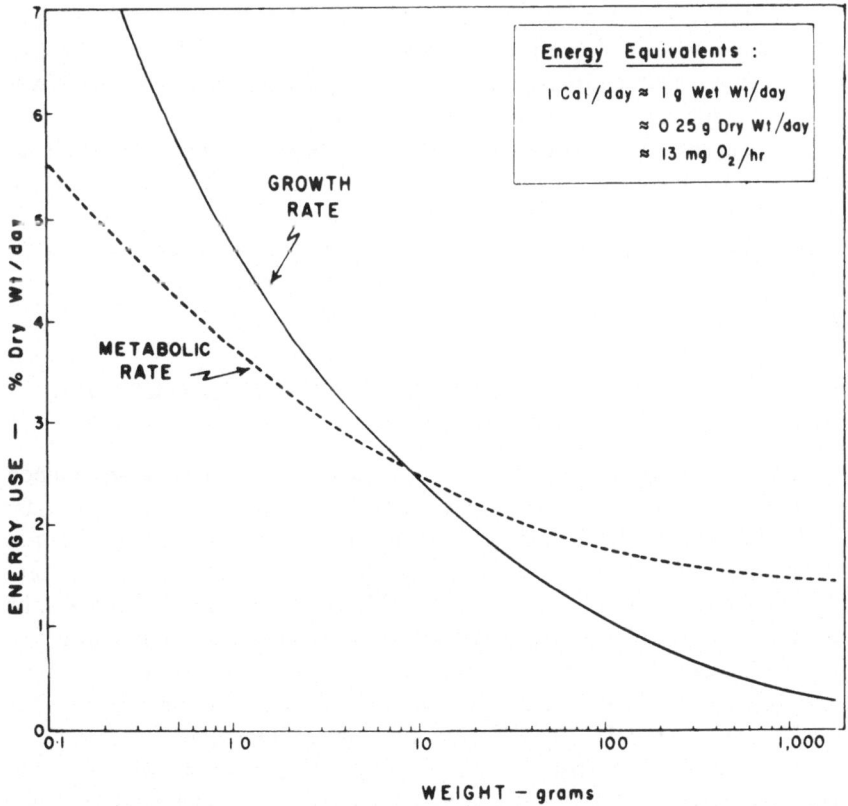

Fig. 10. Relation between energy deposited in growth and that expended in total metabolism, according to size in Pacific salmon. From Brett, 1970*b*.

What emerges from the foregoing discussion of the interaction of quality and quantity of food with rearing temperature is that the feeding strategy is a part of the environment to be manipulated. Feeding must be well managed to take into account interactions with the other variables—photoperiod, salinity, developmental and seasonal differences in physiology, and, of course temperature.

EFFECTS OF WARMING ON DISEASE PROBLEMS
AND THEIR CONTROL

Problems with infectious diseases and parasites are a likely consequence of holding large numbers of the same or related species in close proximity. One group of disease-causing organisms are called opportunistic pathogens. We can assume that these are always present in the fish's environment. Healthy, unstressed fish can usually coexist with these opportunistic pathogens with no ill effects. Brett (1973) sums up this idea as follows: Most healthy stocks of fishes are infected; few are diseased. If our environmental control scheme is effective in optimizing conditions for growth of the species being cultured, the animals will be in the best condition to resist disease outbreaks.

I should like to emphasize that maintenance of suitable environmental conditions will not necessarily be a panacea for potential fish disease problems. Some virulent pathogens strike even healthy stocks. Good management will require the use of disease-free or disease-resistant stocks, sterilization of eggs before shipment, and early diagnosis and treatment of disease.

SEA RANCHING AS AN ALTERNATIVE TO CONTROLLED REARING

The area of greatest expense in an integrated system for rearing large salmon is the period from smolt to adult size. Although equipment and labor costs are reasonable for rearing fish up to the smolt size, these and feeding costs rise sharply after this. Indeed, food accounts for about half the total cost of production. In addition to cost, it is questionable whether there will continue to be adequate supplies of high-quality protein which is an essential component in fish food. Although salmonids, like poultry, are efficient converters of food protein into fish flesh, a rational argument on ecological grounds might be that the available protein should be used directly for human consumption.

A reasonable alternative to controlled rearing is sea ranching, wherein smolts are released to enter the sea, where they forage naturally. Such a scheme takes advantage of salmonid migratory and homing behavior, which takes the fish from their nursery streams to the marine feeding grounds and back to their natal streams. There is growing evidence that salmon are "imprinted" by some chemical characteristic of their natal river, that they follow an inherited migration "program" which takes them to feeding areas at sea and back to the general area of that river where their earlier imprinting allows them to choose a particular tributary.

In our attempts at sea ranching we are using hatchery smolts reared from eggs taken in a river near the intended release and imprinting site. Smolts are tagged or fin-clipped and imprinted using the organic compound morpholine, which has been used successfully for this purpose by Cooper and Hasler (1974). The fish are held in an imprinting pond for several days before release. They are allowed to leave the pond when environmental conditions are suitable and as near as possible to the time of natural smolt emigration. During the summer of expected return, the water in the release pond is dosed with morpholine to attract adults which might have returned to the vicinity.

A sea ranching program is not without problems. First, we must accept that many of the fish released will fall prey to natural and fishing mortality. There is no safeguard that commercial fishermen will not intercept and catch the salmon. The recent ban on much of the Canadian commercial fishery for Atlantic salmon reduces the chances of capture. By using stocks of salmon which mature as grilse or those which do not visit the Greenland area, loss to the international salmon fishery near Greenland can be avoided. Biological studies of the sea ranching concept are being carried out. If sufficient numbers of adult salmon can be made to return to a release and recapture facility to meet the cost of smolt production and to provide a profit, then we shall have a viable alternative to controlled rearing with its high capital and operating costs. More important, we will be making a genuine contribution to increasing protein production rather than merely converting protein that we have already collected into a luxury product with less than half of the original protein remaining.

REFERENCES

Adams, B. L., W. S. Zaugg, and L. R. McLain. 1973. Temperature effect on parr–smolt transformation in steelhead trout (*Salmo gairdneri*) as measured by gill sodium-potassium stimulated adenosinetriphosphatase. *Comp. Biochem. Physiol. 44A:* 1333-1339.

Allen, K. R. 1941. Studies on the biology of the early stages of the salmon (*Salmo salar*). III. Growth in the Thurso River system, Caithness. *J. Anim. Ecol. 10:* 273-295.

Allen, K. R., R. L. Saunders, and P. F. Elson. 1972. Marine growth of Atlantic salmon (*Salmo salar*) in the Northwest Atlantic. *J. Fish. Res. Board Can. 29:* 1373-1380.

Anonymous. 1968. Irish Coastal Pilot, 11th ed. N. P. No. 40. Pub. by Hydrographer of the Navy, London.

Atherton, W. D., and A. Aitken. 1970. Growth, nitrogen metabolism and fat metabolism in *Salmo gairdneri* Rich. *Comp. Biochem. Physiol. 36:* 719-747.

Baggerman, B. 1960*a*. Salinity preference, thyroid activity and seaward migration of four species of Pacific salmon (*Oncorhynchus*). *J. Fish. Res. Board Can. 17:* 295-322.

Baggerman, B. 1960*b*. Factors in the diadromous migration of fish. *Symp. Zool. Soc. London 1:* 33-60.

Belding, D. L., M. J. Pender, and J. A. Rodd. 1932. The early growth of salmon parr in Canadian hatcheries. *Trans. Am. Fish. Soc. 62:* 211-223.

Bergström, E. 1973. The role of nutrition in growth and survival of young hatchery reared Atlantic salmon. *In:* M. W. Smith and W. M. Carter, eds. *Int. Atl. Salmon Symp.* The Int. Atl. Salmon Found., Spec. Publ. *4(1):* 265-282.

Brett, J. R. 1970*a*. Temperature. Pages 513-560 *in:* O. Kinne, ed. *Marine Ecology, Vol. I. Environmental Factors.* Pt. 1.

Brett, J. R. 1970*b*. Fish—The energy cost of living. Pages 37-52 *in:* W. J. McNeil, ed. *Marine Aquaculture.* Oregon State University Press, Corvallis.

Brett, J. R. 1971*a*. Energetic responses of salmon to temperature. A study of some thermal relations in the physiology and ecology of sockeye salmon (*Oncorhynchus nerka*). *Am. Zool. 11:* 99-113.

Brett, J. R. 1971*b*. Growth responses of young sockeye salmon (*Oncorhynchus nerka*) to different diets and planes of nutrition. *J. Fish. Res. Board Can. 28:* 1635-1643.

Brett, J. R. 1973. Marine aquaculture in Canada—The practice and the promise, p. 150-196. Govt.-Industry Policy Development Seminar, Freshwater Institute, Winnipeg, Manitoba. May 31-June 1, 1973.

Brett, J. R., J. E. Shelbourn, and C. T. Shoop. 1969. Growth rate and body composition of fingerling sockeye salmon, *Oncorhynchus nerka*, in relation to temperature and ration size. *J. Fish. Res. Board Can. 26:* 2363-2394.

Brown, M. E. 1946. The growth of brown trout (*Salmo trutta* L.). II. The growth of two year old trout at a constant temperature of 11.5°C. *J. Exp. Biol. 22:* 130-144.

Byrne, J. M., F. W. H. Beamish, and R. L. Saunders. 1972. Influence of salinity, temperature, and exercise on plasma osmolality and ionic concentration in Atlantic salmon (*Salmo salar*). *J. Fish. Res. Board Can. 29:* 1217-1220.

Canagaratnam, P. 1959. Growth of fishes in different salinities. *J. Fish. Res. Board Can. 16:* 121-130.

Conte, F. P., and H. H. Wagner. 1965. Development of osmotic and ionic regulation in juvenile steelhead trout *Salmo gairdneri. Comp. Biochem. Physiol. 14:* 603-620.

Conte, F. P., H. H. Wagner, J. Fessler, and G. Gnose. 1966. Development of osmotic and ionic regulation in juvenile coho salmon *Oncorhynchus kisutch. Comp. Biochem. Physiol. 18:* 1-15.

Cooper, J. C., and A. D. Hasler. 1974. Electroencephalographic evidence for retention of olfactory cues in homing coho salmon. *Science 183:* 336-338.

Doudoroff, P. 1969. Developing thermal requirements for freshwater fishes: Discussion. *In:* P. A. Krenkel, and F. L. Parker, eds. *Biological Aspects of thermal pollution.* Vanderbilt University Press, Nashville, Tenn.

Eales, J. G. 1963. A comparative study of thyroid function in migrant juvenile salmon. *Can. J. Zool. 41:* 811-824.

Eales, J. G. 1965. Factors influencing seasonal changes in thyroid activity in juvenile steelhead trout, *Salmo gairdneri. Can. J. Zool. 43:* 719-729.

Eisler, R. 1957. The influence of light on the early growth of chinook salmon. *Growth 21:* 197-203.

Epstein, F. H., I. A. Katz, and G. B. Pickford. 1967. Sodium- and potassium-activated adenosine triphosphatase of gills: Role in adaptation of teleosts to salt water. *Science 156:* 1245-1247.

Fry, F. E. J. 1947. Effects of the environment on animal activity. *Univ. Toronto Stud. Biol. Ser. 55:* 1-62.

Gaucher, T. A. 1970. A technological perspective. *In:* T. A. Gaucher, ed. *Aquaculture: A New England perspective.* The Res. Inst. of the Gulf of Maine. Proc. from a conference on aquaculture in northern New England.

Gray, J. 1929. The growth of fish. III. The effect of temperature on the development of the eggs of *Salmo fario. J. Exp. Biol. 5:* 125-130.

Gross, W. L., E. W. Roelofs, and P. O. Fromm. 1965. Influence of photoperiod on growth of green sunfish, *Lepomis cyanellus. J. Fish. Res. Board Can. 22:* 1379-1386.

Gudjonsson, T. 1973. Smolt tagging techniques, stocking, and tagged adult salmon recaptures in Iceland. *In:* M. W. Smith and W. M. Carter, eds. *Int. Atl. Salmon Symp.* The Int. Atl. Salmon Found., Spec. Publ. *4* (1): 227-235.

Hayes, F. R., D. Pelluet, and E. Gorham. 1953. Some effects of temperature on the embryonic development of the salmon (*Salmo salar*). *Can. J. Zool. 31:* 42-51.

Hoar, W. S. 1965. The endocrine system as a chemical link between the organism and its environment. *Trans. Roy. Soc. Can.* Ser. IV, Vol. III, *Sect. 3:* 175-200.

Hughes, J. T., J. J. Sullivan, and R. Shleser. 1972. Enhancement of lobster growth. *Science 177:* 1110-1111.

Hurley, D. A., and W. L. Woodall. 1968. Responses of young pink salmon to vertical temperature and salinity gradients. Int. Pac. Sal. Fish. Comm., Prog. Rep. No. 19.

Johnston, C. E., and J. G. Eales. 1968. Influence of body size on silvering of Atlantic salmon (*Salmo salar*) at parr–smolt transformation. *J. Fish. Res. Board Can. 27:* 983-987.

Kamiya, M., and S. Utida. 1969. Sodium-potassium-activated adenosine triphosphatase activity in gills of fresh-water, marine and euryhaline teleosts. *Comp. Biochem. Physiol. 31:* 671-674.

LaRoche, G. 1950. Résistance des saumoneaux á l'eau salée. *Ann. l'Acfas 17:* 125-128.

Lee, D. J., and R. O. Sinnhuber. 1972. Lipid requirements. Pages 145-180 *in:* J. E. Halver, ed. *Fish Nutrition.* Academic Press, New York.

Leggett, W. C., and G. Power. 1969. Differences between two populations of landlocked Atlantic salmon (*Salmo salar*) in Newfoundland. *J. Fish. Res. Board Can. 26:* 1585-1596.

Marcus, H. C. 1962. Hatchery reared Atlantic salmon smolts in ten months. *Prog. Fish. Cult. 24:* 127.

Matthiasson, M. 1970. Beneficial uses of heat in Iceland. Pages 139-184 *in: Proc. Conf. on Benefic. Uses of Thermal Discharges.* N. Y. State Dept. Inviron. Cons., Albany.

McCauley, R. W., and F. Trimborn. 1968. Incubating rainbow trout eggs in heated, recirculated water. *Prog. Fish Cult. 30:* 64.

McInerney, J. E. 1964. Salinity preference: An orientation mechanism in salmon migration. *J. Fish. Res. Board Can. 21:* 995-1018.

Møller, D. 1973. Norwegian salmon farming. *In:* M. W. Smith and W. M. Carter, eds. *Int. Atl. Salmon Symp.* The Int. Atl. Salmon Found., Spec. Publ. *4* (1): 259-263.

Paloheimo, J. E.. and L. M. Dickie. 1966. Food and growth of fishes. III. Relations among food, body size, and growth efficiency. *J. Fish. Res. Board Can. 23:* 1209-1248.

Peters, D. S., and M. T. Boyd. 1972. The effect of temperature, salinity, and availability of food on the feeding and growth of the hogchoker, *Trinectes maculatus* (Bloch and Schneider). *J. Exp. Mar. Biol. Ecol. 9:* 201-207.

Peterson, H. H., O. T. Carlson, and S. Jonasson. 1972. The rearing of Atlantic salmon. Copyright Astra-Ewos AB, Sodertälje. Sweden. Bröd Ljungberg Tryckeri AB, Sodertälje, Sweden. 39 p.

Phillips, A. M., Jr., D. L. Livingston, and H. A. Poston. 1966. The effect of changes in protein quality, calorie sources, and calorie levels upon the growth and chemical composition of brook trout. Cortland Hatchery Rep. No. 34 for 1965. *Fish. Res. Bull. No. 29.*

Phillips, A. M., Jr., H. A. Poston, and D. L. Livingston. 1967. The effect of calorie sources and water temperature upon brook trout growth and body chemistry. Cortland Hatchery Rep. No. 35 for 1966. *Fish. Res. Bull. No. 30.*

Poston, H. A., D. L. Livingston, and A. M. Phillips, Jr. 1969. The effect of source of dietary fat, calorie ratio, and water temperature on growth and chemical composition of brown trout. Cortland Hatchery Rep. No. 37 for 1968. *Fish. Res. Bull. No. 32.*

Saunders, R. L. 1973. Salmonid aquaculture in Norway. *The Atlantic Salmon Journal, 1973, No. 1:* 8–13.

Saunders, R. L., and E. B. Henderson. 1969*a*. Survival and growth of Atlantic salmon fry in relation to salinity and diet. *Fish. Res. Board Can. Tech. Rep. No. 148.*

Saunders, R. L., and E. B. Henderson. 1969*b*. Growth of Atlantic salmon smolts and post-smolts in relation to salinity, temperature and diet. *Fish. Res. Board Can. Tech. Rep.* No. 149.

Saunders, R. L., and E. B. Henderson. 1969*c*. Survival and growth of Atlantic salmon parr in relation to salinity. *Fish. Res. Board Can. Tech. Rep.* No. 147.

Saunders, R. L., and E. B. Henderson. 1970. Influence of photoperiod on smolt development and growth of Atlantic salmon (*Salmo salar*). *J. Fish. Res. Board Can. 27:* 1295–1311.

Siginevich, G. P. 1967. Nature of the relationship between increase in size of Baltic salmon fry and water temperature. *Gidrobiol. Zhurn. 3:* 43–48. Fish. Res. Board Can. Translation Series No. 952.

Stickney, R. R. 1972. Effects of dietary lipids and lipid–temperature interactions on growth, food conversion, percentage lipid and fatty acid composition of channel catfish. Doctoral thesis. Florida State University, Tallahassee (1971) Dissertation Abs. Int. 32, 6545B.

Swift, D. R., and G. E. Pickford. 1965. Seasonal variations in the hormone content of the pituitary gland of the perch, *Perca fluviatilis* L. *Gen. Comp. Endocrinol. 5:* 354–365.

Wagner, H. H. 1970. The parr–smolt metamorphosis in steelhead trout as affected by photoperiod and temperature. Doctoral thesis. Oregon State University. 177 p.

Went, A. E. J. 1971. Salmon of the Foyle system (1970). Foyle River Comm., Nineteenth Ann. Rep. 41–51.

Withey, K. G., and R. L. Saunders. 1973. Effects of a reciprocal photoperiod regime on standard rate of oxygen consumption of post-smolt Atlantic salmon (*Salmo salar*). *J. Fish. Res. Board Can. 30:* 1898–1900.

Zaugg, W. S., and L. R. McLain. 1970. Adenosinetriphosphatase activity in gills of salmonids: Seasonal variations and salt water influence in coho salmon *Oncorhynchus kisutch. Comp. Biochem. Physiol. 35:* 587–596.

Zaugg, W. S., and L. R. McLain. 1971. Gill sampling as a method of following biochemical changes: ATPase activities altered by oubain injection and salt water adaption. *Comp. Biochem. Physiol. 38B:* 501–506.

Zaugg, W. S., and L. R. McLain. 1972. Changes in gill adenosinetriphosphatase activity associated with parr–smolt transformation in steelhead trout, coho, and spring chinook salmon. *J. Fish. Res. Board Can. 29:* 167–171.

DISCUSSION

CONOVER: What is the threshold level of the light required to extend the daylength?

SAUNDERS: Ordinary fluorescent lighting is sufficient. I do not know of any research to establish what the threshold level is.

CONOVER: Is there a relationship in the Atlantic salmon, as there seems to be for Pacific salmon, between the smolt growth and the ultimate size on leaving the river and the successful return of the population and its growth rate in the sea?

SAUNDERS: Swedish experiments have shown that increase in smolt size up to a certain point will lead to an increase in the survival and the recapture rate of tagged salmon. Above a certain smolt size, percentage recapture of adult salmon declines.

CONOVER: Is there a relationship between the return of the salmon, the size of the stock, and the river itself? Some rivers can carry larger sizes of fish than others. Can this be attributed to genetic or environmental factors?

SAUNDERS: It seems to be predominantly an environmental matter, insofar as the river can support so many spawners and there is room for just so many fry and parr. Considering that, in most salmon populations, the fish spend two or three years in the stream before going to the sea, I think population size is limited by the physical and chemical nature of the river. Fish size in a river stock is probably largely controlled by genetics with a lesser influence of the environment.

RAYMONT: How much does territory contribute to this?

SAUNDERS: Salmon are quite territorial and this bears importantly on Dr. Conover's question. The smolt production of a river depends on the area of suitable substrate for the establishment of territories. Salmon held under crowded conditions in hatcheries seem to lose this territoriality.

CONOVER: Do you have any information if demand feeding schemes can be utilized sufficiently in the culture of salmon?

SAUNDERS: A scheme of demand feeding would be quite appropriate, as the feeding routine is very important in achieving good growth. If parr and pre-smolt are fed on a certain schedule–let's say two to three times a day, to satiation–they do not quickly adjust to a schedule of more frequent feeding.

If one uses an automatic feeder which dispenses a small amount of food every few minutes, depending on the temperature, the fish become accustomed to taking just a little food each time and it is difficult to make them take large amounts at less frequent intervals.

DEVIK: We might look at the rearing of salmon as a system, governed by a certain number of stated factors–e.g., food, temperature, salinity, to mention a few. These factors may be subjected to control to a varying degree. We might include in the control factors the genetic variations and possibilities to induce variations. The interesting question for such an ecosystem is how far it is opportune to utilize the natural variations to suit our means, and how much it is worthwhile doing in this context to obtain a certain control of the environment for the particular species.

SAUNDERS: We should choose the stock of salmon very carefully. I mentioned that in some rivers in the western part of the United Kingdom the grilse or one-sea-year salmon seem to be exceptionally large, perhaps two or three times as large as some Norwegian grilse, and certainly two to three times as large as Canadian grilse. This is an obvious example of the sort of stock we should use. I think that the faster growth in these certain stocks must be genetically based, but that there is probably an environmental component as well. For example, in the western part of the U.K. the temperatures are such that the smolts can leave the rivers in March or April, whereas in Norway and Canada the smolts do not go to sea until May or early June. This gives the stocks of early migrating fish quite a head start; two months extra growing time is a very considerable increase. Furthermore, the temperatures in the sea to the west and north of the U.K. are quite favorable for growth, even during winter. Now if it were entirely an environmental question, one might ask why all European stocks of salmon are not fast growing. It has probably to do with their migratory behavior. Some of the stocks from Europe go to west Greenland waters, where the temperatures are just barely adequate for winter feeding. Others do not seem to go so far afield, but stay where growth is faster. There is increasing evidence that migratory behavior is inherited.

MATTHEWS: It seems that different salmon stocks show different growth rates at different stages in the life cycle, with some fishes only needing one year to reach a considerable size before returning to the river. It probably does not matter whether the salmon are away one, two, or three years in the sea, so long as the fish have gained sufficient weight. Can one select for nongrilse?

SAUNDERS: Yes, I think one can select for nongrilse (larger salmon), and I believe it would even be possible to produce grilse or larger salmon through environmental manipulation. I believe that grilse or nongrilse production has a lot to do with genetic background but that it can also be influenced by the environment. In Canadian practice, it might be advisable to select a grilse stock because these, we know, do not contribute to the Greenland fishery. This suggests that grilse do not go so far afield, and probably would not be cropped so heavily by commercial fisheries. Under some conditions it might be reasonable to have a stock of salmon that produce both grilse and larger fish. Total biomass production, as grilse, might be heavier than production of larger salmon, because the longer salmon stay at sea, the more their numbers are reduced by fishing and natural mortality. A choice between grilse and larger salmon would depend, not only on which size gives the greater total return in weight, but also on the comparative values of these two sizes of salmon.

RAYMONT: What is the rate of growth of grilse as opposed to older salmon? Is it more rapid in the growth stage? Does the speed of growth fall off when the age goes up?

SAUNDERS: As fish reach maturity their growth rate decreases and the fish which are maturing as grilse level off in weight from the general growth curve. We believe this is owing to their sexual development and cessation of feeding. It has been observed that for salmon of a given smolt class, those that are mature and have returned as grilse are considerably smaller than the fish of the same smolt class which have not matured and are still feeding at sea. Spawning grilse are commonly over 1 kg lighter than nonmature individuals of the same smolt class but which are still feeding at sea. These comparisons were made between spawners in eastern Canada and tagged fish captured near Greenland.

NYMAN: I can add some information that indicates the growth potential of salmon. During the winter 1972–73 Atlantic salmon were kept in floating cages in the hot-water effluent of a nuclear power plant on the east coast of Sweden. The fish were released in the early spring of 1973 as smolts, two years of age; less than five months later upon recapture some specimens weighed five and a half kilos, the average weight of the tagged returns being three kilos after five months at sea. At release, the smolts weighed 156 to 300 grams. This experiment also proved that sudden drops in temperature did not affect the salmon, even when they were subject to a drop in temperature of more than 10°C over a two-hour period. This excellent growth potential will hopefully be exploited on a more commercial scale in the future in fish farms receiving their water from hot water discharges. The mortality was mainly caused from supersaturation of oxygen of the water. Because of the heating, this could increase to 130% saturation.

SAUNDERS: Thermal conditions in eastern Canada are not always as adverse as indicated in my presentation. Freezing water temperatures in winter are common. There are, therefore, difficulties in holding salmon under natural conditions all the year round. In many places along our east coast, conditions are excellent for summer growth. Accordingly, we have started research on seasonal production wherein we plant smolts in net enclosures in May or early June and harvest them as late as December. The favored market size is from 225–340 g (the so-called portion size). That size can be reached as early as July or August, so the production period would be the months July through December. We are investigating the possibilities of extending this kind of production to a year-round basis, for which we would have to have a source of heated water.

CARSTENS: You mentioned imprinting with a chemical to improve the return. How does this work? Do you have to continuously release the same chemical for ever after, or can you spike your release?

SAUNDERS: Hasler and his associates have used the compound morpholine in the Great Lakes. They subject premigrants to this compound to imprint them, and when they expect adult returns they dose the river or the recapture facilities with some of this compound to decoy the fish when they want them back. This technique has been used quite successfully in the Great Lakes, particularly in Lake Michigan.

GJEDREM: You mentioned the importance of photoperiods. What is the factor of the twenty-four-hour rate for growing salmon? And you also mentioned that parr were successfully reared at $26^0/oo$ salinity.

SAUNDERS: I understand that it is common practice in Norway to rear salmon parr in brackish water and to gradually increase the salinity as the fish approach the smolt stage. It has been shown that hatching, early development, and growth to the smolt stage can be hastened by using appropriate temperatures. Production of yearling smolts is improved considerably by appropriate manipulation of photoperiod, which will stimulate growth of salmonids. But I question whether it is a good practice to use continuous daylight if you are going to move the fish into seawater. There is some evidence that continuous daylight will interfere with the smolting process. It has been suggested that photoperiod is a stimulatory factor acting through the pineal gland to influence hormone production in the pituitary. Pituitary histology changes during normal smolting development. We can further influence these pituitary changes by manipulating the photoperiod.

You asked about the observations that $26^0/oo$ salinity seemed to favor growth of parr. In our early experiments we tested various salinities with salmon alevins or fry after they completed absorption of the yolk sac. We found that they could withstand salinities up to $12^0/oo$, but not 15. Larger parr can stand quite high salinities as long as the temperature is at a reasonable level, say $10-15°C$. Low temperatures seem to potentiate the harmful effect of a high salinity. There are numerous examples of parr surviving quite high salinities. The range $25-30^0/oo$ is critical for parr survival.

RAYMONT: You mentioned briefly that in the use of heated effluent there was some danger of toxic materials. Are you thinking of chlorine, or trace metals, or both?

SAUNDERS: I am thinking of both chlorine and trace metals. In the application we are considering in New Brunswick, it is suggested that we might use the untreated effluent. Some power plants use mechanical rather than chemical treatment to prevent accumulation of fouling organisms. In this case we need only consider heavy metals.

RAYMONT: Do you chlorinate over the year in your power station? In Britian the practice is to chlorinate for six months of the year, roughly when the fouling organisms are active.

SAUNDERS: Some power plants in New Brunswick do not use chlorination at all. This depends on the intake temperatures and the fouling problem in the particular area.

DEVIK: With regard to the ranching of fish, the potentialities of this form of management will depend upon the productive capacity of the areas where the fish feed. Are there any estimates of the potential production in the areas of the sea that are affected?

SAUNDERS: I cannot offer any estimates about the potential production of salmon, only to say that the total number of Atlantic salmon in the world now is many times lower than it was 100 or 200 years ago. The oceanic niche that salmon occupied 100 years ago is probably not vacant. Other organisms have probably taken their place. However, we have no indication of density dependence on salmon growth at sea.

CHRISTENSEN: In your paper there was a reference to a new power station at New Brunswick which is to be started in 1976. I understand the intention is to install a hatchery in conjunction with the power plant, and I wonder if it is possible to give some detail of the installations. For instance: will you install a hatchery for smolt and also for larger fish? What will be the capacity, the economics of the operation, and how large is the temperature rise expected in the cooling cycle?

SAUNDERS: The proposed hatchery will operate on water that, after passing through the condensers, will have a temperature of about 16°C. A feasibility study has recently been conducted to consider engineering, biological, marketing, and economic aspects of operating a hatchery and production plant. The outcome was discouraging in that it would cost a lot of money to use the hot water which is essentially free. Part of the reason is that planning was not done early enough. The hatchery would have to have its own water intake supply rather than sharing the water supply to the power plant. The heated water might have to be put through heat exchangers, which are expensive capital items. Another big cost is in the feed. Forecasts are that the cost of feed may be as high as 40¢ per lb by 1976. Feed could account for half of the total cost of producing salmon. Because of excessive engineering costs at the plant in question, we do not intend to develop a commercial aquaculture station there. Instead, we propose to build a pilot plant in which to transfer our laboratory technology in environmental control of fish growth to a meaningful level of production which, we hope, will be a model for application at some future source of warm water. The pilot study will be multidisciplinary, incorporating economics, engineering, and biological aspects of growing fish and marketing them.

RAYMONT: Yes, this tallies with other experience. The cheap heat is not cheap at all if you are not in at the planning stage, and also at the preplanning stage.

The Ecology of a Texas Bay

C. H. Oppenheimer and W. B. Brogden

Marine Science Institute
University of Texas
Port Aransas, Texas

INTRODUCTION

This contribution is to be related to fringe benefits to long-term localized increases of food supply for mariculture in heated effluents. It is a rather interesting topic that perhaps can be best approached by examining a productive warm water marine natural environment to describe its ecology and the effects of variations of nutrients in the system.

The use of hot water effluents for fish or shellfish productivity in the colder latitudes can be compared in many ways with the warmer waters of the tropical or warm temperate zones of the world. It would seem a most logical approach to select warm water organisms for some northern warm water effluent fish culture rather than to attempt to adapt the colder water species to warm water. To examine this concept, a discussion of the ecology and productivity of warm water coastal environments may be evaluated.

The following is a brief summary of an ecological description of a Texas estuary and its productivity (Collier and Hedgpeth, 1950). The 150,000-acre estuary of Corpus Christi and Nueces Bays was chosen because available ecological information will permit a general discussion leading to its diversity and productivity.

Corpus Christi Bay area is one of five Texas bay systems (Figures 1 and 2), comprising 1.5 million acres of estuarine waters. The bays are shallow. Corpus Christi Bay has a maximum natural depth of three meters, with a lunar tidal amplitude of half a meter and occasional wind or exceptional hurricane tides of up to 3 m.

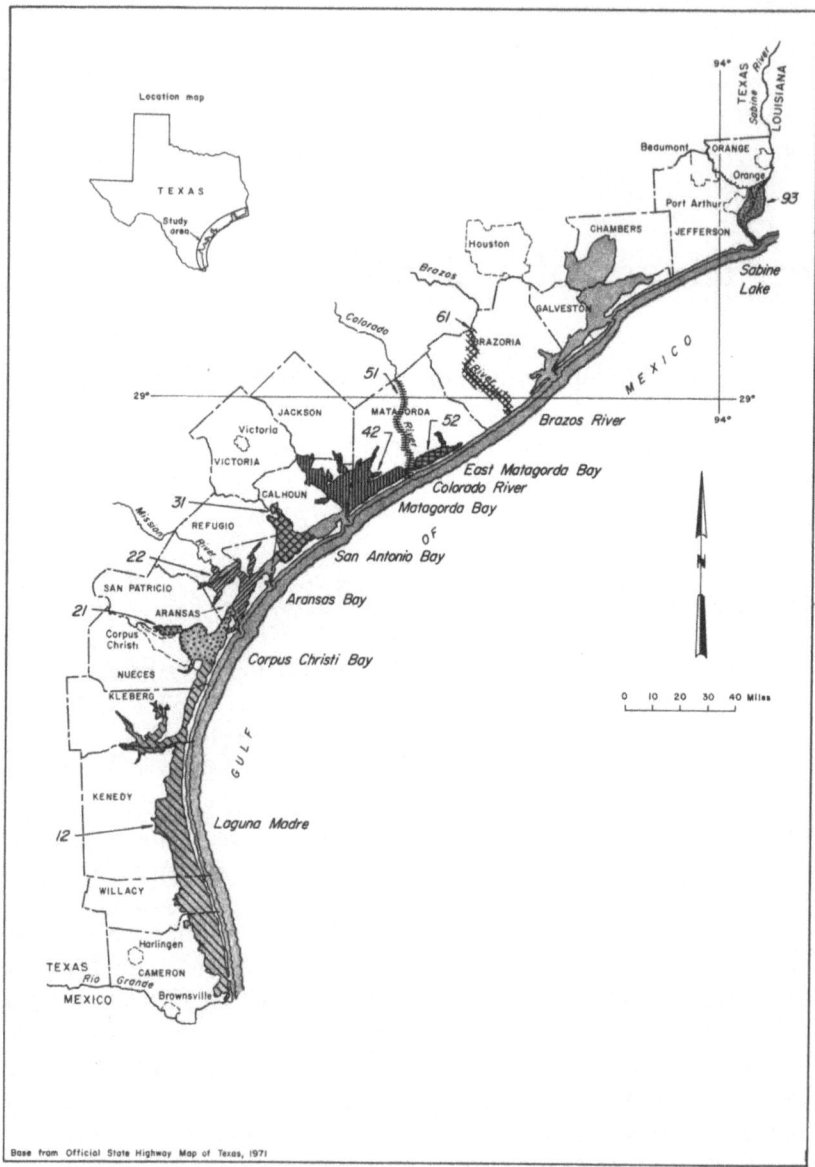

Fig. 1. Locations of bays and estuaries contained in the Coastal Data System.

The bays are in the Texas coastal plain, a vast low, flat expanse surrounded by marshes and grassy vegetation with only a few stunted trees. The barrier islands that separate the bays from the Gulf of Mexico consist primarily of sand with an average elevation of 2 m. The area is constantly subject to winds that are predominantly from either the southeast in the summer or the north in the winter. They average 24 km/hr in the summer with periods up to 64 km/hr. In winter the winds are variable, with a mixture of southeasterlies as in summer and four- to five-day periods of northerly cold winds that may reach 100 km/hr. This constant wind subjects the bays to mixing and produces varying current patterns. As a result, the water is not stratified except for rare periods when the wind ceases for a week or during a heavy rain with overlying freshwater.

The rainfall for the area averages 68 cm/year, usually with short periods of intensive rain followed by long periods of little or no rain. An approximate ten-year drought cycle is evident from past records, at which times the rain may decrease to 25 cm/year. Heavy rainfall, excluding hurricanes, may reach 15 cm in 24 hr. During hurricanes whose frequency is approximately once in 5 years, rains may reach 30 cm in 24 hr.

The variation between rain and drought has been known to change the salinity in portions of the bay from 40 ppt to fresh in one- or two-day periods. The variability in salinity, temperature, and oxygen for 1974 is shown in Figures 3 to 5 (Holland *et al.*, 1974). The water temperature of the shallow bays follows the air temperature in 24 hr. In the summer the shallow edges and sand flats of the bay may reach 40°C. Winter temperatures, at about 3-year intervals, will go below freezing.

As a result of the wind, temperature, rainfall, and shallow waters, the bays are subject to a wide variety of environmental conditions, all of which may cause abrupt changes.

Geologists generally agree that our coast has consisted of bays and estuaries for the past 180 million years. Subsequent changes in water level and changes in contour due to the rise and fall of the sea during the development and decay of ice periods and other geological phenomena have varied during this period. While our coast has long consisted of bays and estuaries, the present pattern was initiated approximately 30,000 years ago when the sea level dropped more than 120 meters due to the last glaciation period. At this time the rivers cut deep into the coastal plain. Then between 18,000 and 4500 years ago the ice retreated. The sea level rose and barrier sand island and overbank mud sheets were deposited as the rivers continued to meander through the river plain. Thus, as the sea level rose and the barrier islands formed, the river valleys were filled with water, creating the present bays and estuaries. Since that time, depositional processes have continued and the bays have become more shallow. Wind and

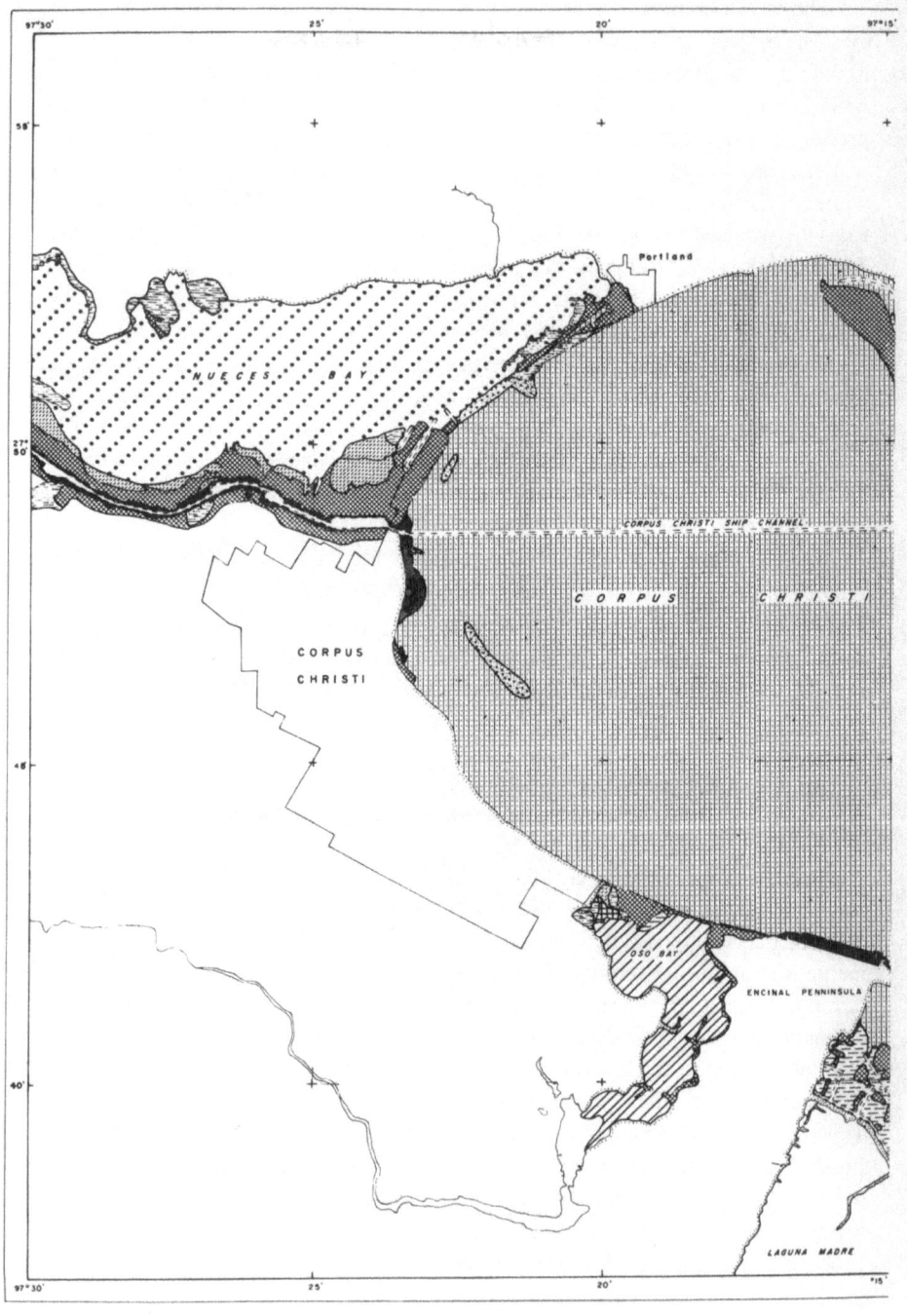

Fig. 2. Biotopes of Corpus Christi Bay.

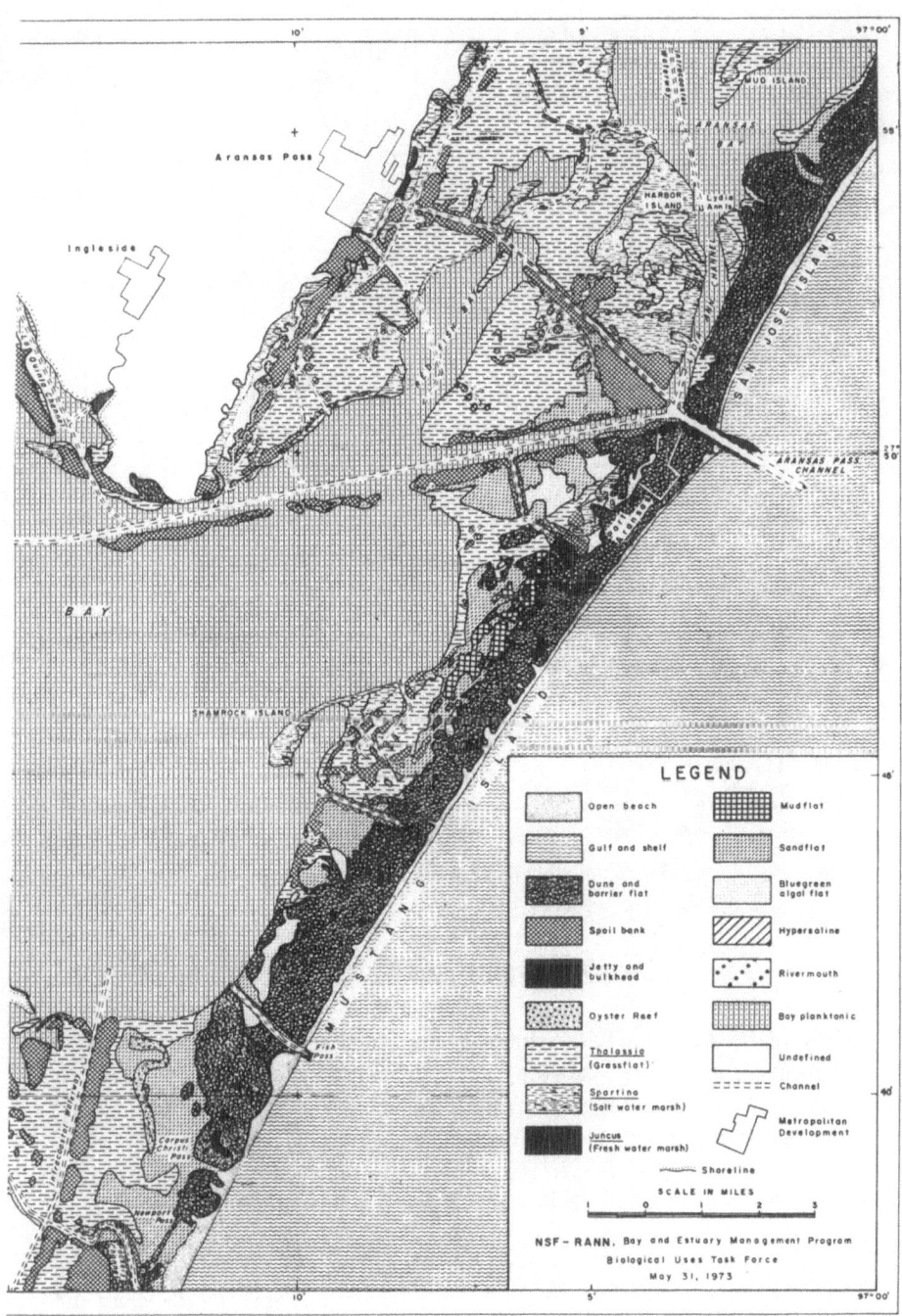

LEGEND

Open beach		Mudflat
Gulf and shelf		Sandflat
Dune and barrier flat		Bluegreen algal flat
Spoil bank		Hypersaline
Jetty and bulkhead		Rivermouth
Oyster Reef		Bay planktonic
Thalassia (Grassflat)		Undefined
Spartina (Salt water marsh)		Channel
Juncus (Fresh water marsh)		Metropolitan Development
		Shoreline

SCALE IN MILES

NSF – RANN, Bay and Estuary Management Program
Biological Uses Task Force
May 31, 1973

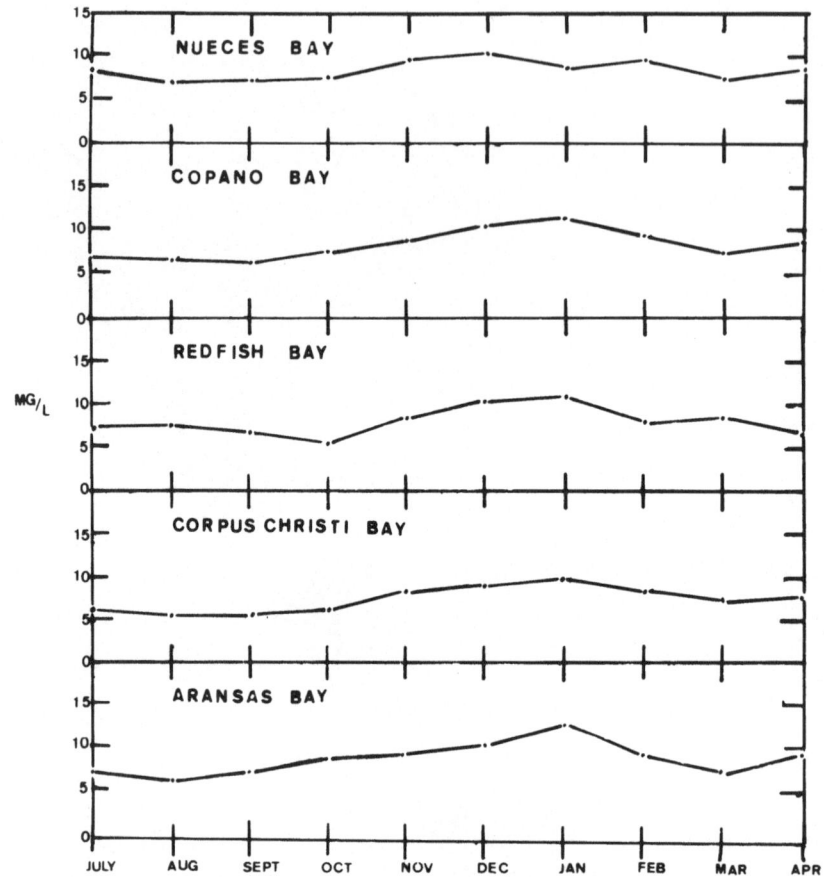

Fig. 3. Monthly mean dissolved oxygen values.

water forces have continually changed the barrier islands, the tidal deltas, and the bays.

During the historical development of the bay systems of Texas and their environmental fluctuations, the living populations have adapted and survived. In general, most of the sport and commercial fishes in the bays have short life cycles and rapid growth characteristics. The oysters will grow to commercial size in 9 to 12 months. The shrimp will grow to prawn size in 12 months. The redfish reach maturity in 3 to 4 years, and at five years they reach a size of 10 to 15 lb.

Most of the commercial and sport fishes in our area have a life cycle between the bays and open Gulf. They spawn in the near offshore area and the larvae and young move into the bays, where they mature into adults to complete the life cycle. A list of significant commercial and sport fishes, found in the bays, are listed with their growth rates in Table I. A typical migration cycle for penaeid shrimp is shown in Figure 6.

The fertility of the bays arises from the adjacent land. The waters entering the Gulf of Mexico through the Yucatan straits are very oligotrophic tropical

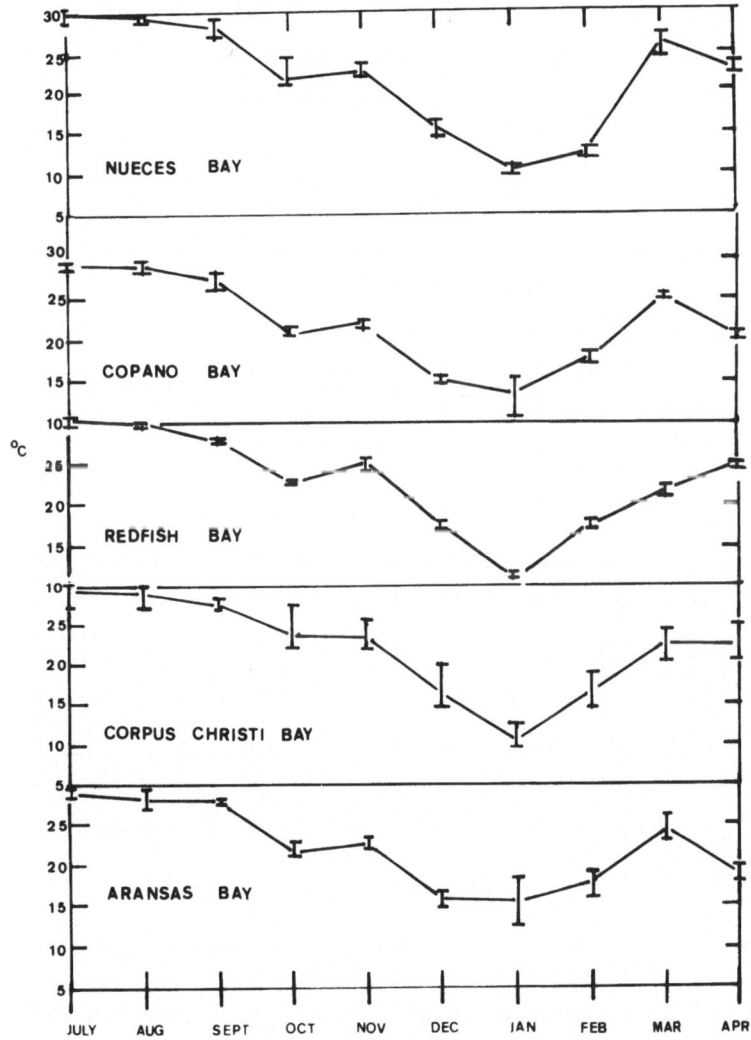

Fig. 4. Monthly mean temperature values, showing minima and maxima.

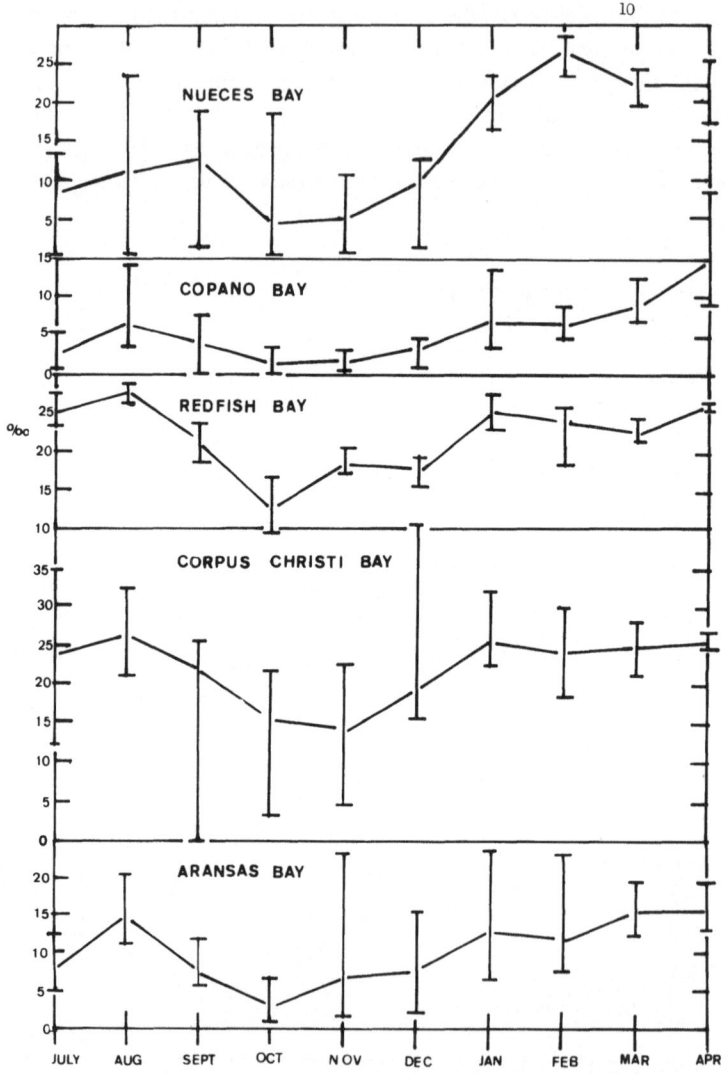

Fig. 5. Monthly mean salinities, showing minima and maxima.

water. While the Gulf produces about 46% of the continental United States fish catch, most of the catch comes from the shelf area. The rivers entering the Gulf drain 75% of the continental United States and approximately 50% of the rain that falls on the continental United States originates from and returns to the Gulf. Therefore the Gulf, which is a semi-inland body of water is continually being fertilized from land.

Table I

Common name	Scientific name	Size/12 months, cm	Years for maturity	Texas, lb caught, 1973	kg
Croaker	Micropogon undulatus	15	3	115,670	52,577.27
Drum black	Pogonias cromis	25	3	1,207,407	548,821.36
Drum redfish	Sciaenops ocellatus	34	4	1,680,147	763,703.18
Flounder	Paralichthys	25	1	354,600	161,181.81
Menhaden	Brevoortia patronus	13	1+	43,059,600[a]	19,572,545.45
Mullet	Mugil cephalus	14	2	168,055[b]	76,388.63
Sea trout, spotted	Cynoscion nebulosus	15	2	1,991,769	905,349.54
Oyster	Crassostrea virginica	7	2 mo.	2,353,785	1,069,902.27
Blue crab	Callinectes sapidus	13	1	6,737,549	3,062,522.27
Shrimp: brown, white, pink	Penaeus sp.	17	9 mo.	80,969,481	36,804,309.54

[a] 1970 values.
[b] Very productive but not a common table fish.

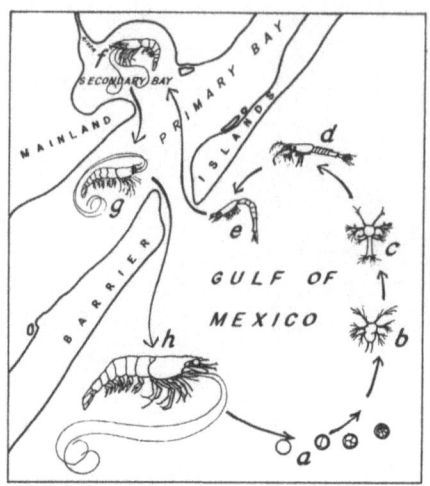

Fig. 6. Life history of a shrimp. (a) shrimp eggs; (b) nauplius larva; (e) protozoea; (d) mysis; (c) postmysis; (f) juvenile shrimp; (g) adolescent shrimp; (h) mature adult shrimp. (a, b, c, d, and e after Heegaard.)

NUTRIENTS

It is generally recognized that the relatively large amount of productivity found in the coastal zone of the Gulf of Mexico has been related to the influence of the adjacent land mass rather than upwelling or some other oceanic nutrient source. A review of the nutrient content of some Texas bays (Table II) reveals that they are similar in nitrogen and phosphorous ratios to other coastal environ-

Table II. Summary of Nutrient Data, Average Concentrations in Milligrams/Liter

Bay system		Nitrogen		Phosphorus	
		Inorganic	Organic	Inorganic	Total
Corpus Christi	1952			0.04	
	1972–73	0.18		0.015	0.04
	1972	0.16	0.62	0.031	0.18
San Antonio	1970–71	0.61		0.13	
	1972–73	0.28		0.16	0.20
Galveston Bay	1968–72				
Trinity Bay		0.39	0.96		0.49
Main Galveston		0.44	0.80		0.55
East Bay		0.16	0.76		0.27
West Bay		0.13	0.60		0.18

Table III. Sources of Nutrient Data

Bay system	Nutrients	Source	Retrieval and analysis

Corpus Christi	PO_4-P
	NH_3, NO_3, NO_2
	PO_4, Total P,
	  Organic N
		Hood, 1953
		J. Holland, unpublished
		  1972–73 surveys
			Manual average
			TWDB Coastal
			Data System

	NH_3, NO_3, NO_2 PO_4, Total P	E. M. Davis, unpublished 1972 surveys	ENVIR Data Bank
San Antonio	NO_2, NO_3, PO_4	Cechova and Davis (1973) 1970–71 survey	Weighted average
	NH_3, NO_2, NO_3 PO_4, Total P	TWDB, unpublished 1972–73 survey	TWDB Coastal Data System

Galveston	NH_3, NO_3
	Total P,
	  Organic N
		Galveston Bay Study,
		  1968–72 surveys
			ENVIR Data Bank

ments (Ryther and Dunstan, 1971; Smith, 1973). The productivity is related primarily to nitrogen, whereas phosphorus is generally in excess over the usual required concentrations found in protoplasm. An analysis of the nitrogen to phosphorus ratios of all available data from the coastal estuaries of Texas indicates that phosphorus is not limiting. Table III shows the data source for the summary in Table II.

Table II shows the relative amounts of nitrogen available to the bays. Such values for nutrients are reflected in the primary productivity for the Texas bays shown in Table IV. Table V shows the sources of nitrogen to the bays. A comparison with Figure 3 indicates that the amount of productivity shown does not materially reduce the oxygen content of the bay.

Table IV. Comparison of Gross Planktonic Production for San Antonio, Corpus Christi, and Galveston Bays

Bay	$g\ C/m^2 day$	lb C/acre day	kg C/year	lb C/year
San Antonio[a]	1.0	8.93	2.12×10^8	4.66×10^8
Corpus Christi[b]	2.52	22.20	3.89×10^8	8.55×10^8
Galveston[c]	5.87	52.46	2.90×10^9	6.37×10^9

[a]Davis, 1971; Jack Nelson, Texas Water Development Board, personal communication.
[b]Davis, 1971; Odum and Odum, 1959; Odum et al., 1958; Odum et al., 1963; Odum and Wilson, 1962; Hellier, 1962; Odum, 1967.
[c]Armstrong and Hinson, 1973; Espey et al., unpublished report; Corliss and Trent, 1971.

Table V. Nitrogen Inflows to Texas Bays

Bay System	Total N input		Relative importance of each source, %		
	10^3 lb/yr		River	Rain	Ind. + Dom.
Corpus Christi avg, 62–71	2,664	19.9	8.5%	12.7%	78.7%
average 3 wet years[a]	4,677	34.9	45.5	9.6	44.8
average 3 dry years[b]	2,419	18.0	4.6	8.6	86.7
San Antonio avg, 61–71	2,610	18.3	83.1%	16.9%	0[c]
Galveston recent?[d]	49,116	144.0	30.7%	3.1%	66.1%
?–74[e]	37,883	111.0	39.8	4.0	56.1

[a]Using 1957, 1958, 1967 with "70" model Industrial and Domestic.
[b]Using 1955, 1962, 1963 with "70" model Industrial and Domestic.
[c]No specific information available, assumed to be very low.
[d]Armstrong and Hinson, 1973.
[e]Galveston Bay Project, unpublished.

The data supplied in these last tables can allow an evaluation of productivity changes that could occur through control of nutrient source. Galveston Bay, with its larger concentration of nutrients, does have the highest commercial fish catch. It does not show any major deterioration of oxygen content and therefore could be used as an example for mariculture practices.

BIOTOPES

During the descriptive evaluation of estuarine environments it is necessary to identify the various types of biological associations in the different areas of the bays. To do this we have utilized a term, *biotopes*, representing recognizable associations of living organisms and their habitats. As shown in Figure 2, nineteen different biotopes have been identified, including both natural and man-produced areas. Several of the biotopes may be of interest when related to possible habitats for warm water mariculture. These habitats can be duplicated in small pilot ponds for experimental use or in larger areas for farms. The following descriptions are provided to show the productive areas in Corpus Christi bay. The organisms shown represent the predominating plants and animals that identify the discrete biotopes. For coastal zone management purposes we have used the biotopes to quantify ecological areas in the bay as shown in Table VI. The following sections and Figures 8 to 13 are descriptions of the biotopes representing the bay planktonic, an oyster reef, *Thalassia* grass flats, *Spartina* salt marsh, *Juncus* fresh water marsh, and a jetty as an artificial reef. Here we must

Table VI. The Biotopes of Corpus Christi and Nueces Bays[a]

Biotope	Acres	Percentage
Open beach	1,980	1.31
Dune and barrier flat	13,358	8.85
Spoil bank	13,327	8.83
Jetty and bulkhead	2,211	1.46
Oyster reef	760	0.50
Thalassia (grassflat)	18,894	12.51
Spartina (saltwater marsh)	7,579	5.02
Juncus (freshwater marsh)	411	0.27
Mudflat	604	0.40
Sandflat	7,348	4.87
Blue-green algal flat	1,208	0.80
Hypersaline	3,033	2.01
Rivermouth	15,755	10.43
Bay planktonic	63,340	41.94
Channel	1,202	0.80
Sum	151,000	100.00

[a]Listed in acres and percentage of the total bays.

explain that the Texas bays do not have rocky environments as found in northern coasts. Jetties, piers, platforms, bulkheads, and marinas then serve as artificial reef structures.

BAY PLANKTONIC

It is difficult, if not impossible, to precisely delimit the geographical boundaries of the bay planktonic biotope because of the spatial and temporal variability exhibited by the plankton. Here the environment is a moving mass of water which may exist at one time as an independent, more or less homogeneous patch, while at other times it may mix indistinguishably into a larger mass. Planktonic organisms, possessing only feeble powers of locomotion, are constrained to travel within these water masses and are restricted from crossing any physical or chemical boundaries. Frolander (1964) shows nine hypothetical positions that might be assumed by an estuarine zooplankton population influenced by tidal phase and time of day while remaining in a given salinity range. These positions are illustrated in Figure 7.

The bay planktonic biotope may vary from a state of great uniformity in chemical and biotic composition to a state in which highly distinctive patches

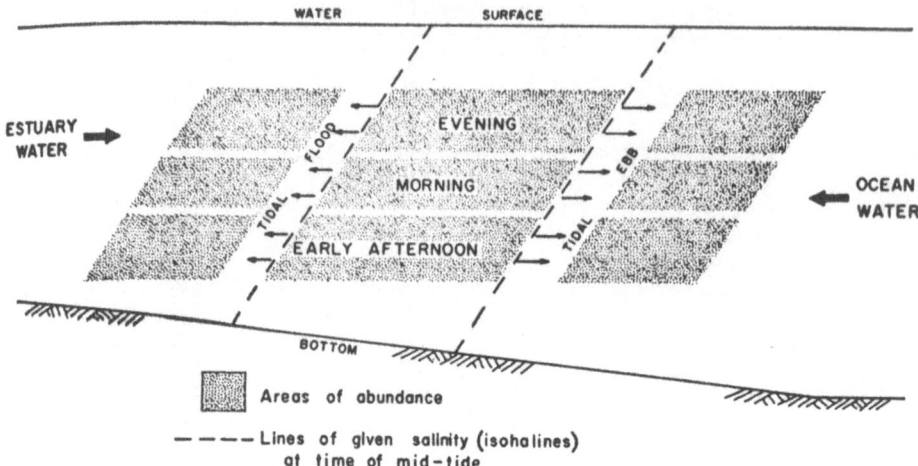

Fig. 7. Nine hypothetical positions that might be assumed by an estuarine zooplankton polpulation influenced by tidal phase and time of day while remaining within a given salinity range.

form a mosaic of different size patches with observable or poorly observable interfaces. An example of a well-defined patch would be a phytoplankton *bloom* (11, in Figure 8).

Phytoplankton are the primary producers within the system and certain plankton associations are the most constant biological feature of the biotope (Fig. 8). Diatoms of the genera *Rhizosolenia* (1), *Asterionella* (2), *Coscinodiscus* (3), *Biddulphia* (4), *Thalassiosira* (17), *Thalassiothrix* (18), *Thalassionema* (19), *Gyrostigma* (20), *Nitzschia* (21), *Skeletonema* (22), and *Actinoptychus* (23), and dinoflagellates of the genera *Ditylum* (6), *Ceratium* (7), and *Peridinium* (8) are microscopic phytoplankton normally present in enormous numbers. Both groups utilize light energy to fix carbon as *food reserves* or incorporate it as integral structural components of the organisms themselves. The fixed carbon of these tiny plants is consumed by barely visible invertebrate zooplankton such as copepods, *Calanus* sp. (24) and *Candacea* sp. (25). In this way organic carbon is moved upward in the food chain as these small copepods (animals) are consumed by even larger animals. Fish and shrimp larvae must have these lower organisms as food sources.

In general, diatoms dominate the winter flora, but share or yield dominance to dinoflagellates during the summer. Nanoflagellates are usually present throughout the year, but may exhibit spring or fall blooms. Higher diversity levels tend to prevail in the lower margins of the bay or estuary, signifying greater variety in

ecological niches. Progressive diminution of diversity up the bay indicates a reduced number of niches resulting from gross pollution or other unfavorable
conditions originating at the end of the bay.

In addition to phytoplankton and zooplankton, larval and postlarval forms
of numerous fish and crustacea (many of commercial importance) contribute to
the total plankton biomass. Depending upon the life history of the species involved, these *meroplankton* may contribute a significant proportion of the primary and secondary consumers in the bay planktonic biotope. It is a well-known
fact that vast numbers of larval and postlarval shrimp, *Penaeus aztecus* (14);
mullet, *Mugil* sp.; spot, *Leiostomus xanthurus* (15); croaker, *Micropogon* sp.;
trout, *Cynoscion arenarius* (13); menhaden, *Brevoortia* sp.; flounder, *Paralichthys*
sp. and *Ancylopsetta quadrocellatus* (16); and redfish, *Sciaenops ocellata* are
found seasonally in this biotope feeding on zooplankton such as *Paracalanus* and
grazing on phytoplankton such as the diatom *Thalassionema* (19) and dinoflagellates such as *Skeletonema* (22) and *Nitzschia* (21).

OYSTER REEF

Wherever currents of sufficient velocity to transport suspended material are
found in combination with solid substrates, sedentary filter-feeding animals tend
to cluster. With time, the hard exoskeletons of these organisms accumulate into
sizable mounds and ridges. Such vertical anomalies formed by the American
oyster, *Crassostrea virginica* (3), and associated organisms constitute the oyster
reef biotope (Figure 9). These reefs occur in all the major Texas bays except
Baffin Bay and Laguna Madre, probably because of a requirement of lower
salinities. In shallow waters the reef may form a low island with a fringe of live
oysters in the intertidal zone, while in deeper waters the reef may form a shoal
rising several feet from the bottom, with live oysters covering its entire surface.
Intertidal oysters will grow at higher salinities than submerged oysters.

Typical associated reef plants in the Texas coastal area are sea lettuce, *Ulva
lactuca* (1); the red alga *Hypnea musciformis* (9); and the green algal genus
Cladophora (8), as shown in Figure 9. Other sessile animals shown in the reef
setting are barnacles, genus *Balanus* (2); anemones, *Bunodosoma cavernata* (4);
various hydroids (25); mussels, *Modiolus americana* (10); and serpulid worms,
genus *Hydroides* (21). Organisms dependent on the shellfish for food include the
Florida rock shell, *Thais haemostoma* (6), a type of oyster drill; and stone crabs,
Menippe mercenaria (15); starfish, *Luidia clathatare* (22); and oyster crabs,
Pinnotheres ostreum (35). Burrowing forms include snapping shrimp, *Alpheus
heterochaelis* (20); boring sponge, genus *Cliona* (19); mud crab, *Panopeus herbstii*

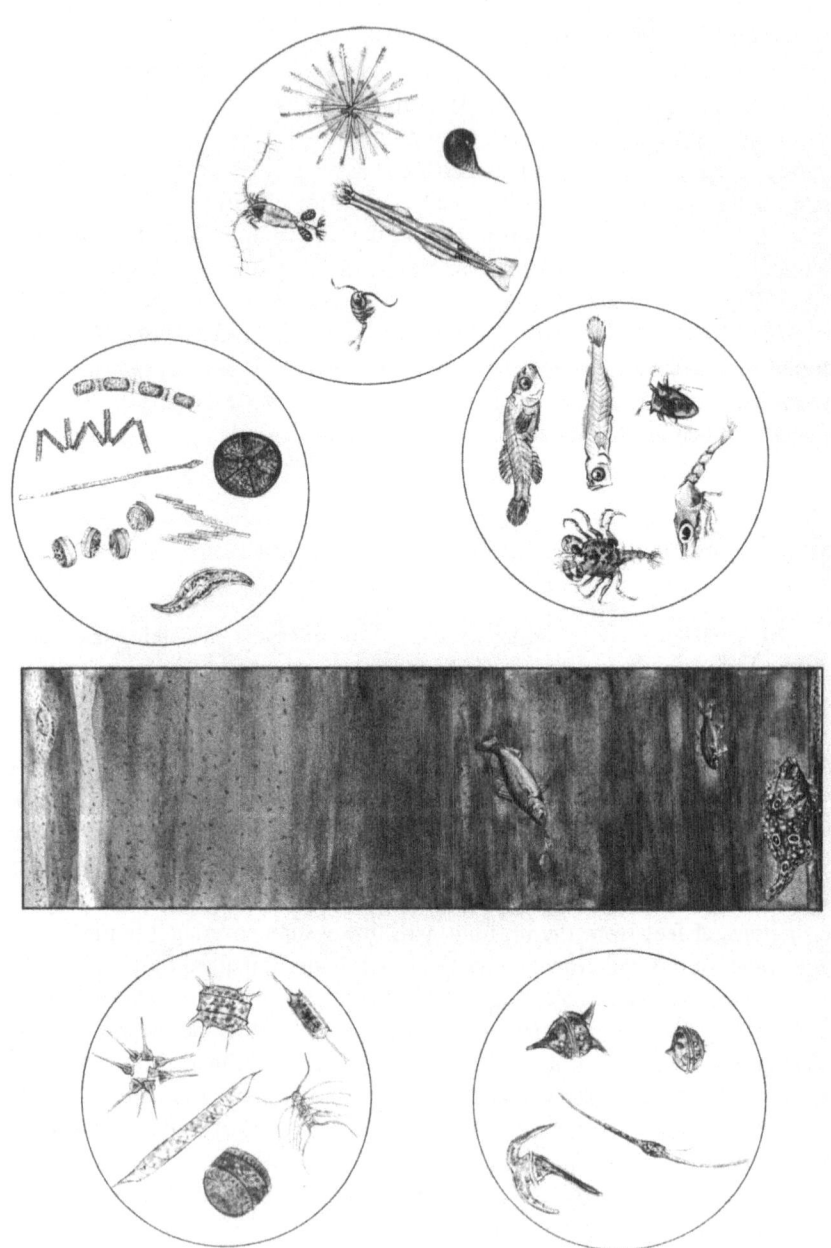

Fig. 8. Bay planktonic.

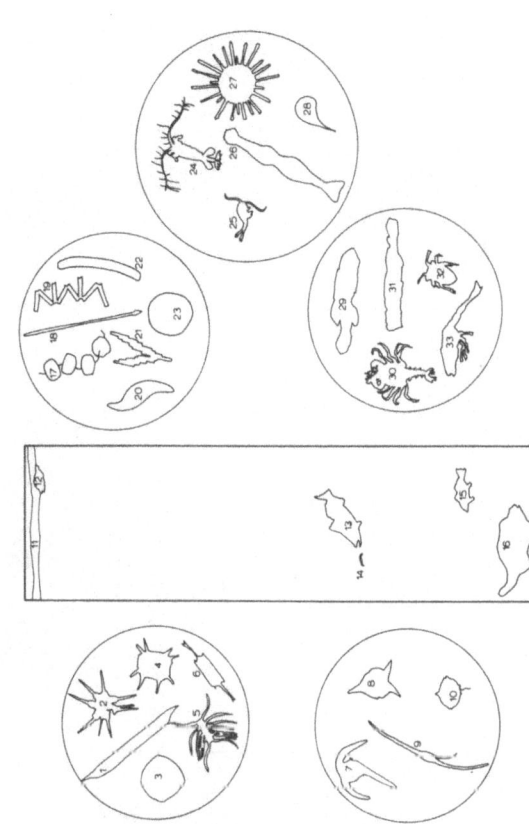

1. *Rhizosolenia styliformis*—Diatom
2. *Asterionella japonica*—Diatom
3. *Coscinodiscus radiatus*—Diatom
4. *Biddulphia mobiliensis*—Diatom
5. *Chaetoceros affinis*—Dinoflagellate
6. *Ditylum brightwellii*—Dinoflagellate
7. *Ceratium tripos*—Dinoflagellate
8. *Peridinium oceanicum*—Dinoflagellate
9. *Ceratium fusus*—Dinoflagellate
10. *Peridinium ornatum*—Dinoflagellate
11. Phytoplankton bloom
12. *Aurelia aurelia*—Jellyfish
13. *Cynoscion arenarius*—Sand trout
14. *Penaeus aztecus*—Brown shrimp
15. *Leiostomus xanthurus*—Spot
16. *Ancylopsetta quadrocellatus*—Flounder
17. *Thalassiosira decipiens*—Diatom
18. *Thalassiothrix longissima*—Diatom
19. *Thalassionema nitzschioides*—Diatom
20. *Gyrostigma* sp.—Diatom
21. *Nitzschia paradoxia*—Diatom
22. *Skeletonema costatum*—Diatom
23. *Actinoptychus undulatus*—Diatom
24. *Catanus* sp.—Copepod
25. *Candacea* sp.—Copepod
26. *Sagitta macrocephla*—Arrow worm
27. *Aulacantha scolymantha*—Siliculose amoeba
28. Foraminifera
29. Larva of *Orthopristis chrysoptera*—Pigfish
30. Megalops stage of *Carcinus maenus*—Crab
31. Larva of *Lagodon rhomboides*—Pinfish
32. Nauplius of *Balanus*—Barnacle
33. Zoea stage of *Pagurus*—Hermit crab

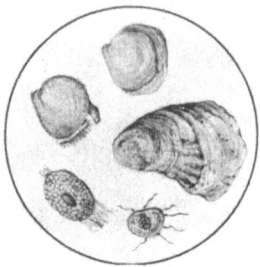 

Fig. 9. Oyster reef.

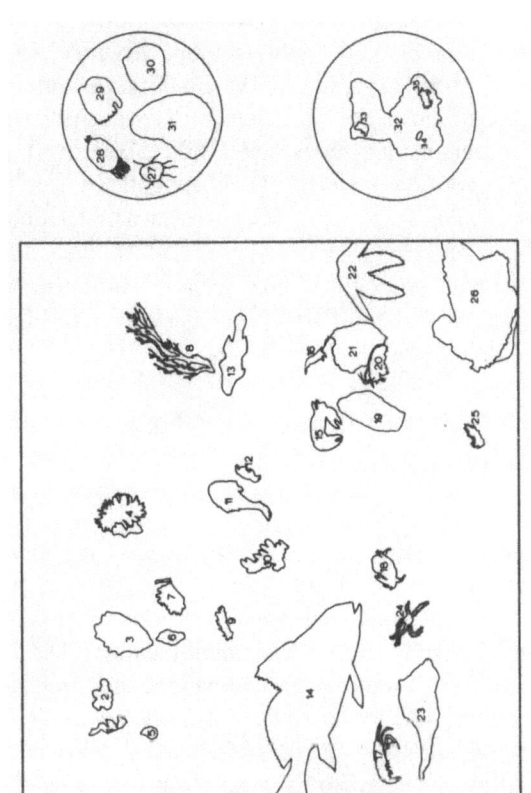

1. *Ulva lactuca* — Sea lettuce
2. *Balanus* sp. — Barnacle
3. *Crassostrea virginica* — Oyster
4. *Bunodosoma cavernata* — Anemone
5. *Ischnochiton papillosus* — Chiton
6. *Thais haemostoma* — Florida rock shell
7. *T. haemostoma* eggs
8. *Cladophora* sp. — Green alga
9. *Hypnea musciformis* — Red alga
10. *Modiolus americana* — Mussel
11. *Gobiesox strumosus* — Skillet fish
12. *Eurypanopeus depressus* — Flat mud crab
13. *Hypleurochilus geminatus* — Crested blenny
14. *Pogonias cromis* — Black drum
15. *Menippe mercenaria* — Stone crab
16. *Palaemontes* sp. — Grass shrimp
17. *Alpheus heterochaelis* — Snapping shrimp
18. *Panopeus herbstii* — Mud crab
19. *Cliona* sp. — Boring sponge
20. *Alpheus heterochaelis* — Snapping shrimp
21. *Hydroides* sp. — Serpulid worms
22. *Luidia clathatare* — Starfish
23. *Busycon contrarium* — Whelk
24. *Ophiothrix* sp. — Brittle star
25. Hydroid
26. *Opsanus beta* — Gulf toadfish
27. Oyster egg undergoing fertilization
28. Beginning of shell formation
29. Last free-swimming stage
30. Spat 5-6 hours after settling
31. Adult *Crassostrea virginica*
32. *Crassostrea virginica* — American oyster

33. *Polydora* sp. — Polychaete
34. *Diplothyra smithii* — Boring clam
35. *Pinnotheres ostreum* — Oyster crab

(27-31) Stages in the development of *Crassostrea virginica*

(18); flat mud crab, *Eurypanopeus depressus* (12); polychaete worms of the genus *Polydora* sp. (33); and the boring clam, *Diplothyra smithii* (34). The chiton, *Ischnochiton papillosus* (5); grass shrimp, genus *Palaemonetes* (16); brittle star, genus *Ophiothrix* (24); and the whelk, *Busycon contrarium* (23) are the predominant grazers shown for this biotope. Several small fish are found associated with the reef, among them skillet fish, *Gobiesox strumosus* (11); crested blenny, *Hypleurochilus geminatus* (13); and gulf toadfish, *Opsanus beta* (26). The black drum, *Pogonias cromis* (14), is known to feed on oysters and other shellfish.

When the reef is exposed, various birds such as white pelicans, *Pelecanus erythrorhynchus;* great blue heron, *Ardea herodias;* and laughing gull, *Larus atricilla*, use it as a resting place.

THALASSIA GRASSFLAT

This extensive and productive biotope is characteristically composed of moderate to dense growths of turtle grass, *Thalassia testudinum* (22); shoal grass, *Halodule wrightii* (20); *Halophila engelmanii* (19); and widgeon grass, *Ruppia maritima*, as shown in Figure 10 (*R. maritima* not shown). The distribution is usually in one to five feet of water along the margins and throughout bays and lagoons. Depths are controlled by turbidity of the water which limits light penetration. Combined with the heavy growths of attached plants and animals, the biomass represented by the grassflats is large. When the plants die back in autumn, the leaves and stems break off and are distributed among the other biotopes, where the material, whether grazed or decomposed, makes significant contributions to the food chain. The growth offers protection and is generally thought of as the major nursery area for the young of many species of fish and crustaceans.

The grass acts as a surface for many invertebrates and microalgae such as diatoms. This adds to the productivity of the area. The sediments, because of the quieting action of the grasses, are generally soft and anaerobic due to entrapment of organic matter.

Because of the seasonal and diurnal fluctuations in temperature and migratory habits, few highly motile animals are found in this biotope on a permanent basis. Among the sedentary species found are large numbers of bryozoans (not shown), hydroids (4), and serpulid worms of the genus *Spirorbus* (5, 6). These organisms share the leaves and stems with equally large numbers of sessile diatoms such as *Cocconeis* sp. (not shown).

Many of the motile forms in this biotope are omnivores, which function as both scavengers and grazers. These include the horn shell, *Cerithidea pliculosa*

(8); olive nerite, *Neritina reclivata* (9); and a small gastropod, *Odostomia gibbosa* (15), as shown; as well as *Melampus* sp. and *Modulus* sp., among the gastropods. Crustacean members shown for this group are the grass shrimp, *Palaemonetes vulgaris* (7); hermit crab, *Clibanarius vittatus* (16); mud crab, *Neopanope texana* (17); blue crab, *Callinectes sapidus* (18); a crab known as *Rhitropanopeus harrisii* (24); the brown and pink shrimps, *Penaeus aztecus* (2) and *P. duorarum* (27); as well as the white shrimp, *Penaeus setiferus,* which is not shown. The shrimp appear in the grassflats as early larval stages and use the cover and food of this biotope as a nursery, migrating offshore to spawn upon maturity. Many larval fish species develop in the protection of this biotope, as well. Final members of this group, as shown, are the sea cucumber, genus *Thyone* (13); the brittle star, genus *Ophiothrix* (14); and the mud worm, *Phascolosoma gouldii* (28).

The burrowing forms of this biotope are the razor clam, *Ensis minor* (23); Venus clam, *Chione cancellata* (25); and Lucina clam, *Phacoides pectinatus* (26), as shown; as well as those of the genera *Tellina, Tagelus,* and *Laevicardium.*

Many fish frequent the grassflats. These include pinfish, *Lagodon rhomboides* (1); spotted sea trout, *Cynoscion nebulosus* (3); tidewater silversides, *Minidia beryllina* (11); redfish, *Sciaenops ocellata* (12); as well as golden croaker, *Micropogon undulatus*; mullets, *Mugil cephalus* and *M. curema*; and menhaden, *Brevoortia patronus.*

Several algae are represented from this biotope in addition to those mentioned as epiphytes. These include the large red alga, *Gracilaria* (19); the diatoms *Nitzschia* (30) and *Cymbella* (31); the dinoflagellate *Ceratium* (29); the euglenoid *Dunaliella* (33); the blue-green *Oscillatoria* (32); and the colonial green alga, *Microcystis* (34, 35).

SPARTINA (SALT WATER MARSH)

This biotope (Figure 11) is subjected to intermittent inundation due to tidal action. Fluctuations in temperature, salinity, water depth, and sediment have exerted a strong selective effect, limiting the numbers of organisms found. The dominant grass in this biotope is smooth cordgrass, *Spartina alterniflora* (11). Like the grassflat biotope, the plant material produced in this biotope, mostly *S. alterniflora* (11), makes a large contribution to the food chain of the estuarine ecosystem. The sediments may range from fine anaerobic silt to sand or shell. Occasionally oyster reefs are found in this biotope. The productivity of the area is high and the grass blades offer protection and attachment for many organisms below and above water. The decayed grass adds to the fertility of the surrounding water areas.

Fig. 10. *Thalassia* grassflat.

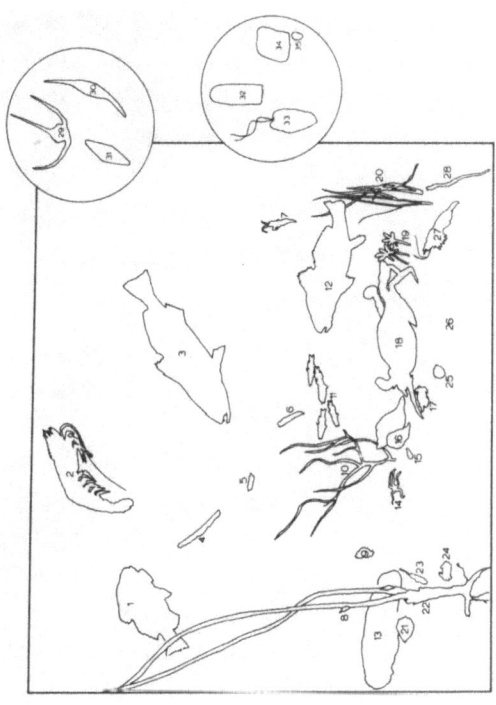

1. *Lagodon rhomboides*—Pinfish
2. *Penaeus aztecus*—Brown shrimp
3. *Cynoscion nebulosus*—Spotted sea trout
4. Hydrozoan
5. *Spirorbus* sp.—Serpulid worm
6. *Spirorbus* sp.—Serpulid worm
7. *Palaemonetes vulgaris*—Grass shrimp
8. *Cerithidea pliculosa*—Horn shell
9. *Neritina reclivata*—Olive nerite
10. *Gracilaria* sp.—Red alga
11. *Minidea beryllina*—Tidewater silverside
12. *Sciaenops ocellata*—Juvenile redfish
13. *Thyone* sp.—Sea cucumber
14. *Ophiothrix* sp.—Brittle star
15. *Odostomia gibbosa*—Small gastropod
16. *Clibanarius vittatus*—Hermit crab
17. *Neopanope texana*—Mud crab
18. *Callinectes sapidus*—Blue crab
19. *Halophila engelmannii*—Sea grass
20. *Halodule wrightii*—Shoal grass
21. *Phacoides pectinatus*—Lucina clam
22. *Thalassia testudinum*—Turtle grass
23. *Ensis minor*—Razor clam
24. *Rhitropanopeus harrisii*—Burrowing crab
25. *Chione cancellata*—Venus clam
26. *Phacoides pectinatus*—Lucina clam
27. *Penaeus duroarum*—Pink shrimp
28. *Phascolosoma gouldii*—Mud worm
29. *Ceratium* sp.—Dinoflagellate
30. *Nitzschia* sp.—Diatom
31. *Cymbella* sp.—Diatom
32. *Oscillatoria* sp.—Blue-green alga
33. *Dunaliella paupera*—Saline euglenoid
34. *Microcystis* sp. (colony)—Green algae
35. *Microcystis* sp. (individual)—Green alga

Fig. 11. *Spartina* (saltwater marsh).

1. *Ardea herodias*—Blue heron
2. *Butorides virescens*—Green heron
3. *Anas discors*—Blue-winged teal
4. *Ajaia ajaja*—Roseate spoonbill
5. *Casmerodius albus*—Common egret
6. *Avicennia germinans*—Black mangrove
7. *Eudocimus albus*—White ibis
8. *Salicornia bigelovii*—Woody glasswort
9. *Procyon lotor*—Raccoon
10. *Distichlis spicata*—Saltgrass
11. *Spartina alterniflora*—Smooth cordgrass
12. *Rallus longirostris*—Clapper rail
13. *Pagurus*—Hermit crab
14. *Telmatodytes pulustris*—Longbilled marsh wren
15. *Croton punctatus*—Beach tea
16. *Sesuvium portulacastrum*—Sea purslane
17. *Batis maritima*—Saltwort
18. *Uca pugnax*—Fiddler crab
19. *Avicennia germinans*—Black mangrove
20. *Littorina irrorata*—Periwinkle
21. *Avicennia germinans*—Black mangrove
22. *Distichlis spicata*—Saltgrass

Other common plants shown in Figure 11 for this biotope are the woody glasswort, *Salicornia bigelovii* (8); and saltwort, *Batis maritima* (17), in the lower areas; and beach tea, *Croton punctatus* (15); saltgrass, *Distichlis spicata* (22); sea purslane, *Sesuvium portulacastrum* (16); and black mangrove, *Avicennia germinans* (6, 19, 21), in the higher, better drained areas.

There are numerous birds that nest or feed in this biotope. Those shown are the great blue heron, *Ardea herodias* (1); green heron, *Butorides virescens* (2); blue-winged teal, *Anas discors* (3); roseate spoonbill, *Ajaia ajaja* (4); common egret, *Casmerodius albus* (5); white ibis, *Eudocimus albus* (7); clapper rail, *Rallus longirostris* (12); and longbilled marsh wren, *Telmatodytes pulustris* (14).

Grazing and scavenging are accomplished by a variety of animals. Those shown include the hermit crabs, *Pagurus* (13); the fiddler crab, *Uca pugnax* (18); and the periwinkle, *Littorina irrorata* (20). The raccoon, *Procyon lotor* (9), is a common visitor, feeding on such shellfish as mussels, cockles, and snails. In the substrate there are untold numbers of annelid and nematode worms, soil arthropods, and bacteria which contribute to final decomposition of detritus.

JUNCUS (FRESHWATER MARSH)

The freshwater marsh biotope is found in permanent freshwater ponding or river areas which are maintained by permanently high water table levels or high rainfall. The dominant vegetation are reeds, genus *Juncus* (4), and bulrush, genus *Scirpus* (5, 12, 20), as shown in Figure 12. Also found here are the cordgrasses, *Spartina alterniflora* and *S. patens* (14), as well as cattails, genus *Typha* (11, 21), and green briars, *Smilax* sp. (10). In areas where there is a salinity gradient, the community composition changes along the gradient into a *Spartina*-dominated salt marsh. The sediments are usually soft mud, often anaerobic due to high organic content. The boundary area is often characterized by the submerged grass *Ruppia maritima* (not shown).

The large amounts of plant material produced annually (estimated at 20,000 lb/acre, Odum and Odum, 1959) provide food and nesting areas for many waterfowl. Among these are the Canada goose, *Branta canadensis* (1); green heron, *Butorides virescens* (2); coot, *Fulica americana* (8); and wood stork, *Mycteria americana* (9). The crustaceans are also represented in the freshwater marsh, with crayfish, *Procambarus clarkii* (7, 17) feeding on the abundant detritus produced. The sheepshead minnow, *Cyprinodon variegatus* (18), also feeds on this material. Common terrestrial vertebrate inhabitants are the western diamondback rattlesnake, *Crotalus atrox* (15); the cottonmouth, *Agkistrodon piscivorus* (19); the opossum, *Didelphis marsupialis* (13); and the Norway rat, *Rattus norvegicus* (6).

With the flushing action due to high tides and heavy runoff, much of the detrital material and bacterial decomposition products are introduced into the economy of the bay. Along drainage channels where there is an intertidal interface, the fiddler crab, *Uca pugnax* (16), predominates along the banks, and the clams, *Mercenaria mercenaria* and *Tagelus divisus* (not shown), on the channel bottoms. Also found, but not shown, is the marsh periwinkle, *Littorina irrorata*, which feeds on the grasses.

JETTY AND BULKHEAD

Jetties and bulkheads are man-made structures of rock, shell, concrete, wood, and steel, placed to restrict sedimentation in channels or to provide docking areas. As a result, these structures are built in areas where there is variable current energy and they offer a surface and protection to a wide variety of organisms. Salinity does control the populations. Therefore, our illustration depicts organisms adapted to salinities above 15 ppt. Thus, most of the forms which inhabit them are either adapted to clinging, physically fixed to the substrate, or free swimming. The flora are predominantly brown, red, and green algae, with some blue-green algae in the splash zone. The fauna represent a wide variety of animals.

The dominant green algae pictured in Figure 13 are of the genera *Ulva* (14), *Enteromorpha* (15), *Cladophora* (13), and *Chaetomorpha* (8). The dominant brown alga is of the genus *Padina* (22) with some *Dictyota* (18). The dominant red alga shown is of the genus *Agardhiella* (21), with *Hypnea* (20), *Gelidium* (9), *Giffordia* (16), *Bryocladia* (6), *Gracilaria* (27), and *Rhodymenia* (24). All of these forms are firmly attached to the rocks and are highly flexible in order to withstand the rigors found on the jetties.

The attached fauna shown are sponges, coelenterates, two mollusks and a crustacean. The sponges are of the genera *Microciona* (25, 26) and *Haliciona* (38). The coelenterates are the anemone, *Bunodosoma cavernata* (23); sea whip, *Leptogorgia setacea* (36); and the remains of an alcyonarian, *Oculina* sp. (37), a sessile anthozoan. The oyster, *Crassostrea virginica* (10); mussel, *Modiolus americanus* (42); and barnacles of the genus *Balanus* (1) complete the range of attached animals shown from this biotope.

Motile forms which cling to the substrate include the gastropods *Thais haemostoma* (41) and *Littorina irrorata* (5); the stone crab, *Menippe mercenaria* (35); hermit crab, *Clibanarius vittatus* (28); the sea urchin, *Arbacia punctulata* (32); and the isopod wharf roach, *Lygia exotica* (4). The crested blenny, *Hypleurochilus geminatus* (11), lives in the sheltered cracks of the jetties.

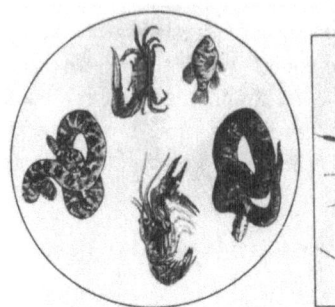
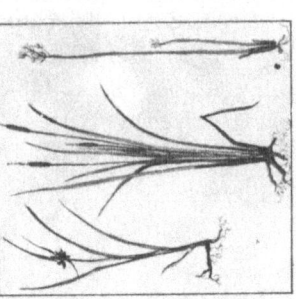

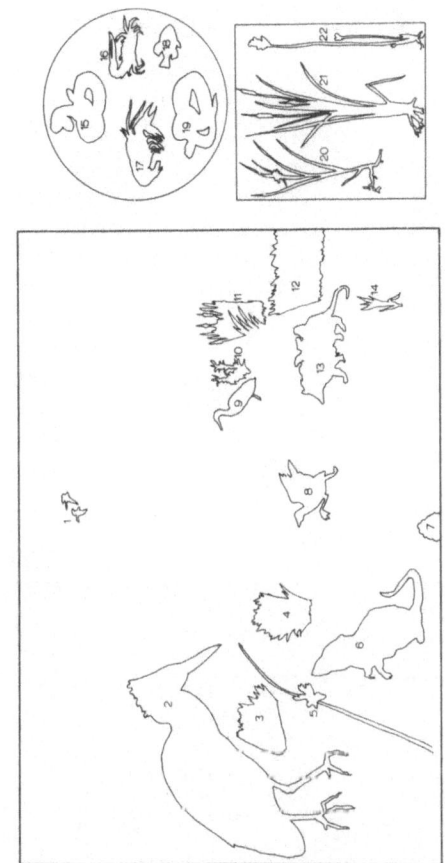

Fig. 12. *Juncus* (freshwater marsh).

1. *Branta canadensis*—Canadian geese
2. *Butorides virescens*—Green heron
3. *Spartina alterniflora*—Smooth cordgrass
4. *Juncus* sp.—Reed
5. *Scirpus* sp.—Bulrush
6. *Rattus norvegicus*—Norway rat
7. *Procambarus* burrow
8. *Fulica americana*—Coot
9. *Mycteria americana*—Wood stork
10. *Smilax* sp.—Greenbriars
11. *Typha domingensis*—Cattails
12. *Scirpus* sp.—Bulrush
13. *Didelphis marsupialis*—Opossum and young
14. *Spartina patens*—Marshhay cordgrass
15. *Crotalus atrox*—Western diamondback
     rattlesnake
16. *Uca pugnax*—Fiddler crab
17. *Procambarus clarkii*—Crayfish
18. *Cyprinodon variegatus*—Sheepshead minnow
19. *Agkistrodon piscivorus*—Cottonmouth snake
20. *Scirpus* sp.—Bulrush
21. *Typha domingensis*—Cattail
22. *Sporobolus virginicus*—Seashore dropseed

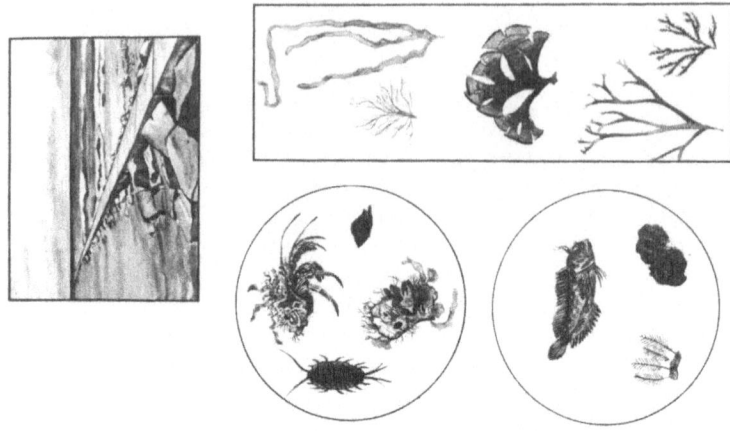

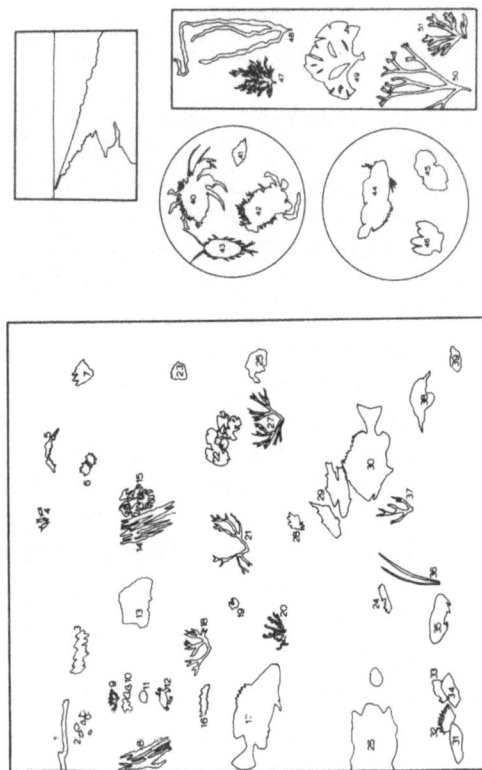

Fig. 13. Jetty and bulkhead.

1. *Balanus* sp.—Barnacle
2. *Thais haemostoma*—Florida rock shell
3. *Enteromorpha flexosa*—Green alga
4. *Lygia exotica*—Wharf roach
5. *Littorina irrorata*—Periwinkle
6. *Bryocladia cuspidata*—Red alga
7. *Ulva lactuca*—Green alga
8. *Chaetomorpha* sp.—Green alga
9. *Gelidium* sp.—Red alga
10. *Crassostrea virginica*—American oyster
11. *Hypleurochilus geminatus*—Crested blenny
12. *Callinectes sapidus*—Blue crab
13. *Cladophora vagabunda*—Green alga
14. *Ulva fasciolata*—Green alga
15. *Enteromorpha lingulata*—Green alga
16. *Giffordia* sp.—Red alga
17. *Promicrops itaiara*—Spotted jewfish
18. *Dictyota dichotoma*—Brown alga
19. *Ovalipes ocellatus*—Swimming crab
20. *Hypnea musciformis*—Red alga
21. *Agardhiella tenera*—Red alga
22. *Padina vickersiae*—Brown alga
23. *Bunodosoma cavernata*—Anemone
24. *Rhodymenia palmata*—Red alga
25. *Microciona* sp.—Sponge
26. *Microciona* sp.—Sponge
27. *Gracilaria prolifera*—Red alga
28. *Clibanarius vittatus*—Hermit crab
29. *Mugil cephalus*—Striped mullet
30. *Archosargus probatocephalus*—Sheepshead
31. White sponge
32. *Arbacia punctulata*—Urchin
33. Hydroid
34. Yellow sponge
35. *Menippe mercenaria*—Stone crab
36. *Leptogorgia setacea*—Sea whip (octocoral)
37. *Oculina* sp.—Hard coral
38. *Haliciona* sp.—Pink sponge
39. *Microciona* sp.—Sponge
40. *Clibanarius vittatus*—Hermit crab
21. *Thais haemostoma*—Florida rock shell
42. *Modiolus* sp.—Mussel and attachments
43. *Lygia exotica*—Wharf roach
44. *Blennius cristatus*—Molly miller
45. *Microciona* sp.—Orange sponge
46. Hydroid
47. *Cladophora vagabunda*—Green alga
48. *Ulva flexosa*—Green alga
49. *Padina vickersiae*—Brown alga
50. *Dictyota dichotoma*—Brown alga
51. *Bryocladia cuspidata*—Red alga

Strongly swimming forms shown include the spotted jewfish, *Promicrops itaiara* (17); sheepshead, *Archosargus probatocephalus* (30); striped mullet, *Mugil cephalus* (29); blue crab, *Callinectes sapidus* (12); and another portunid crab, *Ovalipes ocellatus* (19).

The high productivity and the wide variations in the bay environments have resulted in an interesting relationship between behavior and the food chain. The most productive commercial species—the oyster, blue crab, mullet, menhaden, and shrimp—that make up approximately 90% of the catch, are all omnivores. While these organisms do utilize higher trophic levels, they can sustain themselves on primary producers. This is generally true with an exception, perhaps, for the very early larval stages where some specific food requirements such as rotifers may be needed. The adult organisms then form the food for the other species of carnivores. These data suggest that the environment maintains its high fish productivity, even though it is continually subject to wide changes in temperature and salinity and is in an area of active commercial and urban activity.

A similar effect can be noted for the behavioral patterns of the fish. They appear to be able to evade less desirable habitat changes. An example of the temperature/salinity tolerance of the trout is shown in Figure 14 taken from Copeland and Bechtel (1974).

LIFE HISTORY DATA BANK INFORMATION

During current studies on management criteria for estuarine management, a computerized life history data bank has been assembled from the published literature (Brogden *et al.*, 1974; Oppenheimer *et al.*, 1974). A printout of the food preferences for several of the most significant commercial species of Texas is shown in Table VII. The data bank reviewed 6233 items to provide the list and took approximately 60 seconds of computer time. The large number of references in the bank are not listed in this publication. The data bank includes similar information about all the commercial and many of the noncommercial species of organisms in our area. Other parameters are included, such as shown by Table VIII on the temperature/salinity ranges of the species listed in Table VII. The data bank is also correlated with the biotopes.

The information from Tables I and VII provides a summary of the relative significance of each of the commercial species of marine organisms. Approximately 90% of those listed are omnivores that feed on the primary producers, detritus, and the first trophic level. The croaker and redfish are not included in

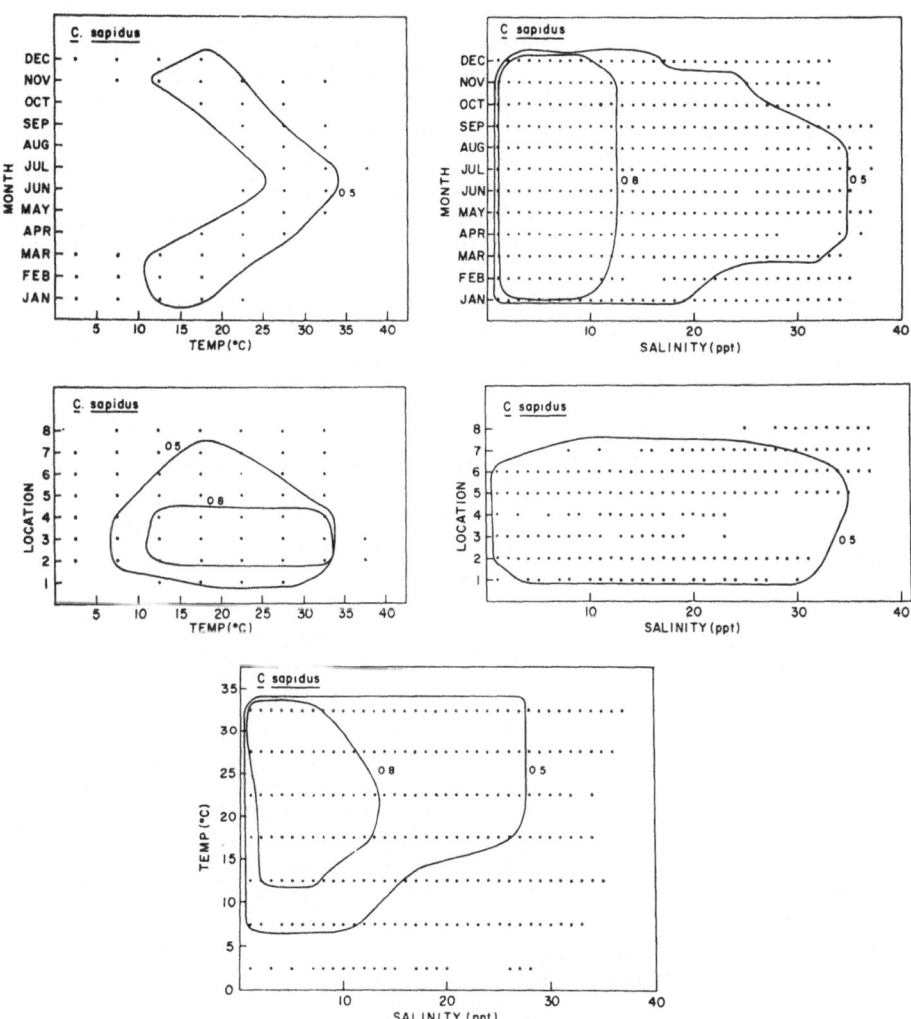

Fig. 14. Catch ratio for the interactions of indicated parameters for *C. sapidus*. The isopleths encompass a catch ratio at or above the indicated values.

Table VII. Food Items of Some Sport and Commercial Species[a]

Pink shrimp (*Penaeus duorarum*)
Adult	Major food	crustaceans, larval fish, polychaetes
	Minor food	dinoflagallates, foraminifera
Juvenile	Other food	algae, amphipods, carideans, caridean eggs, copepods, diatoms, dinoflagellates, foraminifera, isopods, molluscans, mysids, nematodes, ostracods, plant pieces
Larval	Other food	microplankton

White shrimp (*Penaeus setiferus*)
Adult	Major food	detritus, microcrustaceans, Rangia
	Other food	angle worms, blue-green algae, bryozoans, crustacean chitin fragments, fish eggs, foraminifera, green algae, plant fragments, seed pods, shrimp, small gastropods, sponges
Juvenile	Other food	blue-green algae, diatoms, plant pieces

Striped mullet (*Mugil cephalus*)
Adult	Major food	mud, detritus, rooted plants
	Minor food	Biddulphia, diatoms, snails

Blue crab (*Callinectes sapidus*)
Adult and juvenile	Major food	algae, barnacles, clams, crabs, detritus, fish, insects, marsh grass, mussels, snails
Postlarval	Major food	algae, barnacles, clams, crabs, detritus, fish, insects, marsh grass, mussels, snails
Zoea	Major food	dinoflagellates

Southern flounder (*Paralichthys lethostigma*)
Adult	Major food	anchoa, crabs, Micropogon, Rangia
	Minor food	Brevoortia, crabs

Black drum (*Pogonias cromis*)
Adult	Major food	Aquipecten, amphipods, Anomalcardia, branchiodontes, Callinectes, crustaceans, Gobiosoms, Levicardium, molluscans, Mulinia, oysters, penaeid shrimp, Rangia, snake eels, vegetation
	Minor food	annelids, Bulla, clam worms, Crassostrea, cyprinodonts, Donax, fish, grass shrimp insects
Juvenile	Major food	annelids, crustacenas, Callinectes, pelecypods, Penaeus, Tagelus

Gulf menhaden (*Brevoortia patronus*)
Adult	Major food	diatoms, microcrustaceans, organic sediments, phytoplankton, zooplankton
Juvenile	Major food	diatoms, microcrustaceans, organic sediments, plankton

Atlantic croaker (*Micropogon unulatus*)
Adult	Major food	crabs, detritus, fish, molluscans, Rangia, shrimp
	Minor food	Anchoa, annelids, benthos, crabs, fish, isopods, organic detritus, Penaeus, plant pieces, polychaetes, zooplankton
Juvenile	Major food	amphipods, copepods, crustaceans, detritus, isopods

Table VII. (*Continued*)

Sand seatrout (*Cynoscion arenarius*)		
Adult	Major food	Anchoa, Brevoortia
	Minor food	detritus, Penaeus
Juvenile	Major food	copepods, microcrustaceans, schizopods, shrimp, shrimp larvae, small fish
	Minor food	amphipods, crabs, Palaemonetes
Postlarval	Major food	copepods, larval crabs, larval shrimp
Spotted seatrout (*Cynoscion nebulosus*)		
Adult	Major food	Lagodon, Penaeus, Anchoa
	Minor food	Bairdiella, caridean shrimp, Chloroscombrus, Cynoscion, Gobiosoma, Menidia, Mugil Palaeomon, Palaeomonetes
Juvenile	Major food	Angasia, Hippolyte, Periclimenes
	Minor food	ampnipods, Anchoa, Crangon, fish, Palaemon, Palaemonetes, penaeid shrimp
Newly hatched	Major food	copepods
Ripe adult	Minor food	caridean shrimp, mysids
	Major food	Lagodon, Penaeus
Red drum (*Sciaenops ocellata*)		
Adult	Major food	blenny, Callinectes, Crangon, Eucinostomus, Eupagurus, killifish, Libinia, menhaden, Menidia, Menippe, mud crabs, mullet, Palaemonetes, Paralichthys, penaeis shrimp
Juvenile	Major food	ampnipods, copepods, crabs, fingerling mullet, gobies, Palaemonetes, penaeid shrimp, sheepshead minnow, silversides

[a]Partial listing from Computerized Life History Data Bank, University of Texas Marine Science Institute.

the list, even though they are bottom feeders and eat grasses and other detritus (Gunter, 1945). The tables on growth, catch, and food preference could be used to select those marine organisms that are the most efficient for growth and most significant with respect to food requirements. Those species of omnivores would be the most efficient, theoretically, for mariculture development.

A philosophy can be developed for mariculture that is based on the best probable technological transition and utilization of carbon. Normally the omnivore fishes are the least desirable for food. Perhaps the concept can be developed in mariculture to produce these fish on sewage or other waste carbon and then transfer the fish to the food technologist who would strip the protein and retexturize it to some edible material. Thus, esthetic and sanitation problems would be eliminated by the chemist.

C. H. Oppenheimer and W. B. Brogden

Table VIII. Salinity/Temperature Limits for Selected Marine Species in the Gulf Coast[a]

Adult	Salinity				Temperature[b]					
	Range observed, ‰		"Optimum" or preferred range, ‰		Range observed		"Optimum" or preferred range		Lethal under certain conditions	
	Min.	Max.	Min.	Max.	Min.	Max.	Min.	Max.	Min.	Max.
Pink shrimp	0.6	60.0	20.0		10.8	35.5	18.0	26.0		
White shrimp	0.1	57.0			9.0	36.0			4.0°C	
Striped mullet	0.0	75.0			10.7	34.5				
Southern flounder	2.0	60.0	25.0		9.9	30.5				
Black drum	2.6	75.0	15.0–20.0		3.0	35.0	16.0	25.0		
Atlantic croaker	2.0	70.0			3.0	35.0			3.3°C	
Sand seatrout	4.8	45.0	15.0–25.0		13.7	36.7				
Spotted seatrout	2.3	75.0	30.0		4.0	34.9				45°C
Red drum	2.1	50.0	10.0–35.0		8.1	33.0	11.0	30.0		
Blue crab	2.8	60.0			10.0	34.0				
Gulf menhaden	0.5	60.0			9.1	31.6				

[a] Data taken from Life History Data Bank. Literature citation in computer file.
[b] In °C.

SUMMARY

We have described and pointed out some of the major ecological features of a temperate/tropical estuarine environment suitable for northern warm water effluent mariculture. The trophic level significance and rapid rates of growth of many species of commercial marine organisms could be adapted to such mariculture. The described organisms have been shown to tolerate rapid and large fluctuations of temperatures, salinity, and food sources. Detailed descriptions of discrete bay environments are presented that could be helpful in mariculture practice.

REFERENCES

Anderson, W. W. 1966. The shrimp and the shrimp fishery of the Southern U.S. U.S. Dept. of the Interior. Fish and Wildlife Service Fisheries Leaflet 589.

Armstrong, N. E., and M. O. Hinson. 1973. Galveston Bay ecosystem fresh water requirements and phytoplankton productivity. Pages II-1–II-98 in: C. H. Oppenheimer, ed. Toxicity studies of Galveston Bay Project. Report to the Texas Water Quality Board. IAC (72-73)183.

Brogden, W. B., J. J. Cech, Jr., and C. H. Oppenheimer. 1974. A computerized system for the organized retrieval of life history information. Chesapeake Sci. 15:250-254.

Butler, P. A. 1954. Summary of our knowledge of the oyster in the Gulf of Mexico. U.S. Fish Wildl. Serv. Fish. Bull. (89):479-489.

Chechova, I., and E. M. Davis. 1973. Trend surface analysis and seasonal distribution patterns of primary nutrients and chlorophyll in unstratified Gulf Coast estuaries. Water Resour. Res. (9):1543-1554.

Collier, A. and J. W. Hedgpeth. 1950. An introduction to the hydrography of tidal waters of Texas. Publ. Inst. Mar. Sci. Univ. Tex. 1(2):125-194.

Copeland, B. J., and T. J. Bechtel. 1974. Some environmental limits of six Gulf Coast estuarine organisms' contributions in Marine Science. Publ. Inst. Mar. Sci. Univ. Tex. 18: 169-204.

Corliss, J., and L. Trent. 1971. Comparison of phytoplankton production between natural and altered areas in West Bay, Texas. Fish. Res. Bull. 69(4):829-832.

Darnell, R. M. 1959. Studies of the life history of the blue crab (Callinectes sapidus Rathbun) in Louisiana waters. Trans. Am. Fish. Soc. 88(4):294-304.

Davis, E. M. 1971. Report to Texas Water Development Board on development of methodology for evaluation and prediction of the limnological aspects of Matagorda and San Antonio Bays. Contract IAC (70-71)467. University of Texas at Houston.

Davis, E. M. 1973. Assessment of the primary ecological interactions in four Texas estuarine systems. Texas Water Development Board. IAC (72-73)909. University of Texas at Houston. 488 pp.

Frolander, H. F. 1964. Biological and chemical features of tidal estuaries. J. Water Pollut. Control Fed. 36(8):1037-1048.

Galtsoff, P. S. 1931. Survey of oyster bottoms in Texas. U.S. Bur. Fish. Inv. Rept. 6:1-30.

Galtsoff, P. S. 1954. Gulf of Mexico; its origin, waters, and marine life. *U.S. Fish Wildl. Serv. Fish. Bull.* *55*(89). 604 pp.

Gunter, G. 1945 Studies on the marine fishes of Texas. *Publ. Inst. Mar. Sci. Univ. Tex. 1* (1):199–265.

Gunter, G. 1954. Studies on the marine fishes of Texas. *Publ. Inst. Mar. Sci. Univ. Tex. 1*(1):1–190.

Hellier, T. R. 1962. Fish production and biomass studies in relation to photosynthesis in the Laguna Madre of Texas. *Publ. Inst. Mar. Sci. Univ. Tex.* 8:15–21.

Holland, J. S., N. J. Maciolek, D. Kalke, and C. H. Oppenheimer. 1974. A benthos and plankton study of the Corpus Christi, Copano and Aransas Bay systems. Report on data collected July, 1973–April, 1974. Report to Texas Water Development Board.

Hood, D. W. 1953. A hydrographic and chemical survey of Corpus Christi Bay and connecting waters. Tex. A & M Research Project 40. 22pp.

Moffett, A. W. 1970. Shrimp fishery in Texas. *Tex. Pks. Wildl. Bull. No. 50.* 38 pp.

More, W. R. 1969. Contribution to the biology of the blue crab in Texas with description of the fishers. *Tex. Pks. Wildl. Tech. Ser. 1.*

Odum, E. P., and H. T. Odum. 1959. *Fundamentals of ecology.* 2nd ed. Saunders, New York. Pp. 72–73.

Odum, H. T. 1967. Biological circuits and the marine ecosystems of Texas. Pages 99–157 *in:* T. A. Olson and F. J. Burgess, eds. *Pollution and Marine Ecology.* Interscience, New York.

Odum, H. T., and R. P. Cuzon du Rest, R. J. Beyers, and C. Albaugh. 1963. Diurnal metabolism, total phosphorus, Ohle anomaly and zooplankton diversity of abnormal marine ecosystems of Texas. *Publ. Inst. Mar. Sci. Univ. Tex.* 9:404–453.

Odum, H. T., W. McConnell, and W. Abbott. 1958. The chlorophyll "a" of communities. *Publ. Inst. Mar. Sci. Univ. Tex.* 5:65–96.

Odum, H. T., and R. F. Wilson. 1962. Further studies on reaeration and metabolism of Texas bays, 1958–1960. *Publ. Inst. Mar. Sci. Univ. Tex.* 8:23–55.

Oppenheimer, C. H., and K. G. Gordon. 1972. Texas coastal zone biotopes. Interim report to Nat. Sci. Foundation (Grant GI-34870X). University of Texas, Port Aransas, Texas.

Oppenheimer, C. H. 1973. Toxicity studies of Galveston Bay project, Sept. 1, 1971 to Dec. 1, 1972. Final Report to Texas Water Quality Board. IAC (72–73)183.

Oppenheimer, C. H., T. Isensee, W. B. Brogden, and D. Bowman. 1974. Establishment of operational guidelines for Texas coastal zone management. Final report on biological uses criteria for research applied to national needs program, Natl. Sci. Foundation. (Grant GI-34870X). University of Texas Marine Science Institute.

Pearson, J. C. 1929. Natural history and conservation of the redfish and other commercial sciaenids on the Texas coast. *Bull. Bur. Fish.* 44(1929): 129–214.

Ryther, J. H., and W. M. Dunstan. 1971. Nitrogen, phosphorus, and eutrophication in the coastal marine environment. *Science 171:* 1008.

Simmons, E. G., and J. P. Brews. 1950. The Texas menhaden fishery. *Tex. Pks. Wildl. Bull.* 54-A: 1–16.

Smith, D. W. 1973. A preliminary investigation of the nutrient requirements of four Texas bay systems. Unpublished Rept. to Texas Water Development Bd. IAC (72–73)909.

DISCUSSION

CHRISTENSEN: You emphasized the importance of sewage for the aquaculture and for the natural fauna, and among other things said that if the Mississippi River is cleaned, as the engineers would like to do, maybe the productivity of and the fish populations in the

Mexican Gulf will decrease drastically. In such a matter the engineers are often blamed, but mostly the responsibility has to be shared with politicians and biologists. An analogous case from our part of the world is the disposal of sewage from Copenhagen; at present it flows after little more than mechanical purification into Øresund, from where it spreads northward into the Kattegat and then out into the North Sea. Under the pressure of public opinion, politicians in Denmark and Sweden have recently arrived at an agreement between the two countries with the objective of achieving almost complete purification of those effluents before discharge. It remains to be seen what the result will be; maybe one consequence will be a decline of the productivity and fish population in the Kattegat and further out.

My second comment is very brief and possibly a digression. It refers to the discharge of waste heat in some uncertain distant future, when fusion might be realized at an industrial scale. I very much agree with the point of view that also in this case there will be considerable amounts of waste heat to dispose of and *not,* as many politicians or pressure groups will make us believe, that there will be no waste heat associated with the production of electric energy by fusion.

OPPENHEIMER: In reference to your comment, perhaps the elimination of sewage in the passage between Denmark and Sweden may adversely affect the catch of the two million tons of fish that is reported annually for the North Sea. The use of logic may lead to conclusions that are contrary to popular belief, as, for example, the time when I made the statement that by logic I could show that the last 30 years of oil production in the area off Louisiana could be responsible for part of the increase in the fish catch because of the increase in productivity due to the release of oil from the platforms. At least there is a correspondence between the estimated release of oil and the increase of the fish catch in that area. Fishing effort has not increased over these past ten years and it is reasonable to assume that the degradation products from the oil releases enter the food chain to increase the total amount of nutrients available for the food web. We know at least that there is a rapid microbial degradation of the oil released, whereby most of the aromatic and paraffin compounds will be transformed into fatty acids through aldehydes and ketones.

CONOVER: How temperature dependent is the process of oil degradation? We in Canada are of course greatly concerned about oil in the Arctic.

OPPENHEIMER: I have worked with only the temperate microorganisms, but my colleagues in Alaska have demonstrated fairly good rates of degradation, which may be analogous to the fact that the productivity of phytoplankton in the polar regions seems to be as large as in temperate or tropical areas. In this respect it is apparent that we know very little about the initiation of growth, and it may be that these organisms are able to utilize the infrared part of the spectrum for photosynthesis. During light periods, likewise, there is a possibility that molecular activation by infrared accounts for the increase in activity of polar microorganisms, and I would expect that through some research you would find microorganisms that are quite active.

KORRINGA: I may add to this, that Dr. Zo-Bell and his staff have found microorganisms from natural outflows of oil in Alaska that degrade the oil quite rapidly. The rate was not as high as in the Gulf of Mexico, but was sufficiently high to keep pace with the oil seepage so there was no net accumulation of mineral oil.

OPPENHEIMER: This has also been reported by the group at the University of Alaska. Personally, I am not too much concerned about the oil spills as long as they are not too concentrated. Normal photosynthetic activity can be estimated to produce one million barrels of oil a year in the Gulf of Mexico, and for the whole world the estimates are 80 million barrels of oil a year. According to the reports from Woods Hole, the total of man's input including the runoff from roads and land is only six million barrels a year, a very small fraction of what is produced biologically.

So I think the mere fact that in the natural environment where you have oil being produced there are no visible signs of oil accumulations indicates that the biological system is able to cope with the production up to now. As one example we might take the Copano Bay, one of the bays mentioned in my presentation, which is a major oil field as well as an extremely productive bay biologically. There are about fifty producing platforms in that bay and about once a year a pipe break occurs or one of the wells breaks loose. It will produce half an inch of oil on the surface of the bay, without an apparent effect on the productivity of the bay.

DEVIK: In the picture from the oil platforms in the Gulf of Mexico, it was very striking to see the large schools of fish gathering around the foundation of the platforms. In a way it seems that these platforms will serve as shelters for the fish and also to provide them with extra nutrient which leads to a net increase of the total fish population. I would like your comments on whether a similar situation will be created when the oil platforms are erected in the North Sea and on the continental shelf in northern Europe.

OPPENHEIMER: The colonization of the platform area will be very similar to fouling. The sequence of fouling is pretty well known for almost all latitudes. The latest example of such an investigation is the one done on the island of Surtsey, which was formed off Iceland by volcanic action. Below water the island was colonized by the natural fauna and flora within one year, while on the land, it took five to six years before the first plant appeared.

RAYMONT: The pattern of colonization in the North Sea rigs is presumably going to be rather different from the usual fouling pattern, because in the ordinary case we are dealing with rather shallow installations confined usually to harbors or sheltered sites. In the case of oil rigs we are going to have a substantial depth of structures, which might mean that the history of the colonizing organisms is going to be a bit different.

OPPENHEIMER: The organism will be related to the currents bringing the larvae from the nearest populated surface. There will still be a fair amount of nutrients coming from the platform, in spite of the fact that they may take their waste to shore.

KORRINGA: We have good examples of such conditions in the North Sea in the thousands of shipwrecks at different depths. You can observe what is growing there and extrapolate to the reefs. There are a lot of fishes swimming around the shipwrecks because the wrecks give them food and shelter against the current.

PERSOONE: In some of your figures you indicated oxygen contents of about 15 ppm, which would mean about 200% saturation. To which level does the oxygen fall during the night?

OPPENHEIMER: The night time drop is not much; most of the oxygen is derived from mixing due to the wind. The values of 15 ppm are exceptional; on an average, the oxygen values will fluctuate around 9 ppm, which corresponds to about 100% saturation.

RAYMONT: In discussing nutrients you made reference to the phosphorus and the nitrogen cycles, in the case of phosphorus describing solubilization of phosphate during anoxic conditions. You mentioned a similar process for nitrogen and I would like to know how this happens.

OPPENHEIMER: In our area there is a rather significant amount of nitrogen fixation because of the large blue-green algae population. At the same time we have a large anaerobic environment in the sediment, and under anaerobic conditions, decomposition of amino acids and dentrification produces nitrogen gas rather than ammonia. Accordingly, we get quite a loss of nitrogen in the form of gas and an assimilation of nitrogen by the blue-green algae. According to our data, these processes seem to cancel out as shown by calculations on model systems using our collected data. On the basis of data from the bays, we have developed a mathematical model, which gives a good account of the changes in the carbon, nitrogen, and phosphorus balances when the input from the rivers and terrestrial runoff is included.

From this model we are able to determine various mass balances, and to estimate how much carbon, nitrogen, and phosphorus the system is exporting in the way of fish catch, how much goes out through diffusion, and how much comes in. Eventually we hope to utilize this model to predict what will happen if the city of Corpus Christi triples over the next ten years with the subsequent increase of industrial and community sewage effluents.

It seems that in a fairly short time we will obtain this information with the computer system we have developed, but it will still be a long time before we can present the information in such a fashion that we can present print-outs like the one I showed.

KORRINGA: I was very much impressed by the account you gave of experiments with the shrimp to find out what they would like to eat, which put you in a position to shorten the time for developing a complete and palatable food for the shrimp. Is it not so that these shrimp during the day dig into the sand and do not feed? Would it be possible under laboratory conditions to increase the growth by keeping them in darkness longer or in very subdued light. Have you tried this?

OPPENHEIMER: We have only done preliminary experiments, which seem to indicate that when the light is kept off and the animals do not bury, there is an increase in the growth rate.

KORRINGA: For farming this would be very important, because mostly you try to farm an organism in one season and in those countries where there are winter seasons in which they stop growth, it would be very nice if you could speed up growth during summer. Such methods might open the possibility for shrimp farming in several of these areas.

OPPENHEIMER: In our area the shrimp grow to market size in one season. The brown shrimp spawn in about October in the open Gulf, where they escape the winter bay conditions. At times, during the winter, we may have a norther come through which will lower the water temperature very quickly from say, 85°F to as low as 35°F in a matter of one or two days.

In the months of February and March the shrimp larvae return to the bays. They are then about 1 to 4 cm in length and by July and August the same year they will have reached the marketable size of about 25 shrimp to the pound.

KORRINGA: It is remarkable that one could grow shrimp from the egg in the Galveston laboratory, but not larger than to bait size. What could be the reason that you have not managed to grow them to the large size the consumer demands under laboratory conditions?

OPPENHEIMER: I don't know. It might be the food, because the tendency in experimental aspects is to use the shrimp itself for food, which might not produce a sufficiently good diet. I thoroughly agree with the previous speaker that if the animals are provided with a healthy environment and good food, growth rate will be enhanced and disease will drop way down.

PRITCHARD: I was very pleased to hear your remarks on shrimp because this is one of the very exciting areas for research in aquaculture. Your comment on the use of nonprotein nitrogen was most interesting. There seems to be some incompatibility with the very fast rate of food passage and the utilization of nonprotein nitrogen. Usually one would expect some type of a gut fermentation being involved in the utilization of nonprotein nitrogen, but the very rapid rate of food passage seems to rule this out. Are there current theories as to how this comes about?

OPPENHEIMER: As a matter of fact the feather meal undergoes a physical process in its preparation that results in a breakdown of large molecules to smaller ones, which apparently make them more readily available as a nitrogen source. It may be that this physical alteration makes the feathers more palatable and it may be that some of the materials in the feathers have to pass through the gut to be recycled once before they are completely utilized.

PRITCHARD: It may be that the bacteria are not working in the gut but are somewhere else.

OPPENHEIMER: Yes. This is possibly a process going on in the sediment.

CARSTENS: I have a question on the benthos. Yesterday you showed a steel reef which has a very rich community and today you told us about the base which presumably is a muddy bottom not suited for attached algae. What would you think about the walls of Devik's scheme here? They are not solid, they are flexible or rather resilient. Would they be good enough for algae to settle on or do they require a more solid surface?

OPPENHEIMER: I think that the undulation or the pressure of the wave as it passes through the plastic does have some force on the surface of the polyethylene which decreases the fouling on your screens.

DEVIK: My impression was that the flexible walls of the greenhouse showed less growth attached than the rocks around.

CARSTENS: Your answer, then, is that you feel the movement of the plastic sheet would discourage algae from growing there.

OPPENHEIMER: That would be my first guess.

BRATTEGARD: Can you tell us a little about what the main types of food are for the fish coming into the bays?

OPPENHEIMER: The main types of food are, of course, the phytoplankton. In summer there is a very large population of phytoplankton and zooplankton, and in the muddy bottoms the grasses act as a surface for the attachment of organisms. There is a wide species diversity attached to the grasses; the fish and crustaceans eat the surface off the grass. Then there is the detritus of the shoreline grasses like the *Spartina* and *Thalassia*, which during storms are broken off into the water to be degraded by microorganisms into particulate material which is then picked up as food. In addition you have organic material that is brought in by the runoff from the land, pieces of wood and plants, and insect and animal remains.

BRATTEGARD: Stomach analysis of some species of hake and flounder in estuarine areas in Georgia, U.S.A., show that their stomachs at times were almost completely filled with cumaceans and mysids (Crustacea). Have you found something similar in estuaries and lagoons in Texas?

OPPENHEIMER: I have not looked at the stomach content of our fishes, so I could not answer that.

MATTHEWS: I refer to your remarks this morning about the increase in the fish catch being synchronous with oil activity in the Gulf of Mexico and the strong impression I got yesterday of the tremendous opportunities for fishing around these rigs. At the same time the primary productivity of the area must presumably be little changed, except perhaps for a certain slight increase due to degradation of the oil. I wonder whether there is any evidence that the community has shifted more to the pelagic zone from the benthos. Is there a corresponding decrease in the benthos of the Gulf of Mexico which perhaps would be reflected sooner or later in a decrease in the prawn catch?

OPPENHEIMER: There is a transition in the benthos as you go away from the shore. The species diversity does change, but the amount of biomass remains relatively constant. As you go out on the shelf there are some very good reefs, and they are very productive. Studies at the Texas A & M University have shown that there is an amazing diversity of animals on the reef, and a large productivity. In the sandy bottom there is a fairly large population of benthic organisms.

MATTHEWS: I was thinking of the soft bottom. If there were any examination before a rig went down, and after.

OPPENHEIMER: Unfortunately the first rigs were started almost 30 years ago before any studies were initiated. For that reason I am trying at present to promote some interest in doing a survey of the North Sea before oil production starts. I do not know, coming as an outsider, if this effort will ever get off the ground, but we thought we might contribute with our experiences. In our research program in Louisiana, there is a team of 24 scientists working in the offshore area. The group is covering all phases of the circulation productivity, the attached organisms, benthic organisms, the transition from the shore areas in almost every trophic level. We are at the end of a two-year project and we thought that it might be a very interesting experience to make the transition between our temperate environment and the North Sea and use the expertise we had obtained.

PILLAY: In this bay area you have been talking about, is there any ongoing program of aquaculture or any plans to start any aquaculture in the future?

OPPENHEIMER: I do not know of any programs in the Gulf area using wastes. Heat from power plants is used, but so far, not the wastes.

Higher Plants as the Basis for Alternate Food Chains: Their Potentialities in Relation to Mass Culture of Microalgae

Eberhard Stengel

Gesellschaft für Strahlen- und Umweltforschung mbH, München
Abteilung für Algenforschung und Algentechnologie
Dortmund, Germany

INTRODUCTION

Higher aquatic plants have mostly been underestimated with regard to their primary production potential and their importance as the basis of aquatic food chains. This neglect of macrophytes is especially obvious in limnology (Thienemann, 1926), where the research is concentrated almost exclusively on the pelagic vegetation, the phytoplankton. But in marine biology, as well, the investigation of higher plants (angiosperms) has received far less attention than the planktonic algae or the seaweeds. It is only in recent years that a change has come about (Gaevskaya, 1966). From World War II onward a steadily increasing number of publications pointed out that the role of higher plants in primary production in the shallow areas of fresh, brackish, and marine waters can definitely surpass pelagic primary production in such biotopes. Thus the importance of angiosperms for the great productivity of shallow waters has finally been elucidated.

This chapter cannot yet, as would be desirable, report on technologically featured application of submerged higher plants in aquaculture. Instead I shall try to survey in a more general fashion the types of angiosperms occurring in marine waters, and to point out their bearing and functions as the basis of various food chains. I will then formulate what kinds of demands have to be met, until we can consider a marine angiosperm as suitable for aquaculture purposes. Finally, the production rates of submerged higher plants and microalgae shall be compared.

ANGIOSPERMS IN MARINE HABITATS

Only a few angiosperms, i.e., about 45 species, inhabit the sea. However, some of them are able to form impressive vegetations. They are of great importance to the shallow coastal areas, since they supply most of the primary production in these habitats. Nevertheless, most of these plants have been poorly investigated so far, and closer interest in their ecology, physiology, and potential exploitation in marine aquaculture began to arise only during the last several years.

If we leave out for certain reasons the mangrove vegetations, which can be important as a source of detritus (Odum, 1970), the taxonomical monotony of the marine angiosperms vegetation is remarkable. Moreover, all true marine angiosperms are restricted to two families, namely the *Potamogetonaceae* and the *Hydrocharitaceae*. If we include those freshwater and brackish species which can also live in marine habitats—i.e., the *Gramineae* and the *Cyperaceae*—we are dealing with members of four plant families, all of them monocots (Figure 1).

In general, the marine angiosperms occur only in very flat areas (euphotic zone) where the water depth is restricted to a few meters. This holds, for example, for the species of Zostera and Thalassia. Only rarely are marine angiosperms found at greater depths. The maximal depths reported in the literature for *Zostera marina* are 30 m, for *Posidonia oceanica*, 40 m. Only some of the *Halophila* species are able to live in greater depths: *H. Decipiens* was found at 85 m, *H. engelmanni* at 90 m.

In the following, only such genera or species shall be considered as are known to build up dense vegetations and which have been recognized, at least in some cases, as the bases for certain marine food chains. This does not rule out the possibility that other species will eventually become more attractive for marine aquaculture than those particularly mentioned here. Since we know nothing yet of the ability of aquatic angiosperms to make direct use of dissolved organic substances from sewage, a selection of the most promising species would probably be premature.

Marine Species

In the case of those marine angiosperms which form massive vegetations at the borders of the marine habitats, it is interesting to note how restricted to de-

MARINE	BRACKISH
POTAMOGETONACEAE	**POTAMOGETONACEAE**
ZOSTERA (E)	POTAMOGETON pect.(E)
PHYLLOSPADIX	RUPPIA (E)
POSIDONIA (S)	ZANNICHELLIA (E)
HALODULE (E)	
SYRINGODIUM (E)	
	CYPERACEAE
	SCIRPUS (E)
HYDROCHARITACEAE	
THALASSIA (S)	**GRAMINEAE**
ENHALUS	
Halophila* (E)	SPARTINA (E)

(E) = euryhaline   (S) = stenohaline   * = tolerates pollution

Fig. 1. Important genera of angiosperms forming dense vegetations in marine and brackish habitats. [Several authors have divided the family of *Potamogetonaceae* into several families or subfamilies (see chapter titled "Classification," in Den Hartog, 1970; also "Synopsis of families of aquatic vascular plants," in Sculthorpe, 1967). Following the method of Hartog (1970), the old system of Ascherson (1907) is used here.]

fined areas the geographical distribution of the individual species or even of an entire genus is. For example, *Posidonia oceanica* occurs only in the Mediterranean, P. *australis* only along the shores of southern Australia. The *Hydrocharitaceae*, as represented by the genera *Thalassia, Enhalus* and *Halophila*, are confined to the tropical seas.   Only a single species, *Halophila decipiens*, is of pantropical ocurrence. Exceptional to a certain degree is the genus *Zostera*, of which some species have invaded the tropical oceans from the northern or southern colder seas.

With regard to their salinity and temperature requirements, we can distinguish two groups among the marine angiosperms, namely, stenobionts and eurybionts. The most important representative of the latter group is the eurythermal and euryhaline seagrass, *Zostera marina*, known also under the name eelgrass.

It grows in habitats ranging from normal marine salinity down to 6-7‰ Cl^- in the Zuiderzee and to 3‰ Cl^- in the Baltic. Also remarkable is the ability of eelgrass to adapt itself to extreme temperatures. In Alaska it tolerates diurnal temperature changes between 10 and 30°C (Biebl and McRoy, 1971).

Thalassia testudinum, turtle grass, is the only other seagrass besides *Zostera marina* of which the ecology is relatively well known. As already indicated by its name, turtle grass is grazed upon by the sea turtle, *Chelonia mydas* (see also Figure 6). *Thalassia testudinum* is common to the entire Gulf of Mexico and the Carribean in the form of extended, dense vegetations. Contrary to *Zostera marina*, turtle grass is rather stenohaline in that its limits for salinity are narrow, namely 13.5-21‰ Cl^-. For good growth it needs temperatures higher than 20°C. Den Hartog (1970) characterizes *Thalassia testudinum* as follows: "The *Thalassia* association can develop on soft substrates, on coral sand, on dead reef platforms and on rocky substrata. The initial stages in the succession series on these various bottoms are, of course, quite different, but the final stage is in all cases the same. . . . *Thalassia* will arrive sooner or later and will oust the original algal growth."

For the propagation of seagrass populations, it is of some importance that the rhizomes of *Zostera* grow only horizontally. A young population of eelgrass will therefore be confined to the crests of ripple marks, until the interspacial ditches are filled up with sediment. By contrast, the rhizomes of *Thalassia* are able to follow every bend of the substrate surface upward and downward, so that the turtle grass easily covers rather uneven grounds.

Brackish-Water Angiosperms

Besides, the aforementioned examples of genuine marine angiosperms, we have to consider the brackish species of the genera *Ruppia, Zannichellia,* and *Potamogeton* (see Table I). One more quotation from Den Hartog (1970): "The range of salinity which [these members of the *Potamogetonaceae*] can tolerate is even greater than that of the seagrasses, as they are to be found in freshwater, in mixohaline and hyperhaline brackish waters, and in continental saltwaters. As far as the chemical composition of the saltwater is concerned, they are not quite so restricted as the seagrasses. *Ruppia*, for example, occurs not only in waters where sodium chloride is the dominant salt but also in waters where the sulfates of sodium, magnesium or calcium dominate. *Zannichellia* and *Potamogeton pectinatus* have been reported from waters where sodium carbonate is the dominant salt. Furthermore, all these taxa can tolerate very sudden and very large fluctuations in salt content. From an ecological point of view they are in fact ubiquitous. These extremely euryhaline taxa, nevertheless, seldom penetrate into the purely marine environment, although their salt tolerance certainly does not prevent them from

Table 1. Salient Features of Some Aquatic Angiosperms from Marine and Brackish Habitats[a]

Genus species	Geographical latitude (0–70)	Depth (max., in m)	Temperature (°C) min	optimum	max.	Salinity min.	optimum	max.	Direct food for animals	Pollution (sewage)	Mixotrophy proved	Stenobiontic	Eurybiontic	Special remarks
Zostera marina		30	10	(10–20)									+[b]	Needs organic substances in the substrate
Posidonia oceanica		40	10	(17–20)	22							+		Needs well aerated waters
Thalassia testudinum		10	20			20	(24–38)	48	+	–		+		Substrate must be reduced
Halophila ovalis		12	10			18				+			+	Tolerates hyperhalinic conditions
Halophila decipiens		85				18				+			+	
Ruppia maritima		3		Brackish to hyperhaline waters						+				+	Tolerates high contents of NaCl, Na$_2$SO$_4$, MgSO$_4$, CaSO$_4$ and Na$_2$B$_4$O$_7$
Scirpus lacustris										+	+		+	
Potamogeton pectinatus									+	+			+	Tolerates high contents of Na$_2$CO$_3$

[a] Data compiled from Den Hartog, 1970; Sculthorpe, 1967; Seidel, 1967, and others.

[b] + = Tolerated, able, etc.; – = not tolerated; open space: no data available.

doing so. They sometimes occur together with the most euryhaline seagrasses in brackish waters, such as estuaries and lagoons, but are generally restricted to the poikilohaline waters." *Ruppia* has even managed to conquer the Borax Lake in California (Wetzel, 1964).

All of the species mentioned are submerged hydrophytes. In the salt marshes, however, semisubmerged plants, which are only temporarily submerged, play an important role as primary producers. These are members of the marsh grasses of the genus *Spartina* (*Gramineae*) and *Cyperaceae* of the genus *Scirpus* and *Juncus*. The *Spartina* species, especially *Spartina alterniflora*, are common to the east coast of the United States from Texas to Maine. Wherever these plants occur, they are among the most important sources of organic material of their biotopes (Odum, 1970).

THE ROLE OF MARINE ANGIOSPERMS IN THE NITROGEN AND PHOSPHORUS REGIME OF THEIR HABITATS

One of the reasons for the general lack of experimental studies on the physiology of marine angiosperms might be seen in the fact that, up to now, hardly anyone has ever succeeded in cultivating these plants or in keeping them under defined conditions for extended periods of time. It is, therefore, greatly to be appreciated that some important features of the phosphorus and nitrogen metabolism have been elucidated in *Zostera* and *Thalassia*.

In their experiments on the phosphate uptake ($Na_2HP^{32}O_4$) by *Zostera marina*, McRoy and Barsdate (1970) applied a technique which had been developed by Frank and Hodgson (1964) for the measurement of matter transport between the leaves and the roots of freshwater angiosperms. The results obtained by McRoy and co-workers are the following (see also Figure 2):

1. Phosphate is taken up by *Zostera* both through the leaves and through the roots (McRoy and Barsdate, 1970).
2. Depending on the external phosphate concentrations, phosphate is transported within the plant in either direction—from the roots to the leaves and vice versa (McRoy and Barsdate, 1970; McRoy et al., 1972).
3. At low phosphate concentrations in the sediment interstices and sufficiently high concentration in the free water, phosphate is taken up by the the leaves and excreted into the intersticial phase of the sediment surrounding the roots. If the concentration gradient is opposite to the one just described, *Zostera* is able to transport phosphate from the sediment into the open water. This is the normal situation as apparent from the

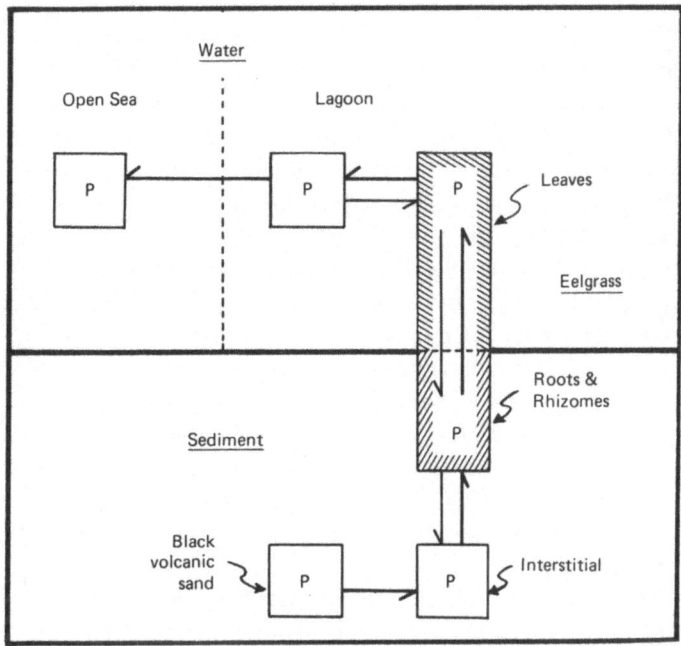

Fig. 2. Pathways of "reactive" phophorus in an eelgrass (*Zostera marine*) ecosystem. Modified after McRoy *et al.*, 1972.

annual cycle of the phosphate regimen in the eelgrass beds of Izembek Lagoon, Alaska (McRoy *et al.*, 1972).

In summarizing these findings one can say that *Zostera* acts either as a phosphorus sink of the ecosystem as long as it absorbs phosphate from the water or as a phosphorus source, which is more frequently the case. In the latter situation a current carries the phosphate released by the *Zostera* beds into the open Bering Sea to such an extent that one can demonstrate the elevated phosphorus concentration in the pelagic zone over long distances. This is a completely new type of phosphorus enrichment of a pelagic zone.

Patriquin and Knowles (1972) have recently studied the ecology and the nitrogen metabolism of *Thalassia testudinum* and some other seagrasses (e.g., *Halodule, Syringodium, Zostera*). From experiments with the sediments and the plant organs, Patriquin and Knowles (1972) obtained evidence for a conspicuous nitrogen fixation in the *Thalassia* beds. This ecologically important activity can mostly be attributed to bacteria which live in the sediment surrounding the rhitomes and roots of the turtle grass, and it is inferred that the bacteria are somehow stimulated by *Thalassia*. The nitrogen fixation was mainly proved by using the acetylene reduction method and corraborated in tests with N-15. From the analyses of ammonia and nitrate it follows that the concentrations of these ions in the interstitial water of the sediment are related to the growth activity of the turtle grass (Patriquin, 1972). Moreover, it is clear that *Thalassia* requires a substrate with a high reduction potential. Organic substances excreted by the roots of *Thalassia* are the source of energy for the nitrogen-fixing bacteria which in turn supply the ammonia needed by the higher plant.

According to the calculation in Patriquin and Knowles (1972), 100 to 500 kg of nitrogen are fixed by the *Thalassia* beds per hectare and year from the atmosphere. In adjacent sediments which are not covered by a turtle grass stand, the corresponding values for nitrogen input are much smaller, namely on the order of 5-10 kg/ha/year. In addition to the nitrogen fixation activity in the rhizosphere of *Thalassia* it was found by Goering and Parker (1972) that blue-green algae living as periphyton on the leaves of *Thalassia* also fix molecular nitrogen. The total nitrogen fixation rate of extremely productive *Thalassia* beds can amount to 350-1700 kg of nitrogen per hectare and year, which is definitely more than maximally reported for legumes (110-220 kg/ha/year, Stewart, 1966; see Figure 3).

Even for the temperate climate, the nitrogen fixation rates expected by Patriquin *et al.* (1972) for *Zostera marina* are likewise remarkable.

In addition to the importance of seagrass vegetations for the community metabolism of their biotopes, the marine angiosperms protect the substratum against erosion and enhance the accumulation of condensed sediments. Their role as shelter for fish and as spawning ground can only be mentioned here.

Plant community	kg/ha x year
Thalassia-bed	
normal average	100 - 500
highest productive	350 - 1700
Sediment adjacent	
without higher plants	5 - 10

after PATRIQUIN & KNOWLES (1972)

Legumes	110 - 220

after STEWART (1966)

Fig. 3. Fixation of molecular nitrogen.

Mammals:
    Trichechus manatus   (Manatee)

Reptiles:
    Chelonia mydas   (Green Turtle)

Fishes: (Herbivorous Reef Fishes)[1]

    Scarus   )
) species   (Parrot-Fishes)
    Sparisoma   )

    Acanthurus sp.   (Surgeon-Fishes)

---

[1]RANDALL (1965): Ecology 46, p. 255-260

Fig. 4. Main grazers on *Thalassia testudinum*.

HIGHER PLANTS AS THE BASIS OF MARINE FOOD CHAINS

The most interesting aspect of marine angiosperms is that they themselves or the periphyton covering their leaves serve as the food of various animals. In addition, the higher plants of the sea are the source of enormous masses of detritus which serves as the trophic basis for shrimp, prawn, and fish. This detritus-bypass of the "classical" food chain, where the primary producers are directly consumed, has recently been reviewed by Quasim (1970). The direct grazing of the mullet (*Mugil cephalus*) on the periphyton of angiosperms has been studied by Odum (1970). This fish also feeds on detritus. Very useful is the extensive review by Hickling (1970), from which one gathers that *Ruppia* seems to be of some importance in the nutrition of the milkfish (*Chanos chanos*). Therefore, we have several examples of animals which feed directly on higher plants or its *Aufwuchs* on the first or second (detritus!) trophic level (see also Figures 4 and 5). The loss of energy through several steps in classical food chains is avoided. Hiatt (1944) created for this the term "telescoping of the food chain." At the moment it seems to be difficult to calculate how "traditional aquaculture systems" such as the tropical milk fish ponds (in which higher plants certainly can play an important role) can serve as a model for culture systems even at higher latitudes. Instead of reporting on these facts, I should like to discuss some of the problems which we will be facing if we want to develop a strategy for the employment of marine angiosperms in the utilization of waste heat and sewage.

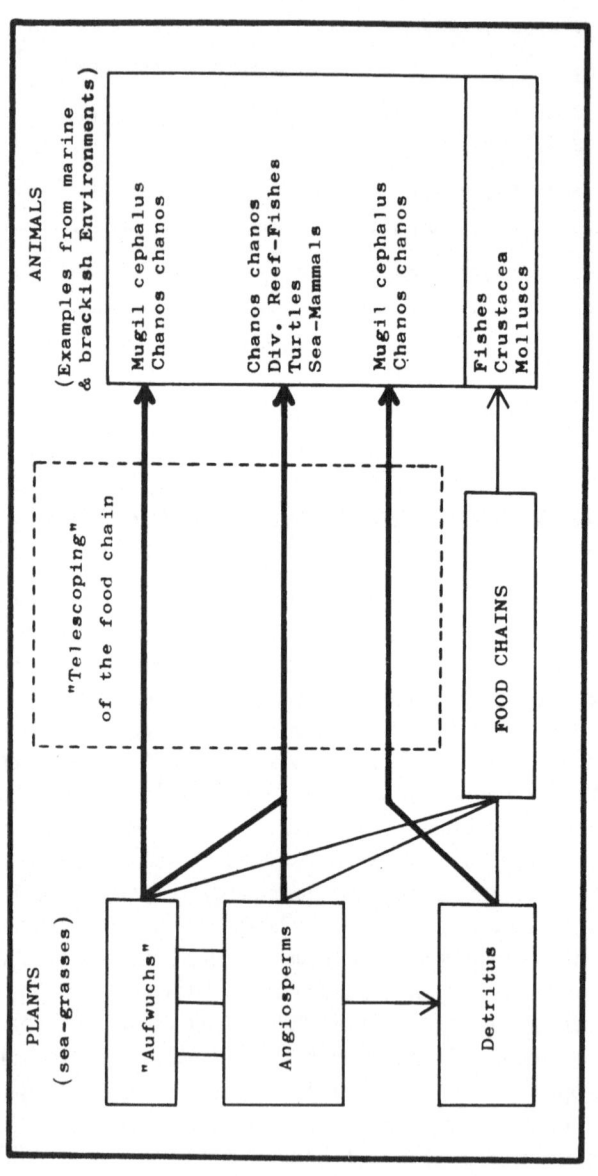

Fig. 5. Higher plants as basis for food chains (scheme).

PREREQUISITES FOR THE APPLICATION OF MARINE ANGIOSPERMS IN AQUACULTURE

What shall we demand from a higher waterplant which we should like to introduce in marine aquaculture? I think such a plant should combine the characteristics shown in Figure 6. Not all of these requirements have to be met in every situation, since a marine aquaculture involving higher plants may be established somewhere between the following two climatical extremes: (a) a constant climate throughout the year and (b) alternation of entirely different seasons. These include every possible combination between constant and variable factors such as temperature, light, salinity, nutrients, and others.

I would expect that a strong seasonal change in incident light energy might create a special difficulty in the cultivation of a higher plant. In the case of microalgae, one can overcome to a certain extent the low productivity during periods of dull weather by the addition of organic substrates such as acetic acid (as carbon and energy source) in open air mass cultures (Stengel 1970). The yields are then mainly dependent on temperature (it seems now feasible to use waste heat from industrial processes for such purposes). Will a higher plant respond in the same way to a supply of dissolved organic material? It seems to me that no

```
1. High productivity

2. High tolerance against changes in
   salinity, temperature & light intensity

3. Tolerance against pollution, diseases
   and plant competitors

4. Ability to utilize organic substrates
   (especially from sewage)

5. Suitability as food for herbivores
   (or as raw material for commercial
   use)*
```

Fig. 6. Characteristics desired for marine aquaculture of a higher waterplant. [*In the literature we can find some notes of the use of seagrasses for economical purposes in earlier times: in the Netherlands *Zostera marina* was widely used as filling material for mattresses and even for road and dike constructions (e.g., Hegi, 1909; Den Hartog, 1970) until the wasting disease had extensively destroyed the seagrass beds around the Atlantic (Renn, 1936). Together with other species of the *Potamogetonaceae* (e.g., *Ruppia maritima*, *Potamogeton pectinatus*), it was sometimes used as manure (either fresh or in the form of ash). The last-mentioned two species were also used as fodder. New investigations on *Thalassia testudinum* (Bauersfeld *et al.*, 1969) showed that this plant can be used as a supplement in a sheep diet with a significant increase in rate of weight gain.]

one has so far tested this experimentally for the true marine angiosperms. Nevertheless Seidel and co-workers (Seidel *et al.* 1967) succeeded in demonstrating that the semisubmerged, sewage-tolerant *Scirpus lacustris* can uptake phenol and indol and metabolize these compounds into substances such as amino acids, peptides, etcetera. In general, it should be possible to find higher aquatic plants which are able to use organic substrates under suitable conditions.

THE PRODUCTION RATES OF HIGHER PLANTS AS COMPARED TO PLANKTONIC ALGAE

Westlake (1965) has compiled all the data available on the net production of plant communities in terms of organic matter production per hectare and year (see Figure 7). From this review, which is still of general validity, we can take the following data: Detailed measurements of the production rates of marine angiosperms are very rare. For *Thalassia* I can quote a figure of 30 annual tons/hectare, and for *Zostera* 3–5 annual tons/hectare (Sculthorpe, 1967; Den Hartog, 1970). The production of *Zostera* has probably been underestimated. If we compare these values for submerged hydrophytes with the corresponding figures for dense

Plant community	t/ha x year
marine phytoplankton	1 – 4.5
lake phytoplankton	1 – 9
limnic submerged macrophytes	4 – 20
marine submerged macrophytes	25 – 40
marine emergent macrophytes (salt marsh)	25 – 80
limnic emergent macrophytes (reed swamp)	30 – 85
sugar cane stands	90

after WESTLAKE (1965)

microalgae in autotrophic open air culture under suboptimal conditions (Bangkok)	55

PAYER et al. (1973)

Fig. 7. Net production of organic matter (dry weight).

	1. microalgae	2. angiosperms
production	highest as possible with the related technical effort	high under favorable conditions, but lower as in 1
area	smaller	big
turbulence needed	high	moderate
treatment (technical)	intensive	extensive or no one
depth	variable	only few meters

Fig. 8. Features of the aquaculture.

cultures of microalgae (Stengel and Soeder, 1975), it appears that the latter are able to give somewhat higher yields even under suboptimal light intensity and carbon dioxide supply, however at a much higher level of capital investment and energy consumption (see Figure 8).

Nevertheless, it should be mentioned that calculations on a protein basis give a somewhat changed view. The crude protein content of microalgae (50-60% of the dry mass) is much higher than in aquatic angiosperms (for *Thalassia testudinum:* 10-13% in the leaves).

CONCLUSIONS

To my knowledge, no one has ever tried to use the biomass of marine angiosperms directly as human food like the seaweeds. Nevertheless, their closer or farther relatives in the inland waters are used as vegetable foodstuffs in Southeast Asia, some of them to a quite large extent: I remember the use of *Ipomoea aquatica* (*Convulvulaceae*), *Wolffia arrhiza* (*Lemnaceae*), *Trapa natans* (*Trapaceae*), and others (Sculthorpe, 1967). A culture system in which the marine angiosperm plants would be attached to strings supported by floating rafts would seem to be a *normal* aquacultural approach corresponding to the techniques used in Japanese seaweed farming (Chapman, 1970). Considering the morphological, ecological, and physiological pecularities of marine angiosperms, I am rather doubtful whether such an approach might be successful, and I am not able to propose any specific technology for the intensive culture of these plants. It is remarkable that transplantation tests with *Thalassia testudinum* were 100% successful when "short-shoots (rhizomes removed) were dipped into a solution of plant hormone (Naphtalene acetic acid) and attached to construction rods for transplanting" (Kelly *et al.*, 1971). However, I am quite sure that until now we have made only very few attempts to cultivate the different species of marine angiosperms because there was no need to do this. I am not convinced that the difficulties which occur in some species would occur in all species.

Of particular relevance may be the observation that in coastal areas which receive a conspicuous load of sewage, *Thalassia* and some other seagrasses are completely replaced by their relatives, *Halophila ovalis* and *H. decipiens* (McNulty, 1961, *Bull. Mar. Sc. Gulf Caribb. 11:* 394–447, cited in Den Hartog, 1970).

Perhaps these species are more able than others to grow in man-influenced fertilized areas even in the higher latitudes when waste heat is implicated, especially because of its probably minor light requirement. You may remember the great depth in which they can live.

As a conclusion it has to be stated that very little is known yet about the productivity, ecology, and physiology of most of the submerged marine angiosperms (except for *Zostera marina* and *Thalassia testudinum*). At present, we can only rely on a very few experimental studies of this interesting group of plants. It may be rewarding to select by proper screeening and breeding strains of marine angiosperms which are suitable to be employed in marine aquaculture.

REFERENCES

Ascherson, P., and P. Graebner. 1907. *Potamogetonaceae. Pflanzenreich 31:* 1–184.

Bauersfeld, P., R. R. Kifer, N. W. Durrant, and J. E. Sykes, 1969. Nutrient content of turtle grass (*Thalassia testudinum*). *Proc. Int. Seaweed Symp. 6:* 637–645.

Biebl, R., and C. P. McRoy. 1971. Plasmatic resistance and rate of respiration and photosynthesis of *Zostera marina* at different salinities and temperatures. *Mar. Biol. 8:* 48–56.

Chapman, V. J. 1970. *Seaweeds and their uses.* Methuen & Co., London.

Den Hartog, C. 1970. The sea-grasses of the world. *Verh. d. Koninkl. Nederlandse Akad. van Wetenschappen, Afd. Natuurkunde Tweede Reeks, Deel 59*, No. 1 North-Holland Publ. Co., Amsterdam and London.

Frank, P. A. and R. H. Hodgson. 1964. A technique for studying absorption and translocation in submersed plants. *Weeds 12,* 80–82.

Gaevskaya, N. S. 1966. *The role of higher aquatic plants in the nutrition of the animals of fresh-water basins.* Vol. I–III. Nauka, Moscow. Transl. by D. G. Maitland Muller. K. H. Mann, ed. National Lending Library for Science and Technology, Boston Spa, Yorkshire, England, 1969.

Goering, J. J., and P. L. Parker. 1972. Nitrogen fixation by epiphytes on sea grasses. *Limnol. Oceanogr. 17* (2): 320-322.

Hegi, G. 1909. 2. Aufl. 1935, K. Suessenguth, ed. *Flora von Mitteleuropa, Bd. I, Pteridophyta, Gymnospermae und Monocotyledones I.* Hanser, Munich (Reprint 1965).

Hiatt, R. W. 1944. Food chains and the food cycle in Hawaiian fish ponds. Part. I. The food and feeding habits of mullet (*Mugil cephalus*), milkfish (*Chanos chanos*) and the tenpounder (*Elops machnata*). *Trans. Am. Fish Soc. 74*(2): 250-261.

Hickling, C. F. 1970. Estuarine fish farming. Sir Frederic Russell and Sir Maurice Yonge, eds. *Adv. Mar. Biol. 8*, 119-213.

Kelly, J. A., Jr., C. M. Fuss, Jr., and J. R. Hall. 1971. The transplanting and survival of turtle grass, *Thalassia testudinum*, in Boca Ciega Bay, Florida. *Fish. Bull. U.S. Dep. Commer. 69*(2): 273-280.

Kirchner, O., E. Loew, and C. Schröter. 1908. *Lebensgeschichte der Blütenpflanzen Mitteleuropas*, I. Band, 1. Abteilung. Allgemeines, *Gymnospermae, Typhaceae, Sparganiceae, Potamogetonaceae, Najadaceae, Juncaginaceae, Alismaceae, Butomaceae, Hydrocharitaceae*. Verlagsbuchhandlung Ulmer, Stuttgart.

McRoy, C. P., and R. J. Barsdate. 1970. Phosphate absorption in eelgrass. *Limnol. Oceanogr. 15:* 6-13.

McRoy, C. P., R. J. Barsdate, and M. Nebert. 1972. Phosphorus cycling in an eelgrass (*Zostera marina L.*) ecosystem. *Limnol. Oceanogr. 17:* 58-67.

Odum, W. E. 1970. Utilization of the direct grazing and plant detritus food chains by the stripped mullet *Mugil cephalus*. *In:* J. H. Steele, ed. *Marine food chains*, University of California Press, Berkeley and Los Angeles.

Patriquin, D. G. 1972. The origin of nitrogen and phosphorus for growth of the marine angiosperm *Thalassia testudinum*. *Mar. Biol. 15*, 35-46.

Patriquin, D., and R. Knowles. 1972. Nitrogen fixation in the rhizophere of marine angiosperms. *Mar. Biol. 16*, 49-58.

Payer, H. D., C. J. Soeder, G. Feldheim, W. Feldheim, U. Gross, and R. Gross. 1973. Dort munder Algen in Übersee. *Umschau Wiss. Tech. 73*(13): 404-405.

Quasim, S. Z. 1970. Some problems related to the food chain in a tropical estuary. *In:* J. H. Steele, ed. *Marine food chains*. University of California Press, Berkeley and Los Angeles.

Renn, C. E. 1936. The wasting disease of *Zostera marina*. *Biol. Bull. 70:* 148-158.

Sculthorpe, C. D. 1967. *The biology of aquatic vascular plants*. Edward Arnold, London.

Seidel, K. 1966. Reinigung von Gewässern durch höhere Pflanzen. *Naturwissenschaften 53:* 289-297.

Seidel, K., F. Scheffer, R. Kickuth, and E. Schlimme. 1967. Mixotrophie bei *Scirpus Lacustris L. 54.* Jg., Heft 7.

Stengel, E. 1970. Anlagentypen und Verfahren der technischen Algenmassenproduktion. *Ber. Dtsch. Bot. Ges. 83*(11): 589-606.

Stengel, E. and C. J. Soeder. 1975. Control of photosynthetic production in aquatic ecosystems. *In:* J. P. Cooper, ed. *Photosynthesis and productivity in different environments*. Cambridge University Press. (Proc. IBP Meeting held in Aberystwyth, April 1973).

Stewart, W. D. P. 1966. *Nitrogen fixation in plants*. Athlone Press, London.

Thienemann, A. 1926. Der Nahrungskreislauf im Wasser. *Zool. Anz. 2* Suppl.

Westlake, D. F. 1965. Some basic data for investigations of the productivity of aquatic macrophytes. Proc. of the I.B.P. Symposium on Primary Productivity in Aquatic Environments Pallanza, Italy, April 1965. *Mem. Ist. Ital. Idrobiol. 18* Suppl.: 229-248.

Wetzel, R. G. 1964. A comparative study of the primary productivity of higher aquatic plants, periphyton and phytoplankton in a large shallow lake. *Int. Revue Gesyamten Hydrobiol. Hydrogr. 49,* 1-61.

Note Added in Proof

Since this chapter was written the following important publications on the topic of this contribution have appeared: Den Hartog, C., ed. 1974. Seagrasses: Transplant experiments, productivity and consumer ecology. Special issue of *Aquaculture* (Vol. *4*, no. 2). Den Hartog, C., ed. 1975. Seagrasses: Responses to ecological factors, community structure and dynamics. Special issue of *Aquatic Botany* (Vol. *1*, no. 2, pp. 93–215).

DISCUSSION

OPPENHEIMER: There has been a very definite move in the United States to use the emergent grasses like *Spartina* to repopulate oil banks from dredging operations. There are at least six projects along the Atlantic seaboard that are doing research right now on transplanting grasses and determining productivity and stabilization of the sands and so forth.

KORRINGA: I would like to refer to investigations about 50 years ago in Denmark on the eelgrass, *Zostera*. In these areas there was an enormous crop of eelgrass growing and few organisms eat the eelgrass directly. To account for the large quantities of biomass in the other parts of the food web, it was then postulated that there was a foodweb (based upon detritus and remains of the eelgrass, which dies off annually) consisting of vertebrates, polychaetes, and others eating the detritus, which were then eaten by the fish, among them the plaice, which was caught in large quantities. This theory was more or less generally accepted until about 1932 when the eelgrass died off completely in all of western Europe. It turned out that this did not affect the productivity of the plaice population.

WALNE: I would want to know how the *Thalassia* fixes nitrogen. Is it similar to the legumes?

STENGEL: The nitrogen is fixed mainly in the sediment around the rhizomes, outside the plant.

WALNE: Would the nitrogen fixing continue if the plant was removed?

STENGEL: I think not, because the lack of energy source for the bacteria. Patriquin has demonstrated, by adding glucose to the sediment, that this stimulates very much the nitrogen-fixing capacity of the system.

OPPENHEIMER: Another mechanism for the production of organic material in our environment is that the *Thalassia* provides a settling effect upon the water. Under one high turbidity condition, this allows materials to settle out, and there is quite a rain of organic materials in the *Thalassia* beds with an attendant population of protozoans, microorganisms, and blue-green algae. Some people think that this is the mechanism whereby the anaerobic muds are primarily produced in our areas.

STENGEL: I think that the experiments from Patriquin are made in regions not so much influenced by rivers. But these studies were carried out in regions of clear tropical water with almost no nutrients, in contrast to the conditions in your lagoons near the Gulf of Mexico.

BRATTEGARD: I would like to refer to some observations on the importance of *Thalassia* beds in Colombia, South America. The observations were made on coral reef flats with steep seaward slope and where dense stands of *Thalassia* partly covered the near-shore parts of the reef flat. The *Thalassia* beds were often interrupted by areas of clean sand. Off the outer parts of the reef flats the gently sloping bottom was sand/silt mixed with mud.

The usual conspicuous fauna associated with coral reefs was present but also a number of smaller, inconspicuous organisms, many of which I believe play an important part in the ecosystem. Among these are mysid crustaceans.

Siriella chierchiae is one which stays close to the substrate during the day and moves up to the surface during the night. Adults and juveniles do not usually mix: The adults are confined to the coral bottom and the juveniles to the *Thalassia* bed.

There are two species of *Mysidium* which occur in large numbers swarming around gorgonarians and *Diadema*. Juveniles tend to keep away from the adults and the subadults and seem to prefer to stay in open sandy areas very close to the sediment surface, especially in depressions where detritus accumulates. In other areas with no corals outside the *Thalassia* beds but with sandy and silty bottoms, adults and subadults of species of the genus *Mysidopsis* and *Brasilomysis castroi* occur in enormous numbers. Juvenile specimens are very common in the *Thalassia* beds. From several local fishmarkets I bought mojarras, small snook, and other fishes that had been fished in such shallow areas. Their stomachs were in most cases filled with remains of mysids.

My point is that adults and juveniles of a specific mysid species in such areas might serve as food for different fish species and/or different age groups of a fish species. Partial destruction of *Thalassia* beds due to harvesting of *Thalassia* might unbalance the fish (predator)/mysid (prey) system.

RAYMONT: Referring to the early work about the *Zostera* forming the base of the food chain eventually going up to plaice, in *Thalassia* beds we have similar studies with mullet feeding on the detritus, particularly from *Thalassia*. What other fishes are known to feed directly on *Thalassia* or on the detritus from *Thalassia*?

STENGEL: I am not familiar with the names, and have no practical experience in this field. Some coral fish (*Acanthurus* spp.) come from the reef, invading the *Thalassia* beds.

PILLAY: There are several species of the fish which feed on detritus. Some of them cannot utilize the algae in their fresh condition; they have to feed on the detritus or at least partially decayed organic material, which only can be digested.

RAYMONT: I was thinking that this is a fair shortening of the food chain again. For aquaculture it might be possible to culture these species of fish which would not feed so much on the living vegetation, but on what has been somewhat decomposed.

PILLAY: Yesterday we discussed the possible role of bacteria as food. It is a little difficult to decide to what extent they contribute to the food of the species, but when you look at the gut contents of most of these detritus feeders, you see enormous quantities of bacteria.

Concluding Discussion
and Index

Concluding Discussion

DEVIK: By now we have covered a lot of ground, from the evaluation of the primary production and its role in the food chain through the discussion of the various food chains that can possibly be based upon planktonic primary production and production from higher plants. In the search for new systems, or additional systems to the one already exploited, we find the example of the integrated aquaculture concept as presented by the Woods Hole group. There seem to be several unexplored possibilities, but there is a shortage of basic data to evaluate these possibilities more closely.

There seems to be no universal approach applicable to all parts of the world; we have to accept regional solutions, but within such solutions there is always room for the inclusion of organisms alien to the particular area or the particular system.

What I would propose to do, then, is to try to sum up my impressions, and try to fit them into the framework of conditions as we meet them, e.g., under Norwegian coastal and climatic conditions. As one way to bring out the specifics of these conditions, one might make an attempt to formulate an experimental system that might have some possibilities of a combined utilization of waste heat and nutrients from domestic effluents within the context of aquaculture. This is presented for discussion, and I would primarily ask the five conveners to comment on such an imagined scheme. From these comments we might then try to draw conclusions on how to proceed and what the most pressing problems are that we can see.

Initially, I think it is worthwhile to consider our frames of reference, which will be several. We have the aquaculture as a very important frame of reference; another frame of reference springs from the differences in geographic locations, which may be referred to by means of the latitude of the location. Other sets of references will be referring to the waste nutrient systems and to the waste heat systems—these frames of reference will overlap to some extent. If we then restrict ourselves to the northern hemisphere in the subarctic or the cold temperate zones. we know that the sunlight and seasonal fluctuations are very important. This will greatly modify the techniques to apply. The overriding problem is how far we can accept the continuous outflow of nutrient-loaded effluents in the winter. when the photosynthetic cycle is barely operating, or operating at a relatively low rate. We may make a check list of problems, and rank these according to priorities. For instance, what kind of organisms should we utilize, what do we know about them from literature, what is the supplemental information to obtain, what will be the technique of handling such large water masses, to mention a few general problems.

At present, the most pressing need is probably for technical solutions that will allow for the concentration and holdup of nutrients until they are assimilated. I believe that the concept of marine greenhouses has arrived at the stage where it should be evaluated in the biological context, and let me try to sketch out one possible approach.

For the sake of argument, we can imagine the use of the marine greenhouse system as an element in the aquaculture system in our fjords. Let us see what the possibilities are for using such a narrow zone in the upper layers of water. This zone can be isolated from the other parts and separated from the productive cycles of the bottom, if desired. In this zone we have an input of waste material, containing carbon. nitrogen, and phosphorus. As a further input we might think of waste heat that will heat the outside leading to a temperature which is higher than in the surroundings. Then we might ask whether we will be able to increase the productivity sufficiently to give living conditions for shellfish like mussels, and the growth rate of these animals might be increased by the application of heat. In the case of mussels or other shellfish, there will be a high fallout of nutrients with the feces. Then we have the question of whether we should try to recycle the part that falls to the bottom.

From the point of view of the recirculation, one might, for a start, be satisfied with a partial recirculation such as can be achieved by the harvesting of the shellfish.

For the next phase we would have to imagine the utilization of the sediments, or seek some means of getting the excretions from the animal mineralized before they reach the bottom.

Another question is whether such a steady rain of nutrients in the form of organic biomass will attract other forms of life to lead to a net increase in biomass and improved changes of an increase in fish catch, as one example.

In such a system a great problem will be to utilize the nutrients supplied during winter. The easy way out is to simply let this part pass out into the sea without being utilized: the other might be to collect it in some kind of mineralizing pond before use. Such a solution might be conceivable for the small communities we find along the Norwegian coast.

Then the other question is whether the productivity of such a system can be made sufficiently high. The objective might be to aim at a productivity five to ten times higher than the average productivity in this area. In such a system where we include the area of photosynthesis in the figure for the area yield, this might mean a yield in wet weight of some 10 tons per hectare, based upon an average productivity of 1 to 1.5 tons per hectare for a Norwegian fjord of average productivity.

Other questions will be whether we can utilize these systems in different ways and are able to devise methods to collect sediments and exploit their productivity.

In such a system we might not necessarily need additional heat to make it function, although a moderate increase in temperature might improve the turnover rate and possibly the total production of biomass. The waste heat would obviously be an advantage and yield a form of combination of resources which may be applicable to our situation, and possibly lead to some kind of control or increase of productivity in a large area. The question is how far we can utilize the greenhouse concept in the formulation of experiments to do from the various problems.

PILLAY: It is difficult to make comments on the summing up of the discussion like the one we had. We covered a lot of ground and many of the papers were of great interest, even if in some cases one might doubt their application to aquaculture development.

The development of solutions to these problems should be based on the experience gained in the use of waste materials and waste heat. There is considerable experience in this field as far as sewage is concerned, as mentioned during the first session, and also described in the presentation of Dr. Korringa. Reference has been made to integrated systems where sewage is treated to remove the solids, or it is digested and the methane produced is used for heating or cooking purposes. The effluent is then passed on to an algae pond, the algae harvested and used for feeding poultry or cattle, and the secondary effluent passed through fish ponds to grow tilapia. This kind of a system has worked very well for centuries in China, Indonesia, India, and other countries.

When trying to apply this type of waste disposal to larger communities, we run into new problems, e.g., the public health hazard. We have to satisfy the public health authorities that such a solution will not affect the health of the people and we have to satisfy ourselves that we are not selling a contaminated product. The heavy metals will, in this context, prove to be the major problem. Can we eliminate industrial waste with their load of heavy metals and other contaminants? This may be possible, provided the operations are properly planned and on a sufficiently large scale so as to make them economically viable.

During the discussions it was concluded that whether we use sewage or whether we use waste heat, the production of algae may not really contribute to the increase of the aquaculture production of, for example, oysters. I think Dr. Walne emphasized the point that the algae have to be selected carefully if you want good growth and not simply survival. It seems that the technology for this purpose is not sufficiently advanced. It might also be that we can develop substitute foods that are easier to produce and just as nutritious as the natural ones. By analogy, at one time it was believed that carp had to be fed with a certain species of algae and the fry and larvae would have to be fed with rotifers. At present, I think there are hardly any hatcheries that use live food for the rearing of carp fry and it does not appear that the production has suffered. So, in the long run, it might be possible for us to utilize the primary production in a different way than we think of at present.

Referring to the use of waste heat, the discussion seemed to indicate that there are many problems in the use of waste heat. Emphasis in future work should be to utilize waste heat for improving primary production or improving the growth and production of aquaculture organisms. From freshwater aquaculture one may consider the examples of carp rearing in cooling water, as practiced in many eastern European countries. The grass carp will usually take 6–8 years to mature in the cold climate of the USSR, but growing these animals under controlled temperature conditions will lead to maturity in about one-and-a-half years' time. It seems, then, that there is a direct use for waste heat in this way, and also for the more rapid production of fry for distribution to various culture stations. It is here that waste heat seems to be used most efficiently. I am trying to underline this aspect, mainly, because it seems that in the case of shellfish like oysters a high temperature easily might result in a poor condition of the oysters with most of the weight increase leading to an increase in the weight of the shells.

Finally, I would like to underline once again the need for a multidisciplinary approach to the problem, especially in relation to the public health aspect. We can, as biologists, declare that there is really no public health hazard in a well-managed system, but unless we can back this up with scientific evidence, whatever technology we might develop and whatever species we might use for aquaculture, it would be extremely difficult to develop it to the proportion of a major industry.

PRITCHARD: I sense in the discussion a good deal of diversity on how to approach the problem of integrated aquaculture. There are some fundamental differences as to what our goals of integrated aquaculture should be. In my bias, I am prone to think that our first concern is for food; our second concern is for environmental matters; and our third concern is for the economy, i.e., financial return. We have to recognize that no matter what approach aquaculturists take in the long run, they are going to have to be able to recoup their investments in the marketplace with a product that can compete with other products. I think that we must really concern ourselves more with the problem of food production. One of the sponsors of this workshop was the Panel on Eco-Sciences, so it is only logical that we have had more input on the environmental concerns. I think that the interplay of both viewpoints has been valuable, but I am not sure that it ever is possible to have the full range of interplay that we need. However, the food or industrialized approach is different from the environmental approach. People approach environmental problems with quite a different urgency and tolerate more inefficiency, then if operating a business. If we are going into aquaculture in an integrated way, then surely we must look at it as an industrialized system where one takes a series of inputs (nutrients from effluents, heat, environment, and technical know-how) and is very pragmatic in order to succeed.

If you conclude that integrated aquaculture is truly industralized chemistry and that what we really need is an enzymatic system to do that job, then the first thing one should do is to find out what enzyme systems are needed to carry out the chemical conversions. On looking over the papers, it struck me that this was the area where we had the least amount of information. The plant/animal relationship in integrated aquaculture is the critical stage of the conversion process. We have to know what the fundamental biochemical processes are if we are going to develop efficient mechanisms and systematic approaches. This is an intriguing area of science and it is one very vitally concerned with the future of food production in controlled ways.

On the other hand, I would hate to think that we would put all our immediate effort into studying enzymes; it is equally important to simultaneously move out of our laboratories into pilot-scale operations. The minute you go into scaling up an operation, a range and magnitude of problems different from the ones met in the laboratory emerge. We know that the scaling up of operations in aquaculture is one of the major problems. One cannot extrapolate from the isolated laboratory situations into very complex environmental situations. However, if you know the biochemistry and can then define precisely what you want your particular culture to do, you can find courses of action that lead to more rapid advance in this area of science.

This is really one of the most exciting areas in biology and I think, above all, it is an area that needs a visibility and a separate identity in research efforts. Frequently, aquaculture is expected to take the spinoff from other biological research, but I do not think that it is good enough just to extrapolate work done with other prime purposes to aquaculture. We have to take a major commitment to aquaculture and there is much to be learned in looking at heat manipulation in this connection. In our Canadian situation, we see heat manipulation as an opportunity to expand a limited season, and as an opportunity to control and culture organisms in an alien environment. We may want to use nonindigenous organisms which could be cultured without jeopardizing the natural environment if they survive only in the heated effluent.

I made a brief comment the other afternoon about experiences in agriculture. For years the plant geneticists developed grasses and legumes with one criterion— yield. They never looked at the nutritional qualities and it was indeed quite a revolution in plant breeding when they began to consider the nutritional quality of pastures. When taking into consideration that the grasses were being grown to feed livestock, they found in many cases that the selections made earlier had no bearing on the productive capacity in the end product. On the basis of such experience, I suggest we should not rush into genetic selection relating to primary productivity with quite the urgency that we might turn to the other end of aquaculture food chain. In other words, I feel that genetic studies are probably a higher priority in fish farming, where protein is recycled to higher valued protein than at the primary production level.

I certainly found a great deal of information in the suggested approaches to develop algal cultures, and I emphasize the serious problems that are encountered here. Certainly such approaches should be encouraged, but I also think that we must recognize that the ideal of a continuous culture may still be beyond our grasp. For this reason it might be simpler to discontinue these cultures at a higher level in the food chain, and it might be easier to make the break after protein has been obtained, and then refine the protein.

I was intrigued by the differences between the amount of water used by North America and Europe in sewage. Conserving water use in sewage is one way to lessen handling of large volumes, with an improvement in economy as a result.

Let it also be mentioned again that if we approach aquaculture simply as an experimental manipulation, we might find ourselves tolerating many inefficiencies that would render the systems noncompetitive in terms of products. There is a real urgency to move into efficient systems; otherwise the industries created will not be able to survive any length of time. If development comes quickly, we might not be justified in applying today's cost benefits analysis, simply because they might be outdated and grossly misleading when extrapolated to the future of aquaculture.

As mentioned by Dr. Pillay, one of the overriding factors is safety. We must recognize that aquaculture can produce foods which are nutritionally desirable, and a variety of foods more suited for balanced human nutrition than any one of the important field.

KORRINGA: I agree with Mr. Pritchard that we are really switching over from one way of thinking to another way of thinking, merely brought about because we are much more numerous at present than we were before. With the increase in numbers, the problems of domestic waste increase. We find, for example, that biodegredation in the water cannot cope with it. Consequently, the water becomes dirty and smelly, fishes die, recreational activities like swimming and even sailing become impossible.

With the increasing amount of waste it becomes increasingly difficult to get rid of it without trouble. And then in rich countries like the European ones which are wealthy enough, there is no direct need for the domestic waste for agriculture, so a solution for these problems is not readily apparent.

My solution would be first to build purification plants on a microbiological basis, which is a partial solution in the sense that there is still too much nutrient in the effluent. We might therefore propose to put this effluent through a third purification stage to avoid blooms in the receiving water. A serious problem is the waste coming from the bioindustries, like chicken farming or pig fattening, which will produce almost as much manure as they ingest feed. Safe disposal of this waste easily becomes too expensive, but we still have the sea, where such unpurified material can be disposed of relatively safely, well offshore. The European countries are rich; they can afford to get rid of waste without too much secondary effect; but peculiarly enough, you can get money for all kinds of things but money for the purpose of getting rid of waste is never readily available.

We are also very worried about the waste heat, with the gradual increase in the water temperature which destroys the fauna and flora in the water where this has not already been done by release of chemicals, as in the Rhine, for example. We have the added complication of mist formation in winter from the warmed-up lakes and even if eels and other fish would grow better, the inconveniences are so great that such an increase is not considered practicable. Then the mist formation has a secondary effect of absorbing light; phytoplankton would not develop so easily. So we cause a lot of problems. Again, one could get rid of this heat if one was prepared to pay for it, and we find that the general solution to these inconveniences is to bring the waste material and waste heat further away from our living areas and our recreation areas, or where we cultivate organisms like shellfish in shallow coastal waters. The engineering problems can be solved if it is politically acceptable.

Moreover, we are now on the verge of another world. We are no longer thinking in terms that we have enough of everything, we have gradually discovered that we cannot go on wasting so much material. And should we go on sending everything to the sea—all the valuable minerals that we need for our plants and other things that we dispose of in the sea? And once it goes into the sea you can never get it back. So should we not be more careful with all the ingredients that we have? Thus we arrive at the philosophy of recycling.

We might study the experiences during the war, which was the last time we had to think about recycling. The solution then was to go down in the food chain by ploughing up many of the pastures to plant potatoes and wheat, and to cut down drastically on the raising of chickens. If we want a substantial increase in the world food production, the best possibilities lie on land. If, by selection, we improve rice and wheat to give only one-tenth of a percent more protein than now, we do not need to consider the other solutions. In this way we can produce vastly more food than we ever can expect to do, for example, by mussel farming. In the mussel farms high productivity is only achieved when the mussels have access to flowing water. In the large production centers of the world we see that a raft will have an immense productivity calculated on the basis of the surface area of the raft. But it should be remembered that for an optimum production on such a raft a supplementary area at least 20 to 30 times as large is needed to produce the necessary amount of feed for the mussels. The mussels need free access to the water around. This means good currents and a rather low population density calculated on the basis including the areas for primary production.

Then comes the possible use of heat. Mussels can live at any temperature from freezing up to about 25-30°C, but for optimal growth the water temperature should be between 10 and 20°C. To get the full use of heat, proper light conditions are necessary, because with no light, no food is produced. So what in theory looks nice, will in practice become very difficult. If you heat the water by a given system, you can only utilize it for mussel farming when you have enough light and enough nutrients to produce the food needed, and one must always be aware of periods with scarcity of food because then the mussels will lose again what they have gained before.

And in the use of sewage, the problems are not any smaller. In my experience, I would be very hesitant to admit any domestic sewage because of the public health hazards and also because the changes of obtaining the right type of plankton for mussels or oysters are limited. Very easily the expense of purification of the marketable product will outweigh the possible benefit of such a nutrient source.

So there still remains the question, can we use heat or waste for aquaculture? And we find that the use depends on the local conditions, and on the situation of the country you live in. Such conditions are present in Indonesia, where they put the domestic waste to use in carp production, likewise in parts of India, where a large acreage is available for pond farming with domestic waste. Such systems are well suited for recycling, because the fish derives part of its food directly from the sewage, part via degradation and carbon dioxide assimilation. The fish produced will be duly boiled or fried before use, so sufficient safeguards against the spreading of viruses or pathogenic organisms are introduced. But all these examples are for the tropics. In the temperate climates things are more difficult, and systems where we have to do it differently in summer than in winter are not very desirable. One might look for alternate solutions. Would it not be possible to lead sewage into fresh water in such a way that in the deposits animals like *Tubifex* can feed? In our normal sewage purification plants we get sludge at the end, which is practically free of organic matter but rich in all kinds of nutrients. I wonder whether one should not instead lead sewage into fresh water and use a certain heating up to grow *Tubifex* in large quantities. There is a great demand for *Tubifex* as food for aquarium fishes. If sea water is the medium,

Capitella might be produced, and *Capitella* also can be harvested mechanically and can be processed to be used, for example, in fish farming.

So to summarize: There is waste heat and domestic waste available. One solution is simply to get rid of it, another is to reuse it for some production of organic material. The alternative of shellfish production will meet many obstacles—apart from securing the right combination of nutrients, light conditions, and temperature, there are the difficulties of creating new markets. So it seems that we have to develop systems that are economical, and we must find uses for the material produced that are acceptable in a wide sense.

CONOVER: I am not a mariculturist nor an agriculturist. I feel more like an ecologist and I would make the point that we are operating with ecosystems. Now the ecosystems which we have around us for the most part are the result of many years of evolution. When we then add heat and domestic waste, we are imposing perturbations on the ecosystem which the system could very well adapt to in some constructive way if given enough time. But we are in a hurry.

There seem to be two choices: One approach is to add heat and an amount of nutrient in the form of waste and see what we get. Then we can try to make the best use of this material. Presumably, we will get production, certainly of algae but also of other kinds of green plants. We are not sure that these algae are going to be particularly valuable to the next level of organism, but we might be able to make a lot of cheap protein this way. We might have to add a little alanine or something else to make the protein palatable to some other organism, so that we can use it as feed. That is one approach.

The other intensive approach would be to pick various components of the ecosystem and develop them, using ecological principles, as far as possible. We have seen that heat will encourage growth, but the resulting quality or condition of the animals might not be so good if grown too fast. What this really means is that we have to optimize the culture conditions. We saw in the graphs which Dr. Saunders showed us the growth curves when salmon were given different rations under different temperature conditions. In this case the results apply potentially to one size class of salmon only. Other size classes of salmon probably respond slightly differently to these factors, so you are rather quickly in a sort of multidimensional scheme when you try to optimize conditions. Because of this complexity we have to think about multifactorial approaches to the optimization of what I would call linear food chains. We can have a few side branches, as demonstrated by Dr. Tenore. Otherwise we must either work monocultures or with situations where there is a beneficial interaction between species. Such interactions are, of course, very prevalent in nature as, for example, the eelgrass and the bacteria and epiphytes associated with it. Here, there is interaction between components in the ecosystem which is apparently mutually beneficial to the system as a whole. So there are probably some situations where we can take groups or complexes of species and manipulate them in some way, but in general I think we are going to be working with a system where the components are related in a linear fashion without cross linkages. In such systems there may be some feedback possible to the first stages of the production process, but in all probability the small amount of waste heat and energy obtained at a lower level of the system is probably not worthwhile trying to recycle. It seems the best idea, then, is to keep the food chain as short as possible, as, for example, the case of the carp which feeds on fecal material.

It is not certain whether we can do something like this in our northern climates. As far as carp go, we don't like carp very well, even if it were quite possible that carp would grow well if they were given some waste heat. It does seem, however, that in northern areas there are a number of species which we don't even think about in terms of aquaculture at the moment. Mention was made of the large numbers of sticklebacks which appeared in the fertilized pond experiments made in the Scottish sea lock during the last world war. I doubt whether anybody would eat the sticklebacks, but there is always the possibility of using such fish as a raw material for protein concentrates that could be put to use in various feed formulations.

There is a difficulty in employing short food chains in such production systems, however. In the perfect food chain, one would like to take plankton algae and make whales from it, by adding waste heat and nutrients. At the moment we are obviously not in a position to aquaculture whales, and if you are going to raise efficient material for export from a planktonic aquacultural system, you start out with the energy in small particles. If you can harvest all those little pieces by farming or some other method, this is obviously the most efficient thing to do to get maximum yield; but if the objective is to raise large organisms, many species will obviously have a hard time getting enough to eat feeding only on planktonic algae. There are a lot of ecosystems which definitely operate more efficiently if the little pieces are fed to slightly larger organisms, and these are in their turn consumed by still larger organisms.

With the short food chain you can make more efficient usage of the initial input, and there is not so much of the material going into the detritus food chain. Is the detritus food chain efficient or is it not? In the rather cold waters off the Canadian east coast, the detritus food chain seems to be important. Recent studies made by Dr. K. H. Mann have indicated the macroalgae produce about 10 times as much organic matter as is presently utilizable directly by the organisms that are members of the seaweed community.

There is reason to suppose that the detrital food chains are rather wasteful. We probably should try to design our aquaculture systems to recycle the feces which, for instance, would be defecated by a suspended mussel population. On the other hand, maybe we should try to keep the amount of such feces down to a minimum by providing the most nutritious algal food source possible.

RAYMONT: Previous speakers have virtually gone over all the ground, with perhaps one or two exceptions. Can I therefore draw on one or two points where I think there is some disagreement?

Dr. Pritchard said there are three main aims in aquaculture—food, environment, and finance—and I think he put food first. Dr. Korringa hopes that the three aims can come together. I am quite sure it is foolish to hope for one solution to be applicable to all regions of the world. In a particular country with a particular type of economy, food may be the dominant objective. Under different circumstances the environment might be polluted, and measures to reduce pollution may be paramount. In both examples the economy will be an important factor, but we should not look for one overall solution.

But we may be emphasizing aquaculture too much. Consider, for instance, domestic waste inland; for example, sewage from a big inland city or from surrounding rural country. Because of the cost, it is obviously not sensible to transport it to the coast for disposal, even if there is an efficient way of getting rid of it there. The solution obviously has to be sought close to the place where the waste is produced. And, even on the coast, I think we ought to look at what we now call pollutants as a possible asset rather than as a danger. It might appear high-sounding, but if we could regulate the disposal of sewage and similar pollutants and monitor its composition to some degree, we could considerably improve the environment. Moreover, we might improve the economy on a limited scale. It is surely no accident that high productivity is generally associated with nearshore areas; estuaries are among the most productive regions. The extra nutrient load from urban regions in addition to the natural shallowness of inshore areas can thus lead to increased productivity, provided the system is not *over*loaded.

I am therefore suggesting that even without full aquacultural practice (which cannot solve all our problems and cover all our needs), proper attention to disposal of so-called pollutants can increase our food supply and maintain our environment.

May I now turn to the suggestion, I believe from Dr. Oppenheimer, that we ought perhaps to start in the opposite direction in utilizing wastes. Should we not take our waste and feed it directly to fish? There are a number of fish that could benefit. The fish could be processed to a protein concentrate which by chemical methods could be modified to fulfill a number of nutritional demands. In this approach we can avoid having second-stage, processed sewage passed into the sea. With processed waste contributed to the sea, there is, of course, the possibility of marine algal production with its large potential growth rates, as, for instance, the Woods Hole and the Dortmund experiments have shown. But I am still not certain whether we are going to be able to effect this except on a limited scale and at relatively high cost. And even this algal material has to be transferred through at least one more stage—mussels or similar grazers—with the accompanying loss in efficiency. So the suggestion of Dr. Oppenheimer that we should look at the possibility of using the particulate waste material directly in feeding fish is significant, perhaps at least in some areas.

So we have to look for solutions for utilizing wastes efficiently that are suited to each country. The solution depends very much on the latitude, not least on whether there is sufficient solar energy over most of the year. This will be one of the main problems in an experiment like the one suggested by Dr. Devik.

For the solution of these problems we do need more biological investigation and more technology. For example, is it certain that at present we cannot get enough algal production in winter at the latitude of Norway? If technically the experiment could be defined to a very shallow layer, even during winter in the marine environment, there would be production, albeit at a lower level than in the summer. I understand there is one area in the Baltic, which Dr. Steeman-Nielsen investigated, and where he showed there was quite a substantial winter production. This related to a shallow layer with a very sharp pycnocline, close to the surface.

Have we also thought sufficiently about the different species of algae for such production? It might be possible to select algal species responsive to the sort of reduced light conditions, and to *seed* these into relatively small marine enclosures. I am reminded of the results of John Bunt, who showed that some of the algae found under the ice in the Antarctic can grow at remarkably low light intensities, much lower than normally expected. It is too early to say whether these organisms can be exploited, but at least we should not utterly reject the possibility of growing algae in winter even at a reasonably high latitude.

In summary, then, we have to seek specific solutions for utilizing waste in aquaculture for the various countries. Depending on the conditions, either food or environment might be the first concern, and labor costs and market opportunities will influence which solution should be adopted. The very interesting experiments on the acceleration of salmon growth is one example. It seems well worth pursuing, but only in a country where you can sell that product relatively easily and command a sufficient price.

OPPENHEIMER: In conjunction with the methods of culture that our chairman has proposed previously, I should like to introduce some thoughts about the type of pond culture using plastic bags and enclosures and sewage nutrients that might be successful and what conditions might be necessary.

First of all, I think the idea has some merit, and that it is a system that can be made to work, even if it may not be economically feasible as yet. This scheme might start with the primary treatment, using some ingenuity to sterilize the material before it is separated into solid and liquid. As one possibility a cobalt-60 treatment could be used for this.

The separation of the material into solid and liquid is one of the aspects that really needs to be looked at, it might be that we have not given this enough attention in the past.

The liquid sewage phase can be used for algal ponds. It may require the algae to be conditioned by chemistry to make them acceptable for whatever species of fish or animal you plan to use in the system. The algae can also be conditioned so that they are acceptable to the larvae. The summer season could then be used to grow larvae. In winter the solid and liquid sewage could be combined to provide a proper nitrogen/phosphorus/carbon ratio, which may be supplemented with a mixture of vitamins and growth factors. By using a pelletizing technique, a food could be manufactured that is completely acceptable to the adults.

The winter part of the cycle might need heat from some outside source to accelerate the growth of whatever fish it is planned to use. Fish could be selected from tropical areas and already adapted to high water temperatures and with a fast-growing cycle; or a fish from a topical area could be adapted to the special conditions.

The system could also include filter feeders such as the oyster. Therefore, the pond or the plastic area should have an open bottom but filled with lattice plastic which would have staggered sections for the oyster culture. In this way the materials raining down as a by-product of the fish activities would be cycled through some second system, and in addition to that, possibly another filter feeder or some other organisms outside of the tank could be included to capture any other organic content that has been filtered through the oysters.

One must then introduce some cost, and assume a plant investment where the heat and unsterilized waste would be supplied at little or no cost. The use of 1000 tons of carbon with a 10% efficiency in the whole system would yield 100 tons at kr. 10.-/kilo to provide an output of one million kroner. Then the question would be whether you can do all this for kr. 1.000.000 per 1000 tons. A scheme like this should certainly take some thought.

Dr. SOEDER: Just one technical side remark. If you were to feed sewage continuously into your delimited greenhouse, it would probably affect the hydrology, and you have to take care of the turbulence created by the continuous inflow of the sewage at the top of the system.

Dr. SAUNDERS: It might be necessary to use artificial light. Perhaps it would be practicable to light the greenhouse in the winter or on dark days at any time of the year.

OPPENHEIMER: The alternative scheme referred to has only a summer light requirement. In the winter you would be feeding the fish to adult size with pellets; in the summer phase the natural light would be used to produce the algae in a pond culture such as suggested by Soeder. Then this material would be conditioned so that it could be fed to the larvae as they develop. As the larvae got old enough to accept solid food they would go into the winter phase and take the pelletized material.

CONOVER: Another possibility in environments with low light condition would be essentially the browsing community. If you could run a constant supply of nutrient over a flat surface, a community of algae would grow there which would support a number of browsing organisms, like the common winkle or limpet, or in special situations abalone, as described by Tenore.

DEVIK: We are getting to the end of the discussion. A number of points have been brought up, and the main conclusion seems to be that we have to grade our effort to the location chosen. For illustration we might consider the conditions in Norway a little more closely. Norway is a country where we have dense populations mainly along the coast. This is in contrast to the Continent, where the big cities are very much inland.

In Norway we are releasing the sewage to the ocean, or rather to one of our fjord systems which are easily thrown off balance. Our labor is scarce, or getting scarce, and our solution then should be to utilize the given resources in the best way possible, with the possible addition of heat in those locations where waste heat is available.

At the moment I think we are in need of technological innovations, particularly when it comes to handling large volumes of water. Handling water is very expensive—we have to look for arrangements that could divert currents, collect the currents in certain areas so that we can concentrate the volumes that are overfertilized.

On the other side, we have to see what we can do with the sewage. We will no doubt have to purify it, and include in the purification a hygienic barrier.

I think the discussion has brought up a number of interesting possibilities. We still have no firm basis to choose between many of them. It might be that the next phase in the development will be one where we should do what you might call some wild experiments, i.e., experiments that are designed to give information about potentialities of various methods, to indicate how to continue the development.

Index